211
Topics in Current Chemistry

Springer-Verlag Berlin Heidelberg GmbH

Bioorganic Chemistry of Biological Signal Transduction

Volume Editor: Herbert Waldmann

With contributions by
G. Dormán, P. J. Hergenrother, C. Herrmann,
J. Kuhlmann, A. Levitzki, S. F. Martin, G. Müller,
M. Thutewohl, H. Waldmann

 Springer

The series *Topics in Current Chemistry* presents critical reviews of the present and future trends in modern chemical research. The scope of coverage includes all areas of chemical science including the interfaces with related disciplines such as biology, medicine and materials science. The goal of eaxch thematic volume is to give the non-specialist reader, whether at the university or in industry, a comprehensive overview of an area where new insights are emerging that are of interest to a larger scientific audience.

As a rule, contributions are specially commissioned. The editors and publishers will, however, always be pleased to receive suggestions and supplementary information. Papers are accepted for *Topics in Current Chemistry* in English.

In references *Topics in Current Chemistry* is abbreviated Top. Curr. Chem. and is cited as a journal.

Springer WWW home page: http://www.springer.de
Visit the TCC home page at http:/link.springer.de/series/tcc/
or http://link.springer-ny.com/series/tcc/

ISSN 0340-1022
ISBN 978-3-662-14697-2   ISBN 978-3-540-45035-1 (eBook)
DOI 10.1007/978-3-540-45035-1

Library of Congress Catalog Card Number 74-644622

© Springer-Verlag Berlin Heidelberg 2001
Originally published by Springer-Verlag Berlin Heidelberg New York in 2001
Softcover reprint of the hardcover 1st edition 2001

Cover design: Friedhelm Steinen-Broo, Barcelona; MEDIO, Berlin
Typesetting: Fotosatz-Service Köhler GmbH, 97084 Würzburg

SPIN: 10649296   02/3020 ra – 5 4 3 2 1 0 – Printed on acid-free paper

Volume Editor

Prof. Dr. Herbert Waldmann
Max-Planck-Institut für Molekulare Physiologie
Universität Dortmund
Otto-Hahn-Straße 11
44221 Dortmund, Germany
E-mail: herbert.waldmann@mpi-dortmund.mpg.de

Editorial Board

Topics in Current Chemistry
Now Also Available Electronically

For all customers with a standing order for Topics in Current Chemistry we offer the electronic form via LINK free of charge. Please contact your librarian who can receive a password for free access to the full articles by registration at:

http://link.springer.de/series/tcc/reg_form.htm

If you do not have a standing order you can nevertheless browse through the table of contents of the volumes and the abstracts of each article at:

http://link.springer.de/series/tcc

There you will also find information about the

– Editorial Board
– Aims and Scope
– Instructions for Authors

Preface

The transduction of signals from the extracellular space across the plasma membrane into the interior of cells and ultimately to the nucleus, where in response to such external signals the transcription of the genetic code is influenced, belongs to the most fundamental and important events in the regulation of the life cycle of cells. During recent years several signal transduction cascades have been elucidated which regulate, for instance, the growth and the proliferation of organisms as diverse as mammals, flies, worms and yeast. The general picture which emerged from these investigations is that nature employs a combination of non-covalent ligand/protein and protein/protein interactions together with a set of covalent protein modifications to generate the signals and transduce them to their destinations. The ligands which are recognized may be low molecular weight compounds like lipids, inositol derivatives, steroids or microbial products like cyclosporin. They may be proteins like, for instance, growth factors or intracellular adaptor proteins which carry SH2 or SH3 domains, and they may be specific DNA stretches which are selectively recognized by transcription factors.

These and other aspects of biological signal transduction provide an open and rewarding field for investigations by scientists from various different disciplines of biology, medical research and chemistry working in academic research institutions or in industry. In particular, it is full of opportunities for bioorganic studies in which the expertises of biologists and chemists have to be merged, and it is highly relevant to medicinal chemistry since "signal transduction therapy" is a new promising approach for the treatment of various diseases including cancer.

This book presents a selection of the most advanced topics from the bioorganic and medicinal chemistry of signal transduction. The authors who have contributed include biologists, biochemists, medicinal chemists and synthetic organic chemists and work in universities and in industry. The topics which are covered address various important events involved in the transduction of chemical signals in biological systems as summarized above. It was my intention to bring together scientists from several subdisciplines and to highlight the progress that has been achieved in this multidisciplinary field. In reviewing the papers which now finally appear in print I am confident that I have succeeded in my attempt.

Dortmund, August 2000   Herbert Waldmann

Contents

Contents of Volume 195

Biosynthesis
Polyketides and Vitamins

Volume Editors: Finian J. Leeper, John C. Vederas
ISBN 3-540-63418-5

Contents of Volume 209

Biosynthesis
Aromatic Polyketides, Isoprenoids, Alkaloids

Volume Editors: Finian J. Leeper, John C. Vederas
ISBN 3-540-66573-0

Protein Tyrosine Kinase Inhibitors as Therapeutic Agents

Alexander Levitzki

Department of Biological Chemistry, The Alexander Silberman Institute of Life Sciences,
The Hebrew University of Jerusalem, Jerusalem 91904, Israel
E-mail: levitzki@vms.huji.ac.il

Protein tyrosine kinases (PTKs) play a key role in cell signaling and regulate biological processes such as proliferation, differentiation, and apoptosis. The malfunctioning of these proteins is the root of many diseases. Over 60 % of all oncoproteins and proto-oncoproteins, which play a key role in cancers, are PTKs. Malfunctioning of PTKs is also the hallmark of other diseases such as psoriasis, Papilloma, Atherosclerosis, pulmonary fibrosis and more. It is therefore logical to target these proteins for drug design aiming at selective and non-toxic drugs. Since the second half of the 1980s we have pioneered and continued to generate tyrosine phosphorylation inhibitors (tyrphostins) as agents against diseases such as cancers, leukemias, lymphomas, psoriasis, restenosis, angiogenesis, and more. Currently, there are already a few tyrphostins in clinical trials. We argue that it is likely that the most effective tyrphostins in the future will be those which target the substrate binding domain of the PTK and not the ATP sub-domain. It is highly likely that tyrphostins against the key PTKs that play a pivotal role in diseases will become an important component in human therapy.

Keywords: Tyrphostins, Protein tyrosine, Kinases, JAK-2 EGF, PDGF BCR-ABL, Psoriasis, Papilloma, Restenosis, Leukemia, Lymphoma

1
Introduction

Discoveries in the area of signal transduction pathways over the past two decades have revealed that many diseases result from the malfunctioning of these pathways. In fact, nearly every disease can be defined in molecular terms and can be characterized as a "fingerprint" of a set of biochemical lesions. As a consequence of this revolution in understanding disease an entirely new field has evolved known as "signal transduction therapy" [1]. Both small molecules and proteins can now be used as correcting agents for the malfunction of signal transduction pathways thus taking on the function of a drug. This review will concentrate on protein tyrosine kinases (PTKs).

The malfunction of protein tyrosine kinases (PTKs) is the hallmark of numerous diseases. Not only do PTKs account for over 60% of oncogenes and proto-oncogenes involved in human cancers but enhanced activity of PTKs has also been implicated in such nonmalignant diseases as psoriasis, Papilloma, restenosis and pulmonary fibrosis. It is therefore not surprising that there has been such a surge of studies attempting to target PTKs for drug development over the past decade (for review see [2]). For many diseases the understanding of molecular pathology has so advanced as to enable identification of molecular aberrations. Among such malfunctioning signaling elements PTKs appear prominently. Table 1 summarizes those PTK signaling molecules whose altered activities have proven to be directly correlated with a human disease. Other PTKs are also implicated but the level of evidence is more correlatory than in the cases summarized in Table 1.

2
Universal Targets and Selective Targets

In most cases of proliferative disease more than one signaling pathway is involved; this is especially true in cancers where many genetic alterations have taken place on the cell's pathway to its transformed state. Even in such benign conditions as psoriasis or restenosis several signaling pathways have been implicated. In all of the above cases more than one PTK is likely to be involved though in many instances it is possible to identify one PTK that plays a key role in the disease. Table 1 lists most of these PTKs and the diseases in which they are involved. These PTKs are candidates for drug targeting. PTKs such as IGF-1R and the Src family kinases have been implicated in many types of cancer thus qualifying as *universal targets* [2, 3] for the development of PTK inhibitors.

Out of those PTKs involved in pathophysiological states a relatively small number show up in cancers and other proliferative states. For example, over-expression of the EGF receptor kinase is the hallmark of most, if not all,

Table 1. PTKs whose enhanced activities is correlated with specific diseases

PTK involved	Type of altered activity	Disease implicated
EGFR, unmutated	Overexpression and autocrine stimulation	Epithelial cancers, Psoriasis, Papilloma
EGFR, truncated	Constitutive activation	Gliomas breast, and ovary cancers
PDGFR	Overexpression and autocrine stimulation. Activation of blood vessel smooth muscle cells in the media	Glioblastomas, Restenosis, Atherosclerosis
TEL-PDGFR fusion protein	Constitutive activation of PDGF receptor	Chronic myelomonocytic, Leukemia
Bcr-Abl	Constitutive activation in the cytoplasm of blood stem cells	Chronic myeloid leukemia (CML)
Jak-2	Persistent activation	Recurrent pre-B acute lymphoblastic leukemia (pre-B ALL); IL-6 dependent multiple myeloma
Her-2/neu, Her-3, Her-4 heterodimers	Overexpression	Breast, ovary, lung, gastric cancers
VEGFR/Flk-1/KDR	Activation of tumor vascularization by the anoxic tumor	All cancers
c-Src, c-Yes	Persistently activated	Cancers of the lung, colon breast, prostate
IGF1-R	Overexpression, autocrine stimulation	Cancers, psoriasis

epithelial cancers. This is usually accompanied by the autocrine or paracrine expression of its ligands, producing persistent enhanced stimulation of EGF dependent pathways. In certain tumors, a truncated persistently active version of the receptor is overexpressed and induces intense signaling. It is therefore little wonder that attempts to generate EGFR kinase directed tyrphostins have been pursued since the search for PTK inhibitors began. The development of an inhibitor for the EGFR kinase which could possibly become a universal inhibitor should be regarded as an important objective.

The involvement of enhanced EGFR signaling is also the hallmark of Papilloma induced by HPV 16/18 [4] and psoriasis [5]. Blockers of the EGFR kinase were therefore suggested as potent anti-psoriasis agents [5–7] and anti-Papilloma agents [4]. Indeed AG 1517(SU 5271 = PD 153035, see below) has been undergoing clinical trials since early 1997 [7]. Similarly, the involvement of HER-2/neu in breast, ovarian, lung and gastric cancers makes this close relative of the EGFR kinase an attractive target for drug design. Overexpression of HER-2 occurs in a significant proportion (10–35%) of the cases. In some instances it is possible to identify one particular PTK whose activity is correlated with the disease. For example, in the chronic phase of chronic myeloid leukemia

(CML) in which the fusion protein Bcr-Abl, the product of the Philadelphia chromosome, is implicated as the cause of the leukemia [8]. TEL-PDGFR [9] and Jak-2 [10] are similarly associated with other forms of leukemia. In these cases highly selective inhibitors can be used to inhibit and even selectively purge the diseased cells. For most tumors, however, one signal transduction inhibitor may not in itself be sufficient to eradicate the disease. Indeed early in vivo experiments show that tyrphostin RG 13022, an EGFR kinase blocker, is capable of inhibiting the progress of tumor growth and prolonging survival of nude mice implanted with a human squamous tumor which overexpresses EGFR. Treatment with RG 13022 is not by itself sufficient to cause total eradication of the disease [11]. It is, therefore, probable that PTK inhibitors will prove most efficient in combination with other drugs such as cytotoxic drugs ([12, 13] and discussion below) or antibodies [11].

3
Design and Synthesis of Protein Tyrosine Phosphorylation Inhibitors

Once it became apparent that the enhanced activity of PTKs was a major contributor to oncogenesis, a search for tyrosine kinase inhibitors began. The systematic synthesis of selective PTK inhibitors (tyrosine phosphorylation inhibitors = tyrphostins) that show selectivity towards the isolated EGFR kinase and do not inhibit Ter/Thr kinases was reported in the late 1980 s [14]. It took a few more years to produce highly potent and selective tyrosine kinase inhibitors, mostly by semi-rational drug design and high throughput screening (for review see [2, 3, 15, 16]). Kinetic analysis of the mode of EGFR kinase action and of pp60$^{c\text{-Src}}$ shows that ATP and the substrate bind independently to the kinase domain and no sequential binding occurs [17]. This property simplifies the kinetic analysis on the inhibitors' mode of inhibition of PTKs [18].

One of the most surprising findings on the selectivity of inhibitors discovered so far is the extent to which ATP-competitive inhibitors can be selective. For example, quinoxalines are highly selective inhibitors of PDGFR kinase [19, 20] and quinozalines for the EGFR kinase ([21] and see below). Further analysis on the mode of tyrosine kinase inhibition reveals that the affinity of the inhibitor and its mode of binding to the kinase domain depend on whether the kinase is in its activated form or basal inactive state. Two examples illustrate this point: Activated Abl kinase like p210$^{Bcr\text{-Abl}}$ and p185$^{Bcr\text{-Abl}}$ possess different affinities to both substrate and inhibitors (tyrphostins) as compared to the proto-oncogenic form p140$^{c\text{-Abl}}$ [8]. For example, p210 $^{Bcr\text{-Abl}}$ and p185 $^{Bcr\text{-Abl}}$ are inhibited by the tyrphostin AG 957 with Ki values of 0.75 µmol/l and 1.5 µµmol/l as compared to Ki = 10.0 µM for the cellular wild type p140$^{c\text{-Abl}}$. In all these cases AG 957 is competitive with the substrate and non-competitive with ATP. The difference between c-Abl and Bcr-Abl is not found in the kinase domain, since the Bcr sequence is fused in-frame with the intact kinase domain. It would therefore appear that tethering the Bcr sequence upstream to the c-Abl alters the conformation of the kinase domain such that it binds the inhibitor more tightly. The same pattern of behavior is observed vis-à-vis the substrates: the oncogenic forms exhibit lower Km values towards the substrates as compared to the proto-oncogenic forms. A similar relationship was found for the mouse proteins where v-Abl(= gagAbl) exhibited higher affinity towards the substrates and inhibitors as compared to the proto-oncogenic form of the protein [8]. In the case of the PDGFR, kinase activation of the receptor also leads to changes in the structure of the kinase domain but the situation is slightly more complex: upon activation the mode of inhibition of the selective inhibitor AG 1296 (or AG 1295) is altered. Whereas the inhibitor is competitive vis-à-vis ATP in the inactive form of the receptor, it binds with higher affinity and becomes mixed competitive vis-à-vis ATP subsequent to receptor activation by PDGF [22]. These two examples suggest that this may be a general pattern of behavior for PTKs [23]. This finding also points to the necessity of examining potential PTK inhibitors not only as blockers of PTK autophosphorylation, which is done routinely, but also as blockers of the PTK action on exogenous substrates.

With the advance of X-ray crystalography and the ability to determine the three dimensional structure of the kinase with the bound inhibitor, drug design has become more precise and rational. The three dimensional structures of a two tyrosine kinases complexed with a kinase inhibitor have already been solved:

1. The structure of the FGF receptor with selective and non-selective inhibitors has already been published [24, 25].
2. The Src kinase Hck has been crystallized in its inactive form with the inhibitor PP1 complexed with it [26]. Similarly, Lck in its active form has been crystallized with the inhibitor PP2 [27].

These structures are currently guiding a number of laboratories in their attempt to design novel, more selective Src kinase inhibitors.

This being the case, the next challenge is to attempt to design compounds which selectively block pp60^{c-Src}, Yes and Fyn are activated in many human malignancies ([28, 29] and references therein) but do not significantly block other Src family kinases involved in other pathways such as p56^{Lck}. Similarly, the availability of the insulin receptor kinase structure in its inactive form [30] as well as in its active form complexed with APPNHP and a peptide substrate [31] has enabled an educated search for inhibitors for the IGF-1R kinase which is highly homologous to the insulin receptor kinase.

This search has resulted in identification of AG 538 as a potent inhibitor of the IGF-1R kinase [32]. The AG 538 family, identified by this type of "educated" search was designed to find inhibitors which mimic the encompassing protein loop between tyrosine 1158 and tyrosine 1162 and actually occupies the substrate site in the inactive state. This then becomes phosphorylated upon activation within the insulin receptor. Indeed AG 538 is capable of inhibiting IGF1-R autophosphorylation with an IC$_{50}$ value of 0.4 µM [32]. This series of inhibitors is different from AG 1024 and some of its analogs, which were found by random screening [33]. It will be most interesting to compare the mode of binding of AG 1024 to AG 538 within the active site of the receptor. It is important to note that some of the inhibitors discriminate between the insulin receptor kinase and the IGF1-R kinase by a factor of up to 8 [33]. The challenge remains to design inhibitors of IGF1-R kinase which differentiate between the insulin receptor kinase and IGF-1R kinase by a wider margin. Table 2 summarizes the main pharmaco-

Table 2. Lead pharmacophores for PTK inhibitors

BENZENE MALONO NITRILES · LAVENDUSTINS · ANILIDO PHTHALIMIDES

PYRAZOLO PYRIMIDINES · PYRIDO PYRIMIDINES · QUINAZOLINES

QUINAXOLINES · BENZO QUINAXOLINES · PYRROLO ISATINES

phores which have proven to be effective PTK inhibitors with no significant effects on Ser/The kinases. Table 3 summarizes the specific compounds which have shown biological efficacy on the targets quoted.

4
ATP Mimics vis-à-vis Substrate Mimics

It was initially argued that the best potential PTK inhibitors would be compounds that compete for the substrate in the kinase binding domain. It was argued that such compounds would be less toxic than ATP mimics since they bind to those domains at the kinase site that are less conserved than the substrate binding domains. Indeed tyrphostins like AG 490 which blocks Jak-2 [10] and AG 556 which possesses anti-inflammatory properties have been shown to be highly non-toxic in vivo [34–37].

Table 3. Tyrphostins with biological activities

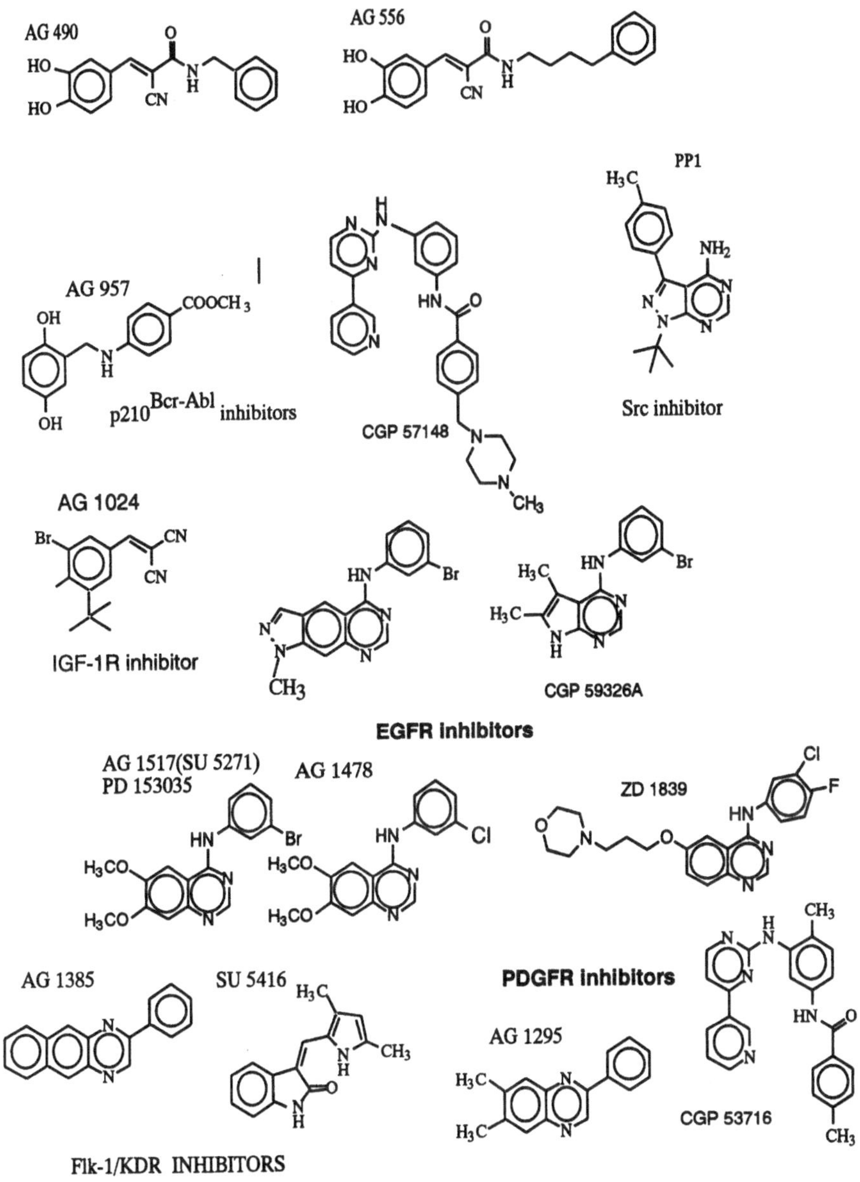

AG 490

AG 556

PP1

AG 957

p210^Bcr-Abl inhibitors

CGP 57148

Src inhibitor

AG 1024

IGF-1R inhibitor

CGP 59326A

EGFR inhibitors

AG 1517(SU 5271)
PD 153035

AG 1478

ZD 1839

AG 1385

SU 5416

PDGFR inhibitors

AG 1295

CGP 53716

Flk-1/KDR INHIBITORS

Over the past decade, however, the PTK inhibitors favored by most investigators have been ATP mimics that compete with ATP at the binding site. Most of the compounds depicted in Table 2 are ATP mimics with the exception of the tyrphostins developed by us. In the case of tyrphostins one can indeed classify compounds which compete with ATP, compounds which compete with the substrate and bisubstrate inhibitors which compete with the substrate and ATP simultaneously [18]. Compounds can also be identified that act as "mixed" competitive inhibitors which bind simultaneously with ATP and/or substrate but decrease the affinity of ATP and the substrate to their respective sites [18, 22]. Among the tyrphostins all classes of compounds can be identified [18] but the real question is which of these is preferable for clinical development.

In recent clinical trials, in which PTK inhibitors are being utilized, doses of 50 – 100 mg/kg have been used. These high doses reflect the relatively low efficacy of these compounds in vivo despite the fact that their IC50 values towards their molecular targets such as the EGFR, VEGFR-2/Flk-1 are in the nanomolar range. The high doses required partly reflect the competitive relationship between the intracellular millimolar concentrations of ATP and the administered drug. It is noteworthy that drugs such as beta-adrenergic blockers are administered at doses which are lower by about 100-fold. In this case a drug possessing an affinity in the nanomolar concentration range has to compete with up to 0.1 µmol/l concentration of the endogenous ligand adrenaline or nor-adrenaline. Thus β-blockers can be administered in this case at doses of 1.0 mg/kg and still be effective. It seems in light of this that the possibility of developing substrate competitive tyrosine kinase inhibitors as opposed to ATP-mimics should be reconsidered. A further consideration is the possibility of ATP-mimics being fairly toxic since the selectivity of newly developed compounds are only tested against a limited number of PTKs and Ser/Thr kinases, where the number of PTKs falls in the 400 – 500 range and the number of Ser/ kinases is numbered in the thousands. It has already been obsereved that the so called selective Src family kinase inhibitor PP1 is in fact equipotent as a PDGFR kinase inhibitor [38]. Similarly the Novartis Bce-Abl kinase inhibitor, CGP57148 [39] currently undergoing clinical trials for chronic myeloid leukemia (CML), is as potent against PDGFR kinase [40]. It is likely that this type of behavior will be encountered as progress continues to be made in developing ATP mimics as therapeutic agents.

5
Screening for Tyrosine Kinase Inhibitors

The involvement of several tyrosine kinases in various cancers requires efficient screening methodologies for the inhibitory compounds. Screening is divided into three steps: (1) *primary screening* against the pure isolated PTK in a cell-free system. The objective is always an ELISA format. The compounds are screened against a battery of PTKs and Ser/Ther kinases in order that the pattern of selectivity can be established quickly [2].

Successful candidates, that show high affinity towards the isolated PTK move to (2) *secondary screening* in which the compound is tested for its potency in inhibiting the PTK in the intact cell. Many compounds fail at this stage either because they are not permeable or do not reach their target in the intact cell. Compounds that pass this test are then examined against other kinases within the intact cell context. This screening is also conducted in a 96 well format and is therefore amenable to high throughput screening. Successful compounds are tested in (3) *tertiary screening* for their ability to block the growth of the cell when such growth is driven by the targeted PTK. This test is in many cases accompanied by an additional test, which examines the ability of the compound to block the growth of these cells on soft agar. Such potency indicates that these compounds could be successful in vivo. Successful compounds undergo (4) *quaternary screening* through testing in animal models. Animal models range from SCID mice or nude mice harboring the human disease to the inhibition of balloon injury induced stenosis in pigs as a model for restenosis.

Due to the availability of chemical libraries, plant and microbial extracts and rapid screening methods lead compounds are rapidly identified. When the structures of PTKs are known optimization by organic synthesis and computer modeling follows. Compounds that successfully pass all the screening tests are ready to be evaluated for clinical trials.

6
Biological Activity of Tyrphostins

A number of families of PTK inhibitors (tyrphostins) have shown efficacy in tissue culture as well as in vivo. A significant number of benzene malono nitriles (BMNs), which belong to the founding family of PTK inhibitors were found to possess biological activity. These include AG 490, AG 126 and AG 556 for which promising in vivo data already exist [10, 11, 34 – 37].

Anilidopthalimides, which inhibit EGFR [41], also show efficacy in vivo. Pyrido pyrimidines like CGP 53716 which inhibit PDGFR [42], have shown efficacy in vivo. A similar compound, CGP 57148B, from the pyrazolo pyrimidines, blocks Bcr-Abl and shows efficacy in intact cells harboring Bcr-Abl [43]. However, this inhibitor also blocks PDGFR and Kit. AG 957, a substrate competitive tyrphostin derived from Lavendustin A, is very selective against Bcr-Abl kinase [8, 44] and has been shown to selectively purge Ph$^+$ cells from blood obtained from CML patients at the chronic phase of the disease [45]. Quinoxalines were found to be very selective PDGFR kinase inhibitors with very high efficacy in intact cells [19, 20]. The quinoxaline AG 1295 has also shown efficacy in vivo by inhibiting balloon injury induced stenosis in pigs ([46] and see below), a predictive animal model for restenosis in humans. In vivo activity of AG 1295 was also shown in the rat model [47]. Quinozalines show good efficacy against the EGFR [7, 13, 48 – 50] and one of them, AG 1517 [13, 49] which is identical to PD153035 [48] is currently in clinical trials for psoriasis [7]. This compound is very effective in blocking the growth of psoriatic keratinocytes [7]. Another quinazoline 2D 1839 (Table 3) developed by Zeneca is currently in clinical trails. A related pharmacophore (see Table 2), which has been shown to be effective against the EGFR kinase is derived from 4-(Phenylamino) pyrrolopyrimidine [51], a compound which was rationally designed on the basis of successful inhibitors from the dianilinphthalimide family. Yet another successful family of EGFR kinase inhibitors was obtained by the fusion of the quinozaline moeity with a third ring [52].

7
The Synergistic Action of Tyrphostins with Cytotoxic Agents Antibodies and Cytokines

The enhanced activity of PTKs generates anti-apoptotic robustness in cancer cells. This is most probably due to the enhancement of anti-apoptotic pathways mediated by BclX/Bcl-2 and c-Akt/PKB as well as by other biochemical elements. Intensification of anti-apoptotic pathways seems to be the predominant factor in the emergence of resistance to cytotoxic drugs and radiation. It is probable that these inhibitory pathways suppress the stress pathways triggered by cyto-toxic agents, thus preventing cell death. This principle has been demonstrated in two instances and the molecular mechanism has been at least partially analyzed. Tyrphostin AG 825 was shown to strongly synergize with *cis*-platin, etoposide and doxorubicin, all DNA damaging agents triggering the stress pathways [12]. A series of non-small lung cancer tumor [NLSC) cell lines with different levels of HER-2/neu overexpression was examined for this effect.

The extent of synergy was found to be dependent on the level of expression of the HER-2/neu receptor. While no synergy was observed, additivity was in fact observed for cells with no overexpression of HER-2/neu, the highest synergy was found in cells with very high expression of HER-2/neu. This was reflected quantitatively in the combination index (CI) whose value was much lower than 1.0 for the high overexpressors and equal to 1.0 for the cell in which HER-2/neu is not overexpressed [12]. It seems that the higher the overexpression of HER-2, the higher the anti-apoptotic shield provided by the receptor tyrosine kinase. It would seem however, that while the sensitivity of the cells to stress signaling is also enhanced upon higher expression of HER-2/neu, this sensitivity is masked by the high expression of HER-2/neu. Thus, when the cells are stripped of the robust anti-apoptotic shield their enhanced vulnerability to stress is exposed.

A very similar result was obtained recently with human glioma cells, which overexpress the truncated form of the EGF receptor Δ (2–7)-EGFR. In this case EGFR kinase selective agent inhibitors such as AG 1478 or AG 1517 synergize with *cis*-platin (CDDP). It has been demonstrated that AG 1478 reduces the level of expression of Bcl-X_L and increases the activity of Caspase 3. The combina-tion of the tyrphostin with *cis*-platin is lethal to the tumor cells since it leads to apoptosis [13]. It seems therefore that ΔEGFR transmits strong anti-apoptotic signals which, when inhibited, sensitize the cells to the stress signaling triggered by *cis*-platin. Such a success would allow clinical application of these tyrphostins as sensitizers of these tumors to cytotoxic drugs, to which they are normally insensitive. Indeed recently the Ludwig Institute for Cancer Research (LICR) has begun preparation to treat patients with glioma multiformis with the combina-tion of AG 1478 and CDDP.

The combinatorial approach, which combines a blocker of anti-apoptotic sig-naling with a stimulator of stress signaling, could prove to be a general approach.

It is therefore probable that blockers of Bcl-2/Bcl-X_L proteins and c-Akt/PKB will have a broad application for cancer therapy when they become available. However, since PTKs, especially in tumors, are frequently, if not always, the source of anti-apoptotic signals, tyrphostins may actually fulfill this task. In some cases such as Pre-B ALL the selective tyrphostin AG 490 was found to be sufficient by itself to induce apoptosis [10]. It seems that in this case stress signaling is so strong that dismantling of the cells from the anti-apoptotic signals is sufficient to induce apoptosis. This point, however, is still under investigation. In IL-6 dependent multiple myeloma cells, significant cell death can be attained by the Jak-2 inhibitor AG 490, but its effect is enhanced dramatically by the Fas receptor antibody CH-11[53]. Similarly, in CML one finds very siginificant synergy between the Bcr-Abl kinase inhibitor AG 957 and the Fas receptor antiboby CH-11 in selectively purging Ph+ cells [45].

The combination of tyrphostins with cytotoxic drugs can be followed by immunotherapy in order to eliminate residual disease. Though such combinations have not yet been examined, the combination of AG 490 and IL-12 against IL-6 dependent multiple myeloma recently showed impressive tumor suppressive effects [54], suggesting that the general idea may indeed be correct.

8
Success of Tyrphostins In Vivo

In many cases the hypothesis that PTK inhibitors provide an effective block to proliferative disease or tumor growth in vivo has in fact been proven in preclinical animal models. Such cases are listed below.

8.1
PDGFR Kinase Directed PTK Inhibitors

Tyrphostin AG 1295 and its close analog AG 1296, both quinoxalines (see Table 2), have been shown to selectively block PDGFR kinase with insignificant inhibitory effects on EGFR, Src, Flk-1, HER-2/neu and IGF1-R. These tyrphostins have also been shown to reverse the transformed phenotype of *sis* transformed NIH 3T3 cells [19] and slow C6 glioma induced tumors in nude mice (Shawver, L., unpublished). In recent experiments on pigs, AG 1295 was shown to block balloon injury induced stenosis in the femoral artery [46]. The extent of inhibition is ~50–60% compared to the treatmemt of the balloon-injured femoral artery with the vehicle alone. Similar results were obtained for rats. The pig model, which is much more faithful to the human situation and a better predictor for humans, tends to indicate AG 1295 and its more potent analogs AGL 2034 and AGL 2044 (Gazit, Banai and Levitzki, in preparation) as candidates for clinical testing. A quinoxaline developed by Rhône-Poulenc-Rorer RPR has also shown efficacy in inhibiting stenosis in balloon injured minipigs fed on a high fat diet [55]. The ability of very selective PDGFR kinase inhibitors to block balloon injury-induced stenosis supports Russel Ross' (1976) "PDGF hypothesis" [56, 57] which implicated this growth factor and its receptor in the process of atherosclerosis and its accelerated form-restenosis. The finding that PDGFR kinase directed tyrphostins block ~60% stenosis strongly suggests that, in order to block the other proliferative signals, other signaling pathways should be targeted. These include those signaling pathways elicited by FGF and TGFα. Since accelerated atherosclerosis in transplanted hearts is the major cause of death among heart transplant patients, AG 1295 and its analogs are currently being tested in an animal heart transplant model with very promising results (Karck and Levitzki, unpublished experiments).

8.2
Jak-2 Inhibitor

Among the many tyrphostins, it has been shown that the AG 490 family are potent inhibitors of Jak-2 and to a lesser extent Jak-3 [58]. The high potency of AG 490 in blocking Jak-2 may well explain these agents' ability to block recurrent pre-B acute lymphoblastic leukemia (Pre-B ALL) when engrafted in SCID mice [10]. AG 490 induces apoptotic death in Pre-B cells taken from patients suffering from this disease. Jak-2 is persistently activated in these Pre-B cells whereas in normal B cells this is not the case. It is still not known whether the persistent activation of Jak-2 is due to an activating mutation or to a robust autocrine stimulation. The efficacy of AG 490 in vivo accompanied by a complete absence of toxicity in normal blood tissue is gratifying [10]. This agent is currently undergoing evaluation for clinical trials. AG 490 also induces apoptosis in IL-6 dependent multiple myeploma [53] and synergizes with the Fas antibody CH-11 and with IL-12 to induce the eradication of the IL-6 dependent tumor [54].

8.3
EGFR Kinase Inhibitors

The eficacy EGFR kinase blockers from the early tyrphostin family was demonstrated in 1991. They were the first PTK blockers to show efficacy in vivo [11]. Due to their unfavorable pharmacokinetic properties, however, they did not make further progress in pre-clinical animal models and were later abandoned as possible candidates for the clinical set up. These experiments nonetheless constituted a milestone in the sense that they provided proof of the principle. These experiments showed for the first time that the activity of a tyrphostin identified in a cell-free system resulted in in vivo success. A few heterocyclic EGFR kinase inhibitors such as DAPH [41] and CGP 59326 A [59] have shown efficacy in vivo against tumors that overexpress EGFR. Furthermore, the combination of this inhibitor with cytotoxic drugs such as adriamycin and vinblastine induced tumor regression. One quinazoline, AG 1517 (SU 5271), is currently being tested as an anti-psoriatic agent using topical application. Since no animal model is available for psoriasis, the potency of AG 1517 (SU 5271) in inhibiting the growth of psoriatic cells and the absence of adverse toxic effects has enabled the beginning of clinical trials. Since not only psoriatic cells [6, 7] but also Papilloma (HPV 16) infected keratinocytes are driven by the EGFR-kinase system [4] success may lead to the testing of EGFR-directed PTK inhibitors as anti-Papilloma agents [4]. As with psoriasis, there is no accepted animal model for Papilloma. It is therefore likely that EGFR kinase inhibitors will go straight to the clinic, skipping the pre-clinical animal model. In the case of Papilloma, topical application is intended to prevent the transformation of the HPV 16 infected tissue into cervical cancer [4].

A quinazoline developed by Zeneca, ZD 1839 (Table 3) has shown efficacy in the clinic against EGFR overexpressing tumors. AG 1478 [3, 29] is effective against glioma multiformis in cells [13] and in vivo (Cavenee W., unpublished experiments) only in combination with the cytotoxic drug CDDP (*cis*-platin). In 50% of these tumors the EGFR is truncated and is persistently active. We have shown that this truncated receptor emits very strong anti-apoptotic signals which result in the tumor's drug resistance. Upon blockade of the truncated receptor by AG 1478, the tumor is re-sensitized to the drug and its growth is severely inhibited. Therefore, the combination of AG 1478 and CDDP is currently being developed for clinical trial for patients with glioma multiformis.

8.4
Tyrphostins as Anti-Inflammatory Agents

A number of tyrphostins such as AG 126 [34, 37] and AG 556 [34 – 37] have shown high efficacy as anti-inflammatory agents. The most prominent is AG 556, which was found to be effective as anti-sepsis agent in mice [36, 37] and dogs [35] and was very effective in alleviating the symptoms of experimental autoimmune encephalitis, even with late administration [60]. AG 126 was found to be effective in preventing LPS-induced mortality in mice [37] and to alleviate cirrhosis-like symptoms in rats [34]. The search for such compounds within the tyrphostin family was based on scattered reports in the literature that the activation of PTKs is on the pathway of LPS induced inflamatory response. Similarly, scattered reports suggest that the action of TNFα involves PTK activation at some point on the overall signaling network activated by the cytokine (see, for example, [61 – 65]). Since no specific molecular target has been defined, screening for active tyrphostins was based on the potency of compounds tested to block LPS induced TNFα production and TNFα action [37]. It remains to be seen whether the biological activity of this family of tyrphostins is exclusively due to the inhibition of PTKs.

9
References

1. Levitzki A (1994) Eur J Biochem 226:1
2. Levitzki A (1996) Curr Opin Cell Biol 3:239
3. Levitzki A, Gazit A (1995) Science 267:1782
4. Ben-Bassat H, Rosenbaum-Mitrani S, Hartzstark Z, Shlomai Z, Kleinberger-Doron N, Gazit A, Plowman G, Levitzki R, Tsvieli R, Levitzki A (1997) Cancer Res 57:3741
5. Ben-Bassat H, Vardi DV, Gazit A, Klaud SN, Chaouat M, Hartzstark Z, Levitzki A (1995) Experimental Dermatology 4:82
6. Dvir A, Milner Y, Chomsky O, Gilon C, Gazit A, Levitzki A (1991) J Cell Biol 113:857

7. Powell TJ, Ben-Bassat H, Klein B, Chen H, Shenoy N, McCollough J, Narog B, Gazit A, Harstark Z, Chaouat M, Tang C, McMahon J, Shawver L, Levitzki A (2000) Br J Dermatol 14: 802–810 (1999)
8. Anafi M, Gazit A, Gilon C, Ben-Neriah Y, Levitzki A (1992) J Biol Chem 267:4518
9. Carroll M, Tomasson M, Barker G, Golub T, Gillialand D (1996) Proc Natl Acad Sci USA 10:14845
10. Meydan N, Grunberger T, Dadi H, Shahar M, Arpaia E, Lapidot Z, Leader S, Freedman M, Cohen A, Gazit A (1996) Nature 379:645
11. Yoneda T, Lyall RMAM, Pearsons PE, Spada AP, Levitzki A, Zilberstein A, Mundy GR (1991) Cancer Res 51:4430
12. Tsai CM, Levitzki A, Wu L-H, Chang K-T, Cheng C-C, Gazit A, Pemg R-P (1996) Cancer Res 56:1068
13. Nagane M, Levitzki A, Cavenee WK, Su Huang H-J (1998) Proc Natl Acad Sci USA 95:5724
14. Yaish P, Gazit A, Gilon C, Levitzki A (1988) Science 242:933
15. Levitzki A (1992) FASEB J 6:3275
16. Groundwater PW, Solomons KRH, Drewe JA, Munawar MA (1996) Progr Med Chem 33:233
17. Posner I, Engel M, Levitzki A (1992) J Biol Chem 267:20638
18. Posner I, Engel M, Gazit A, Levitzki A (1994) Mol Pharmacol 45:673
19. Kovalenko M, Gazit A, Bohmer A, Rorsman C, Ronnstrand L, Heldin C, Waltenberger JFD, Bohmer F, Levitzki A (1994) Cancer Res 54:6106
20. Gazit A, App H, McMahon G, Chen J, Levitzki A, Bohmer FD (1996) J Med Chem 39: 2170
21. Ward WHJ, Cook PN, Slater AM, Davies DH, Holdgate GA, Gree LR (1994) Biochem Pharmacol 48:639
22. Kovalenko M, Romstrand L, Heldin C-H, Loubochenko M, Gazit A, Levitzki A, Bohmer FD (1997) Biochemistry 36:6260
23. Levitzki A, Bohmer FD (1998) Anticancer Drug Des 13:731–734
24. Mohammadi M, McMahon G, Sun L, Tang C, Hirth P, Yeh BK, Hubbard SR, Schlessinger J (1997) Science 276:955
25. Mohammadi M, Froum S, Hamby JM, Schroeder MC, Panek RL, Lu GH, Eliseenkova AV, Green D, Schlessinger J, Hubbard S (1998) EMBO J 1720:5896
26. Schindler T, Sicheri F, Pico A, Gazit A, Levitzki A, Kuriyan J (1999) Mol Cell 3:639
27. Zhu X, Kim JL, Newcomb JR, Rose PE, Stover DR, Toledo LM, Zhao H, Morgenstern KA (1999) Structure Fold 7:65,125
28. Levitzki A (1996) Anti-Cancer Drug Design 11:75
29. Osherov N, Levitzki A (1994) Eur J Biochem 225:1047
30. Hubbard SR, Wei L, Ellis L, Hendrikson W (1997) Nature 372:746
31. Hubbard SR (1997) EMBO J 16:5572
32. Blum G, Gazit A, Levitzki A (1998) (submitted)
33. Parrizas M, LeRoith D (1997) Endocrinology 138:1355
34. Lopez-Talavera JC, Levitzki A, Martinez A, Gazit A, Esteban E, Guardian J (1997) J Clin Invest 100:664
35. Sevransky JE, Shaked G, Novogrodsky A, Levitzki A, Gazit A, Hoffman Z, Quezado BD (1997) J Clin Invest 99:1966
36. Vanichkin A, Palya M, Gazit A, Levitzki A, Novogrodsky A (1996) J Infect Dis 173:927
37. Novogrodsky A, Vanichkin M, Patya A, Gazit N, Osherov N, Levitzki A (1994) Science 264:1319
38. Waltenberger J, Uecker A, Kroll J, Frank H, Mayr U, Bjorge JD, Fujita D, Gazit A, Hombach V, Levitzki A, Bohmer FD (1999) Circ Res 85:12
39. Druker BJ, Tamura S, Buchdunger E, Ohno S, Segal GM, Fanning S, Zimmermann J, Lydon NB (1996) Nat Med 2:561
40. Buchdunger E, Zimmermann J, Mett H, Meyer T, Muller M, Druker BJ, Lydon NB (1996) Cancer Res 56:100
41. Buchdunger E, Mett H, Trinks U, Regenass U, Muller M, Meyer T, Beilstein P, Wirz B, Schneider P, Traxler P (1995) Clin Cancer Res 1:813

42. Buchdunger E, Zimmermann J, Mett H, Meyer T, Muller M, Regenass U, Lydon N (1995) Proc Natl Acad Sci USA 92:2558
43. Druker BJ, Tamura S, Buchdunger E, Ohno S, Segal GM, Fanning, S, Zimmermann J, Lydon NB (1996) Nature Med 2:561
44. Kaur G, Gazit A, Levitzki A, Stowe E, Cooney DA, Sausville EA (1994) Anti-Cancer Drugs 5:213
45. Carlo-Stella C, Regazzi E, Sammarelli G, Colla S, Garau D, Gazit A, Savoldo B, Cilloni D, Tabilio A, Levitzki A, Rizzoli V (1999) Blood 93:3973
46. Banai S, Wolf Y, Golomb G, Pearle A, Waltenberger J, Fishtein I, Schneider A, Gazit A, Perez L, Huber R, Lazarovichi G, Levitzki A, Gertz SD (1998) Circulation 19:1960
47. Fishben I, Waltenberger J, Banai S, Rabinovich L, Chorny M, Levitzki A, Gazit A, Huber R, Gertz DS, Golomb G (2000) (in press)
48. Fry DW, Kraker AJ, McMichael A, Ambroso LA, Nelson JM, Leopold WR, Connors RW, Bridges AJ (1994) Science 265:1093
49. Gazit A, Chen J, App H, McMahon G, Hirth P, Chen I, Levitzki A (1996) Biorg Med Chem 8:1203
50. Bridges A J, Zhou H, Cody DR, Rewcastle GW, McMichael A, Showalter HDH, Fry DW, Kraker AJ, Denny WA (1996) J Med Chem 39:267
51. Traxler PM, Furet P, Mett H, Buchdunger E, Meyer T, Lydon N (1996) J Med Chem 39:2285
52. Rewcastle GW, Palmer BD, Bridges AJ, Showalter HD, Sun L, Nelson J, McMichael A, Kraker AJ, Fry DW, Denny WA (1996) J Med Chem 39:918
53. Catlett-Falcone R, Landowski TH, Oshiro MM, Turkson J, Levitzki A, Savino R, Ciliberto G, Moscinski L, Fernandez-Luna JL, Nunez G, Dalton WS, Jove R (1999) Immunity 10:105
54. Burdeyla L, Carlett-Falcone R, Levitzki A, Coppola D, Dalton WD, Jove R, Yu H (2000) Cancer Res., in press
55. Bilder G, Wentz T, Leadley R, Amin D, Byan L, O'Conner B, Needle S, Galczenski H, Bostwick J, Kasiewski C, Myers M, Spada A, Merkel L, Ly C, Persons P, Page K, Perrone M, Dunwiddie C (1999) Circulation 99:3292–3299
56. Ross R, Glomset JA (1976) N Ing J Med 295:369
57. Ross R (1986) N Ing J Med 314:488
58. Sharfe N, Dadi HK, Roifman CM (1995) Blood 86:2077
59. Lydon NB, Mett H, Mueller M, Becker M, Cozens RM, Stover D, Daniels D, Traxler P, Buchdunger E (1999) Int J Cancer 81:669
60. Brenner T, Poradosu E, Soffer D, Sicsic C, Gazit A, Levitzki A (1998) Exp Neurol 154:489
61. Orlicek SL, Hanke JH, English BK (1999) Shock 12:350
62. Han Y, Rogers N, Ransohoff RM (1999) Interferon Cytokine Res 7:731
63. Fuortes M, Melchior M, Han H, Lyon GJ, Nathan C (1999) Clin Invest 3:327
64. Yan SR, Novak MJ (1999) Inflammation 2:167
65. Hanna AN, Chan EY, Xu J, Stone JC, Brindley DN (1999) J Biol Chem 274:12,722

Peptidomimetic SH2 Domain Antagonists for Targeting Signal Transduction

Gerhard Müller

BAYER AG, Central Research ZF-WFM, Building Q18, 51368 Leverkusen, Germany
E-mail: gerhard.mueller.gm1@bayer-ag.de

This contribution is dedicated to Prof. Dr. Horst Kessler on the occasion of his 60th birthday.

Within the last few years, therapeutic intervention into signal transduction cascades attracted the interest of pharmaceutical research. Apart from selective inhibition of phosphorylating and dephosphorylating enzymes, medicinal chemistry has focused on an SH2 domain-targeted drug design approach. This review will briefly summarize the role of adaptor proteins in signaling networks, followed by an introduction on the functional and structural details of SH2 domains in terms of protein architecture and mode of ligand binding. Major emphasis is laid on medicinal chemistry programs that aim to develop peptidomimetic SH2 domain inhibitors. Different strategies for phosphotyrosine-replacement will be presented, followed by a detailed discussion of the iterative optimization cycles revealing new peptidomimetic and non-peptide SH2 domain antagonists. Finally, an outlook is given on further applications of the modular chemistry concept and the privileged building blocks, that have been developed in SH2 domain inhibitor programs, onto a new class of adaptor proteins, notably the PTB domains.

Keywords: SH2 domains, PTB domains, Peptidomimetics, Phosphotyrosine, Structure-Based Drug Design, Adaptor Proteins, Privileged Structures

List of Abbreviations

Abu	α-aminobutyric acid
Ac$_6$c	1-amino-cyclohexanecarboxylic acid
Achec	2-amino-cyclohex-3-enecarboxyxlic acid
Aib	α-amino-isobutyric acid
APP	amyloid precursor protein
CARD	caspase recruitment domain
chmF	4-(carboxyhydroxymethyl)phenylalanine
cmF	4-(carboxymethyl)phenylalanine
cmT	4-(carboxymethyl)tyrosine
DEATH	domains involved in cell death (apoptosis)
F$_2$cmF	4-(carboxydifluoromethyl)phenylalanine
FOMT	O-(2-(2-fluoromalonly))tyrosine
FPmp	4-(phosphonofluoromethyl)phenylalanine
F$_2$Pmp	4-(phosphonodifluoromethyl)phenylalanine
F$_4$Pmp	2,3,5,6-tetrafluoro-4-(phosphonomethyl)phenylalanine
Hcy	homocyclohexylalanine
HGF	hepatocytic growth factor
HPmp	4-(phosphonohydroxymethyl)phenylalanine
IRS-1	insulin receptor substrate 1
ITAM	immune receptor tyrosine activation motif
Jak	Janus kinase
MDCK	Madin-Darby canine kidney
OMT	O-(2-malonyl)tyrosine
PDGF	platelet-derived growth factor
PDZ	domain found in PSD-95, Dlg, and ZO-1/2 proteins
PH	pleckstrin homology domain
PI	phosphotyrosine-interaction domain
Pmf	p-(2-malonyl)phenylalanine
Pmp	4-(phosphonomethyl)phenylalanine
PTB	phosphotyrosine binding domain
PTK	protein tyrosine kinase

PTP	protein tyrosine phosphatase
ptyr	phosphotyrosine
RTK	receptor tyrosine kinase
SF	scatter factor
SH2	src homology 2 domain
SH3	src homology 3 domain
SMART	simple molecular architecture research tool
Stat	signal transducers and activators of transcription

1
Introduction

Over the past few years, novel opportunities for pharmaceutical research emerged from a deeper understanding of intracellular signal transduction pathways together with the involved molecular components and intermolecular interaction events [1 - 6]. By revealing the complex networks that generally underlie the extracellular ligand-induced regulation of cell function, numerous novel molecular targets have been suggested for a therapeutic intervention into signal transduction-related disorders. Diseases resulting from aberrant signal transduction processes range from cancers, atherosclerosis, or psoriasis as examples of proliferative disorders, over inflammatory conditions such as sepsis, rheumatoid arthritis, multiple sclerosis, or tissue rejection, to various immunological disease states, and metabolic disorders[1, 4, 6]. Consequently, current medicinal chemistry is in the process of refocusing its efforts in seeking new approaches to disease management by controlled intervention into signal transduction pathways. Apart from the more classical approach of medicinal chemistry to interfere in signal transduction cascades by low-molecular weight enzyme inhibitors of, e.g., receptor tyrosine kinases (RTKs), cellular tyrosine kinases, farnesyl transferases, or phospholipases, the controlled modulation of non-catalytic adaptor protein domains emerges as a novel research area [2, 5, 7].

Signaling cascades are choreographed to a large degree by exactly those non-catalytic signaling domains such as SH2, SH3, PTB, PDZ, PH, WW, CARD, DEATH, 14-3-3, etc., which mediate the propagation of a receptor activation event through the cytoplasm by establishing highly specific protein-protein interactions [7 - 10]. In this context, phosphorylation of particular tyrosine residues serves as an on-off switch for, e.g., SH2 domain binding, while the sequential peptide environment around a phosphotyrosine (pTyr) confers a secondary level of specificity. Therefore, the pTyr determines a critical pharmacophore in cellular signaling [7, 11]. Its inappropriate expression, either through generation by protein tyrosine kinases (PTKs) or cleavage by protein tyrosine phosphatases (PTPs), contributes to the aforementioned diseases. Agents that selectively modulate pTyr-based molecular recognition events through interference into, e.g., the SH2 domain-mediated protein-protein interactions are not only considered as useful biochemical tools, but are believed to yield new medicines.

The design of such low-molecular weight compounds that inhibit or mimic the action of a naturally occurring protein epitope or peptide ligand is a continuing challenge for modern medicinal chemistry [12–15]. Peptides generally lack the appropriate physicochemical properties and metabolic stability to be useful therapeutic agents. This has led to the well-established concept of peptidomimetics that comprises a stepwise de-peptidization approach based on the knowledge of the structural requirements that govern the corresponding ligand-target binding event.

This contribution will summarize the currently pursued approaches towards non-peptide SH2 domain antagonists and the development of promising lead structures for SH2 domain-associated disease processes. It is clearly beyond the scope of this review to provide an in-depth overview of the entire research area.

After introducing functional and structural aspects of SH2 domains and the mode of ligand binding, special emphasis will be laid on lead finding attempts that utilize a modular, peptidomimetic-based design concept for the development of non-peptide SH2 domain inhibitors, rather than discussing any projects that remain in the realm of linear or cyclic oligopeptides. The corresponding section will be divided into two distinct parts: first pTyr mimics are introduced, followed by the discussion of promising oligopeptide mimetics. Instead of listing the distinct research programs pursued in different research institutes chronologically, a conceptional classification of the lead structures of different generations is preferred, thus accounting for the steady decrease in peptide character achieved throughout an iterative optimization process. This should allow the reader to appreciate fully the distinct steps that finally led to a significant structural departure from the original peptide lead sequence. Finally, an opportunity for extrapolating the elaborated peptidomimetic building blocks as well as the modular chemistry concepts from the SH2 domain target family onto a second class of adaptor proteins, notably the PTB domains, will be highlighted.

Even though adaptor proteins do not function as classical receptor proteins or catalytically active enzymes, the terms "antagonist" and "inhibitor" will be used interchangeably throughout the text for reasons of convenience.

2
SH2 Domains

2.1
Adaptor Proteins in Signal Transduction

A signal is defined as "any event or action that causes some general activity" [16]. According to that definition, the function and activity of cells is primarily controlled by external signals that modulate intracellular events upon binding to transmembrane cell surface receptors. The subsequent activation of the cell surface receptor by extracellular factors is transduced over the cellular membrane into the cytoplasm, causing the intracellular receptor domains to establish contact with "downstream" signal transducing proteins. This initial intracellular protein-protein interaction event then stimulates a cascade of non-catalytic as well as catalytically modifying protein-protein contacts, by which the biological signal is disseminated throughout the cell along multiple intracellular pathways. Most generally, the term signal transduction refers to a controlled flow of biological information by a sequential or parallel engagement of molecular regulators with the aim to elicit a cellular response [17].

The biochemical paradigm of signal transduction pathways is the ordered alteration of the phosphorylation state of surface-exposed tyrosine (Tyr), threonine (Thr), or serine (Ser) residues on target proteins. Thus, the molecular recognition of phosphorylated proteins, together with enzymatic phosphorylation and dephosphorylation is the primary mechanism of cellular signal transduction. Apart from kinases and phosphatases, numerous cytoplasmic proteins involved in signal transduction pathways are not active in the classical fashion of enzymes. These non-catalytic adaptor proteins or protein domains confer a remarkable specificity to the distinct signal transduction pathways in which they recruit enzymes into signaling networks by establishing spatial proximity between the biochemical reaction partners [18]. This fundamental process of scaffolding and anchoring pursued by adaptor proteins governs where and when corresponding kinases, phosphatases, and other enzymes involved are activated. In general, the enzymatic activity of a catalytically active signaling protein is encoded on a distinct protein domain separate from their adaptor region. These binding regions are localized on modular protein domains demonstrating a core recognition ability which is conserved throughout the entire homology family a particular adaptor module belongs to. By the repeated use of adaptor proteins and the associated "lock-and-key" recognition events, a complex and diverse regulatory network of molecular interactions can be assembled [18]. In several cases multiple and different adaptor protein domains and enzymes are encoded on a single polypeptide chain, thus allowing a single multiple-domain protein to bind multiple targets. Within signaling cascades, this is used to couple activated receptors to several downstream targets and associated pathways or, alternatively, to increase affinity and specificity by simultaneously employing more than one adaptor for a single binding event [18].

Consequently, nature utilizes a combination of adaptor protein domains, to ensure specificity by forming signaling complexes. The adaptor protein-mediated assembly of signaling components into pathways is typified by the association of phosphorylated or autophosphorylated receptor tyrosine kinases (RTKs) with cytoplasmic multi-domain proteins. Currently, numerous adaptor protein modules are known to be involved in orchestrating the correct repertoires of enzymes into individual signal transduction pathways [8–10], the most prominent representations being the SH2 [19, 20], SH3 [21–25], PTB [26–28], PDZ [29–32], DEATH [33–36], WW [37–41], or 14-3-3 [42–44] domains. As of 18 March 2000 the SMART (simple molecular architecture research tool) database at the EMBL in Heidelberg [9, 10] annotated the amazing huge number of, e. g., 154 distinct SH2 domains in 143 human signaling proteins, 359 SH3 domains in 268 distinct human proteins, 55 PTB domains in 51 human proteins, 282 PDZ domains in 157 human proteins, 79 WW domains in 45 human proteins, and 10 14-3-3 domains in 10 human proteins. This impressive list of currently known signaling domains together with associated background information on functional and structural aspects can be accessed via the World Wide Web under http://smart.embl-heidelberg.de/smart/domain-table.cgi [9, 10].

2.2
Function of SH2 Domains

Cellular signal transduction is initiated by the extracellular binding of an agonist to its receptor which is followed by phosphorylation of regulatory tyrosines
on the cytoplasmic receptor domains. Phosphorylation is achieved either by
receptor tyrosine kinases (RTKs) which are integral part of the receptor protein
itself, or by non-receptor tyrosine kinases that are recruited to the receptor
protein from the cytoplasm. This cytoplasmic receptor autophosphorylation
or phosphorylation acts as a switch to induce physical association between
the activated receptor and the adaptor protein modules. Most generally, an
adaptor is defined as a "device that connects pieces of equipment that were not
originally designed to be connected" [16]. SH2 domains act as an adaptor
by connecting activated receptors with downstream signaling proteins such
as kinases, phosphatases, phospholipases, or GTPases, respectively (Fig. 1)
[45–47]. The SH2 domain was first identified on the oncoproteins Src [48] and
Fps [19]. Functionally, SH2 domain-containing signaling proteins can be classified in two groups, multi-domain enzymes and catalytically inactive adaptors.
The family of enzymatically active SH2-containing proteins comprise Src, Lyn,
Hck, Syk, ZAP70, Syp, Btk, Tec, Abl, Itk, Csk, PLCγ, and GAP. Among the en-

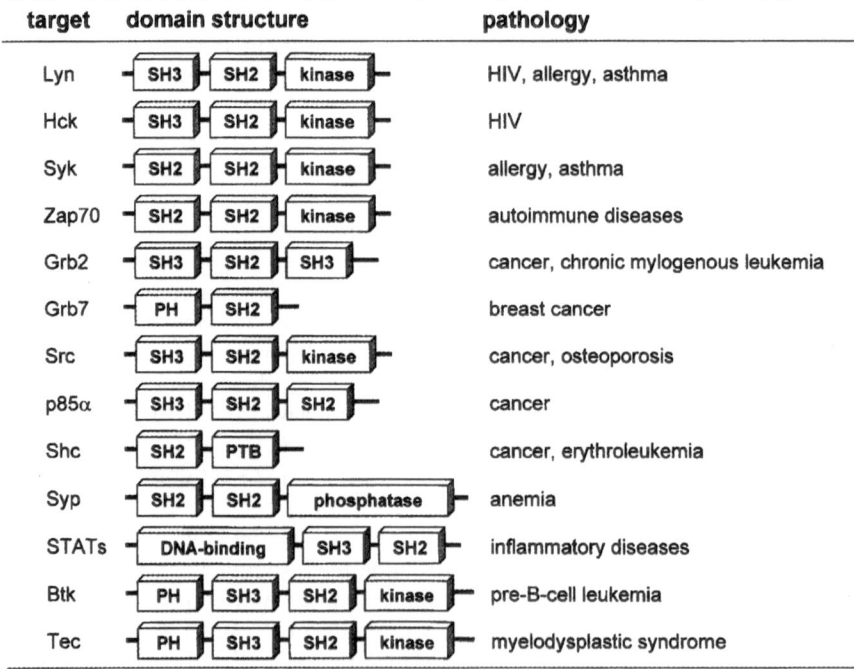

target	domain structure	pathology
Lyn	SH3 – SH2 – kinase	HIV, allergy, asthma
Hck	SH3 – SH2 – kinase	HIV
Syk	SH2 – SH2 – kinase	allergy, asthma
Zap70	SH2 – SH2 – kinase	autoimmune diseases
Grb2	SH3 – SH2 – SH3	cancer, chronic mylogenous leukemia
Grb7	PH – SH2	breast cancer
Src	SH3 – SH2 – kinase	cancer, osteoporosis
p85α	SH3 – SH2 – SH2	cancer
Shc	SH2 – PTB	cancer, erythroleukemia
Syp	SH2 – SH2 – phosphatase	anemia
STATs	DNA-binding – SH3 – SH2	inflammatory diseases
Btk	PH – SH3 – SH2 – kinase	pre-B-cell leukemia
Tec	PH – SH3 – SH2 – kinase	myelodysplastic syndrome

Fig. 1. Schematic representation of the modular architecture of selected pharmaceutically
relevant SH2 domain-containing targets involved in signal transduction. For further details
refer to http://smart.embl-heidelberg.de [9, 10]

zymatically inactive adaptor proteins are Grb2, Grb7, p85α, Shc, Nck, or Tensin (Fig. 1).

SH2 domains specifically recognize phosphorylated tyrosine (pTyr) residues on cytoplasmic receptor domains or soluble signaling components [49–51]. The pTyr-flanking amino acid residues constitute the SH2 domain docking site and determine which SH2-containing signaling proteins associate with the phosphorylated protein surface [52–56]. Multiple pTyrs on, e.g., a single cytoplasmic receptor domain can recruit numerous different SH2 domain proteins as exemplified for the PDGF (platelet-derived growth factor) receptor (Fig. 2) that attracts 8 different SH2 domain proteins in 12 distinct binding events.

Upon recognizing a pTyr-presenting ligand, SH2 domain proteins can be phosphorylated, thus recruiting other signal transducers to the activated signaling complex, or altering the intrinsic enzymatic activity of an SH2-containing enzyme. Apart from the pure adaptor function and the enzyme activation, SH2 domain proteins have been shown to translocate to the cell nucleus where they

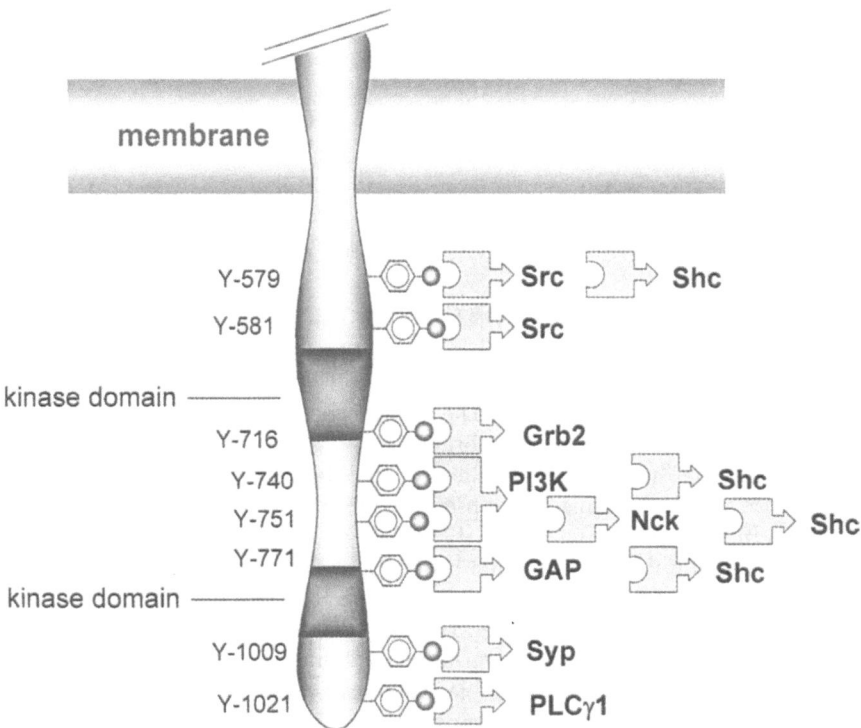

Fig. 2. Presentation of the intracellular domain of the PDGF (platelet-derived growth factor) receptor. The two catalytically active kinase domains are highlighted. Docking sites for SH2 domains are symbolized with a pTyr sidechain. SH2 domains identified to bind to the corresponding phosphorylated tyrosines (numbered on the *left*) are schematically depicted and labeled on the *right*

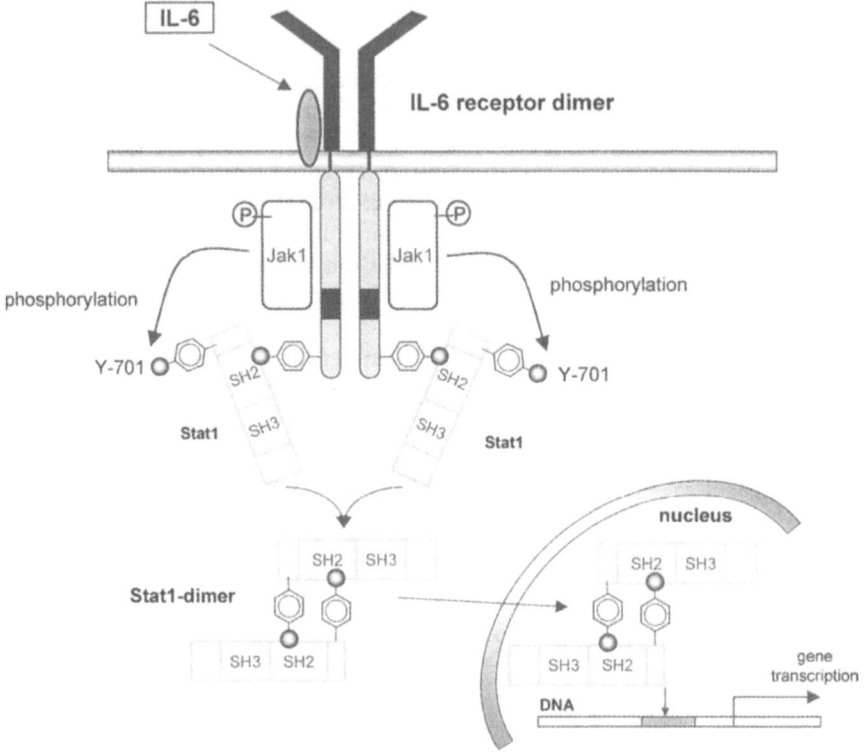

Fig. 3. IL-6 receptor-mediated Jak-Stat pathway [57–59]

regulate gene transcription upon DNA binding. The Jak-Stat pathway (Jak: Janus kinase; Stat: signal transducers and activators of transcription) (Fig. 3) is an illustrative example for a straightforward signal transduction cascade involving only a few distinct steps [57–59]. The non-receptor tyrosine kinase Jak associates with an activated transmembrane receptor, e.g., the IL-6 receptor, and subsequently phosphorylates members of the Stat protein family. Due to the occurrence of SH2 domains in Stat proteins, they associate to dimers, stabilized by the reciprocal interaction of the SH2 domain of one Stat protein with the pTyr ligand epitope on the opposing Stat protein and vice versa. The dimers translocate from the cytoplasm to the nucleus where they specifically bind to DNA and activate gene transcription (Fig. 3) [57–59].

The potential binding sites for various SH2 domain proteins on cytoplasmic receptor domains as well as on soluble proteins have been precisely mapped for several pTyr-containing target proteins. Apart from the impressive diversity of SH2 domain function, the fundamental property of all SH2 domains refers to the specific recognition of pTyr epitopes.

2.3
Structure of SH2 Domains

SH2 domains generally consist of approximately 100 amino acid residues [49, 50, 60, 61] and have been first identified as a conserved sequence region between the oncoproteins Src and Fps [19, 48]. By means of sequence homology, SH2 domains have been uncovered in numerous other intracellular signal transduction proteins (Figs. 1 and 4) [20].

Due to the ready accessibility of SH2 domains by molecular biology techniques, numerous experimentally determined 3D structures of SH2 domains derived by X-ray crystallography as well as heteronuclear multidimensional NMR spectroscopy are known today. The current version of the protein structure database, accessible to the scientific community by, e.g., the Internet (http://www.rcsb.org/pdb/) contains around 80 entries of SH2 domain structures and complexes thereof. Today, the SH2 domain structures of Hck [62], Src [63–66], Abl [67], Grb2 [68–71], Syp [72], PLCγ [73], Fyn [74], SAP [75], Lck [76,77], the C- and N-terminal SH2 domain of p85α [78–80], and of the tandem SH2 domains Syk [81, 82], ZAP70 [83, 84], and SHP-2 [85] are determined. All SH2 domains display a conserved 3D structure as can be expected from multiple sequence alignments (Fig. 4). The common structural fold consists of a central three-stranded antiparallel β sheet that is occasionally extended by one to three additional short strands (Fig. 5). This central β sheet forms the spine of the domain which is flanked on both sides by regular α helices [49, 50, 60].

	helix 1	strand 1	strand 2
LCK	WFFknlSRkd AErqLLapgN	ThGsFLIRES	ESta.GsfsL SVRDFDqNqG
SRC	WyFGKItRrE sErlLLnaeN	prGtFLvRES	Ettk.GaycL SVsDFDNaKG
SYK	WFhGKISREE sEqivLigsk	TnGkFLIRar	dnn..GsyaL cll....heG
GRB2	WFFGKIpRak AEe.mLskqr	hDGaFLIRES	ESaP.GDfsL SVk.....fG
SYP	WFhpnItgvE AEn.LLltrg	vDGsFLaRpS	kSnP.GDftL SVR.....rn
P85α	WywGdISREE vnekL..rdt	aDGtFLvRda	stkmhGDytL tlRkggNNKl
Consensus	WFFGKISREE AE--LL---N	TDG-FLIRES	ES-P-GD--L SVRDFDNNKG

	strand 3	helix 2
LCK	evVKHYKIRn LDnGGfYISp	ritFpgLhEL VrhYtNasDG LCtRLsrpC~
SRC	lnVKHYKIRk LDsGGfYIts	rtqFNSLqqL VayYSkHaDG LChRLTtVC~
SYK	k.VlHYrIdk dktGklsIpe	gkKFdtLwqL VehYSykaDG LlrvLTvpC~
GRB2	ndVqHfKvlr .DGaGkYflw	vvKFNSLnEL VdyHrstsvs RnqqifLrd~
SYP	gaVtHiKIqn .tGdyydlyg	geKFatLaEL VqyYmeHhgq LkekngDVi~
P85α	ikifH.... .rdGkygfSd	pltFsSvvEL inhYrNesla qynpklDVk~
Consensus	--VKHYKIR- LDGGG-YIS-	--KFNSL-EL V--YSNH-DG LC-RLTDVC-

Fig. 4. Multiple sequence alignment of six distinct SH2 domain sequences. The SH2 domain notifiers are given on the *left*, the *bottom line* represents the consensus sequence as analyzed from the alignment. Highly conserved sequence positions are *marked with capital letters*. The sequential position of the major secondary structure elements (helices and sheets) are *indicated by boxes*

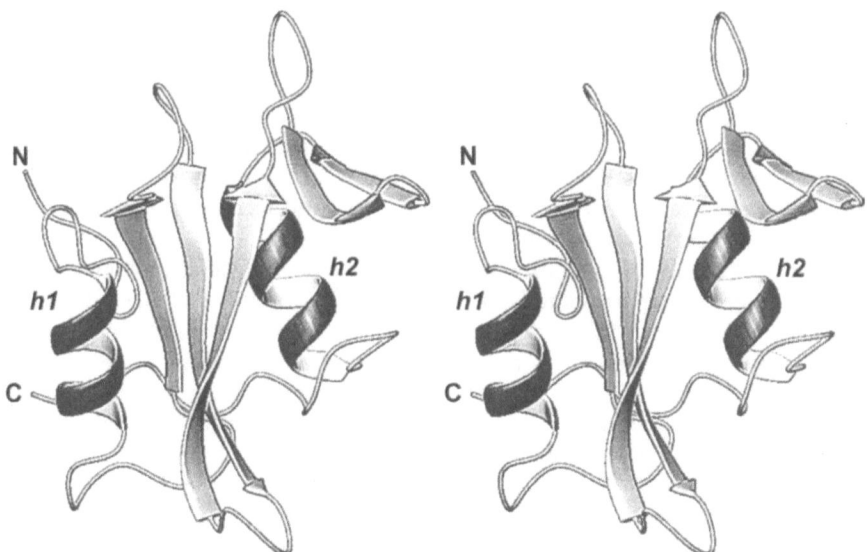

Fig. 5. Side-by-side stereo presentation of a ribbon diagram of the Lck SH2 domain, Protein Databank entry code: 1BHF.PDB [118]

In most SH2 domains the peptide ligands bind in an extended backbone conformation towards the edge strand (strand 3) of the central β sheet in an orthogonal orientation. The protein part accommodating the N-terminal α helix is primarily concerned in providing the binding determinant for the pTyr residue within peptide or protein ligands, whereas the opposite compartment formed by the C-terminal α helix provides interaction partners for ligand residues downstream of pTyr. The protein architecture is tailor-made for an "end-on" recognition mode, since both termini of SH2 domains are located opposite to the ligand binding site, thus not affecting ligand recognition upon covalent N-, C-, or N- and C-terminal incorporation of the SH2 domain into a multi-domain protein context [49, 50, 60].

2.4
Mode of Ligand Binding

The majority of pTyr-containing peptide or protein ligands bind in an extended conformation [49, 50, 60]. However, significant deviations are observed for, e.g., Grb2-peptide complexes, in which the ligand is bound in a tight reverse turn conformation of type βI [68–70]. In all SH2 domains, the pTyr sidechain projects into a defined binding pocket, aligned with residues capable of forming an extended network of hydrogen bonds with the phosphate groups (Fig. 6). Highly conserved Arg residues coordinate the phosphate by bidentate hydrogen bonds reinforced by additional charge-charge interactions.

Apart from the hydrogen bond potential at the bottom of the pTyr pocket, the phenol ring of pTyr makes a number of hydrophobic contacts with the SH2

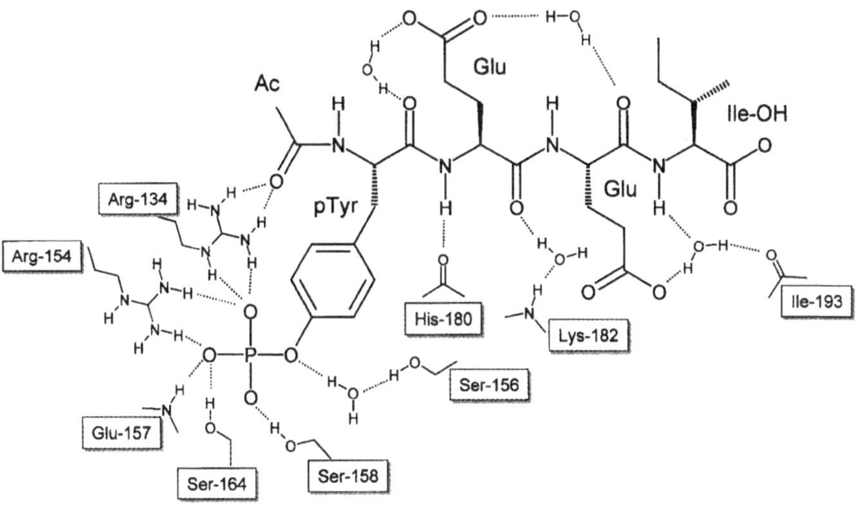

Fig. 6. Schematic diagram of the peptide-protein interaction mode as seen in the crystallo-graphically refined structured of the Lck SH2 domain-peptide complex, Protein Databank entry code: 1LKK.PDB. The residues directly engaged in intramolecular hydrogen bonds (*dotted lines*) are labeled explicitly

domain. A second pronounced pocket, primarily lined with hydrophobic resi-dues, binds the amino acid sidechain three positions towards the C-terminus relative to the pTyr residue (pTyr+3). Due to the observation that pTyr peptides bind to most of the SH2 domains by addressing these two binding pockets (pTyr-pTyr+3), the binding mode has been described as a "two-pronged plug engaging a two-holed socket" (Fig. 7) [64].

In a variety of SH2 domains the "two-holed socket" template for binding the "two-pronged plug" is less pronounced than in others, and in some cases ligand residues further downstream than pTyr+3 are bound in hydrophobic pockets. The pTyr-pTyr+3 connecting residues are generally surface exposed with sol-vent-accessible sidechain functionalities and, apart from a hydrogen bond involving the pTyr+1 backbone N-H, no specific direct intermolecular contacts are conserved [49, 50, 60, 64].

This specific interaction mechanism whereby the binding is hypothesized to resemble a "two-pronged plug" (peptide, protein ligand) inserting into a "two-holed socket" (SH2 domain) predicts predominantly a hydrophobic interaction for high-affinity binding. This is mainly determined by the level of insertion of the pTyr+3 residue into a hydrophobic pocket (Figs. 7 and 8).

However, thermodynamic investigations of this binding event employing, e.g., isothermal titration calorimetry revealed the "two-pronged plug – two-holed socket" model to be an oversimplification of the details underlying the true binding mechanism [86–97]. No significant differences within the spe-cific thermodynamic signature were detected for binding of different peptides with altered sidechains of variable size and hydrophobicity at the pTyr+3 position.

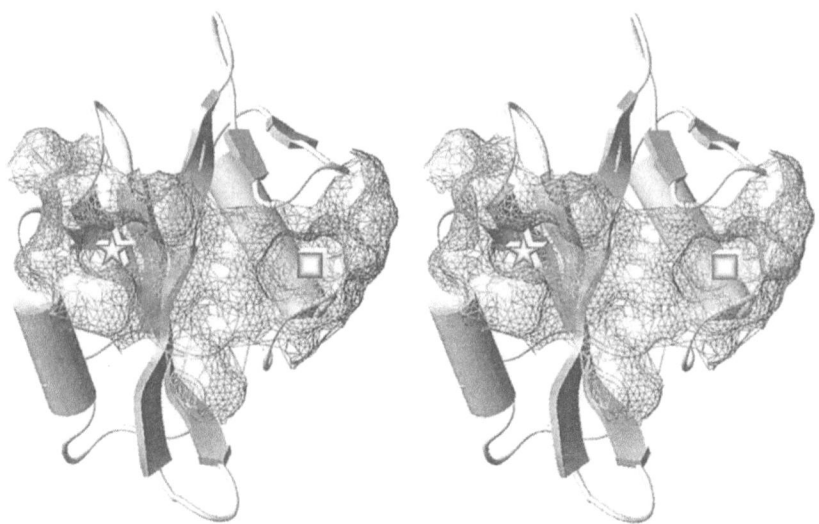

Fig. 7. Side-by-side stereo presentation of the Lck SH2 domain (1BHF.PDB). The protein is shown as a ribbon diagram, while the ligand binding site is highlighted by a Connolly surface in line-mode. The two interaction-mediating ligand sidechains bind into the "two-holed socket", the binding pockets of which are marked by an asterisk (*left*, pTyr-binding pocket), and a cube (*right*, pTyr+3 sidechain)

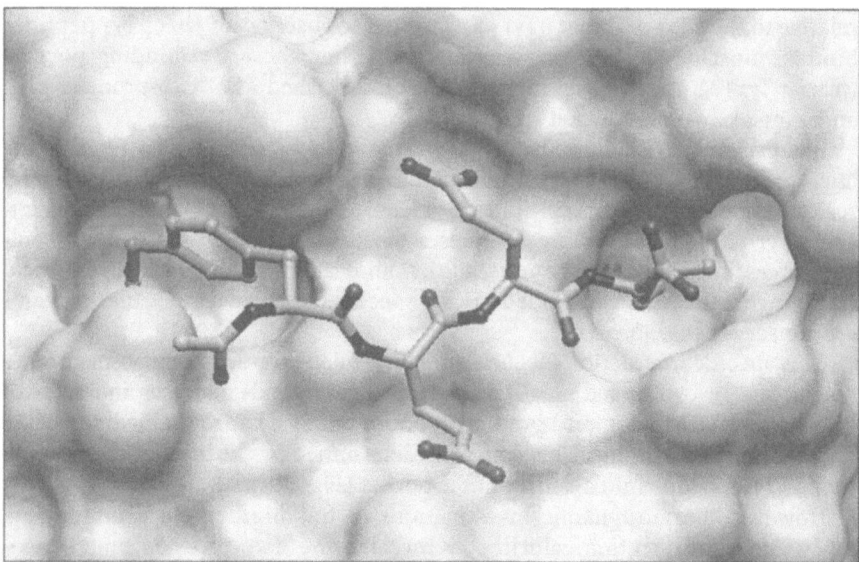

Fig. 8. Lck SH2 domain-peptide complex (Ac-cmF-Glu-Glu-Ile-OH, **12**) revealing the "two-pronged plug engaging a two-holed socket" binding mode, reminiscent of the majority of SH2 domains (Protein Databank entry code: 1BHF.PDB [118]). The protein is depicted in a Connolly surface mode, the ligand is given in a ball-and-stick representation. The cmF residue is deeply buried in its binding pocket (*left*)

In these experiments, only moderate heat capacity changes upon binding of high- and low-affinity pTyr peptides to SH2 domains were observed, suggesting that high-affinity binding may not be determined solely by a hydrophobic effect centered at the pTyr+3 position. In the case of the Src SH2 domain-pTyr-Glu-Glu-Ile peptide interaction, it could be calorimetrically shown that the pTyr+2 position, in this case Glu, was nearly as critical for high-affinity binding as the pTyr+3 located Ile [94]. In the light of numerous high-resolution structures, this finding is rationalized with a water-mediated hydrogen bonding network emanating from the pTyr+2 Glu sidechain to polar and charged surface-exposed residues of the Src SH2 domain [95].

Following a combined spectroscopic and calorimetric study on the binding mechanism of the tandem SH2 domain of Syk and ITAM (a nanomolar-binding dually phosphorylated peptidic immune receptor tyrosine activation motif), an even more complicated binding mechanism characterized by a ligation-coupled conformational change of the tandem SH2 domain was observed [97]. The studies further revealed the single SH2 domains to be rather rigid locks or sockets when binding their keys or plugs, respectively. No remarkable induced-fit effects were observed upon ligand binding [97].

2.5
Specificity of SH2 Domains

The specificity profile of individual SH2 domains was determined in a series of studies employing degenerate phosphopeptide library screens [53, 98] from which two consensus motifs emerged for SH2 domain binding [52–55, 99, 100]. One group of SH2 domains preferentially binds to a pTyr-Glu-Glu-Ile motif that defines a generic recognition sequence:

pTyr-Xaa-Xaa-Yaa,

Xaa being hydrophilic and/or negatively charged, and Yaa being bulky and hydrophobic.

The second group of SH2 domain proteins discriminates less and selects mainly hydrophobic residues with a consensus sequence:

pTyr-Yaa-Any-Yaa,

Yaa being hydrophobic.

Based on these sequence-affinity relationships it was assumed that ligand selection has to be performed by SH2 domain residues strategically aligning the binding pocket which specifically interact with pTyr-exposing tetrapeptide epitopes on the protein binding partner. Consequently, the paradigm of most SH2 domain-targeted inhibitor design attempts relies on the assumption that utilizing non-natural amino acids and peptidomimetic building blocks will permit the synthesis of de-peptidized SH2 domain antagonists with improved affinity, specificity, stability, and pharmacokinetic properties.

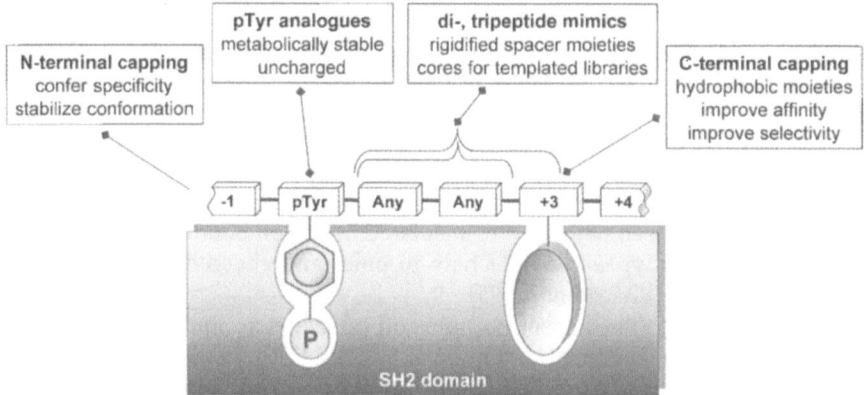

two-pronged plug engaging a two-holed socket

Fig. 9. Peptidomimetic design principle of SH2 antagonists derived from a fragmentation of the native peptide ligand. The lead finding efforts discussed in Sect. 3 are classified according to the modules given above the peptide lead sequence

3
Peptidomimetic SH2 Domain Antagonists

Numerous efforts towards the design and synthesis of promising lead structures specifically targeting the ligand binding site of distinct SH2 domains have been described mainly within the last five years, highlighting the novelty of this research area within current medicinal chemistry (for reviews see [7, 11, 101]). The chemistry programs pursued in different research groups in industry and academia were immensely stimulated, if not initiated, by the experimentally determined 3D structures of SH2 domains and their complexes with peptide ligands. Since the targeted protein-protein interaction event involves a sequentially continuous binding epitope, comprising the pTyr 1 on the SH2 domain ligand side, a modular peptidomimetic-based strategy was chosen by most research groups (Fig. 9).

According to the peculiarities of the ligand binding mode and the specificity profiles described in Sects. 2.4 and 2.5, respectively, the lead design problem can be separated into individual subtopics as shown in Fig. 9.

Prior to the discussion of biologically active peptidomimetic and non-peptide SH2 domain antagonists (Sect. 3.2), the following section will introduce a variety of attempts towards the replacement of pTyr 1 (Scheme 1).

3.1
Phosphotyrosine Mimetics

The hydrolytic accessibility of the ester bond in pTyr 1 (Scheme 1) to cellular protein-tyrosine phosphatases motivated numerous research groups to develop phosphatase-resistant bioisosteric analogues. It was soon recognized that the

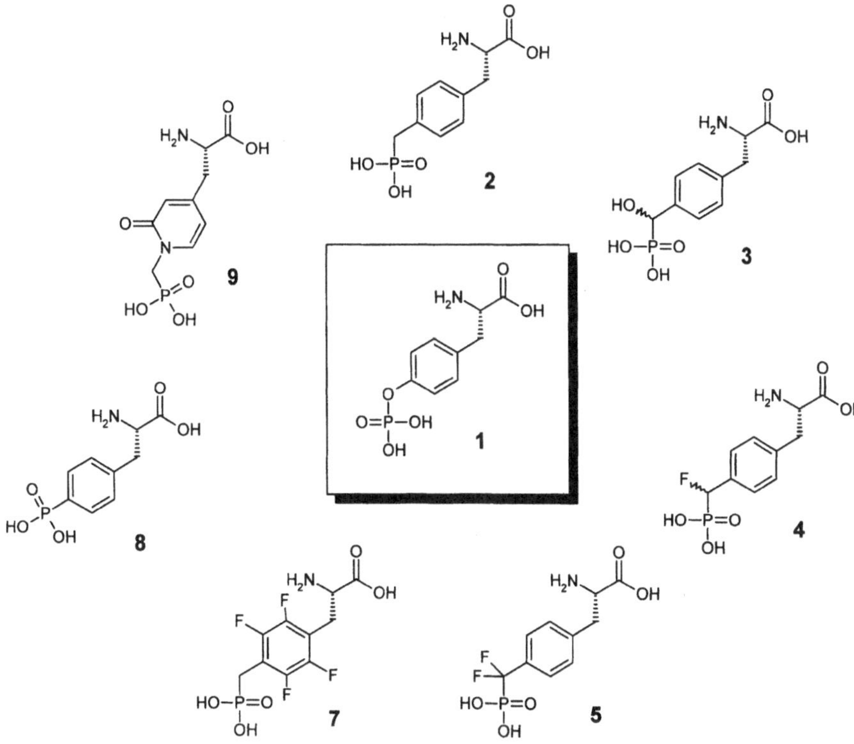

Scheme 1. Phosphorus-containing pTyr mimics

phosphate structure itself imparts critical functionality to the tyrosine. Therefore the search for phosphate mimics that can be introduced onto, e.g., phenylalanine platforms has finally attracted the interest of classical peptidomimetic research. In the following five sections (Sects. 3.1.1–3.1.5) representative examples will be given that are classified into phosphorus-containing and non-phosphorus-based pTyr analogues, the latter being roughly grouped into different classes according to the "spatial density" of oxygen atoms.

3.1.1
Phosphonates

The phosphonate analogue Pmp **2** [4-(phosphonomethyl)phenylalanine] (Scheme 1) can be considered as one of the closest pTyr mimics, since the only structural modification is the replacement of the phosphate ester oxygen by a methylene group [102–107]. Pmp-containing peptides generally display decreased binding affinities for a specific target SH2 domain when compared to the pTyr-containing parent peptide. This was attributed to an increased pK_{A2} of the phosphonate **2** (7.1) relative to the pTyr **1** (5.7) and to the loss of hydrogen bonding capabilities [63, 108, 109]. A decrease of the corresponding pK_A was achieved by introducing electron-withdrawing groups onto the methylene bridge of Pmp 2. In a comparative study using the C-terminal SH2 domain of p85α and a standard peptide ligand, different modified Pmp analogues 2–5 (Scheme 1) were investigated [110]. It turned out that only with F$_2$Pmp 5 [4-(phosphonodifluoromethyl)phenylalanine] affinities comparable to that of the pTyr peptides could be achieved, while FPmp 4 [4(phosphonofluoromethyl)-phenylalanine], Pmp 2, and HPmp 3 [4-(phosphonohydroxymethyl)phenyl-alanine] produced peptides with decreased affinities. The determined binding potencies increased in the order: HPmp 3 < Pmp 2 < FPmp 4 < F$_2$Pmp 5 ≈ pTyr 1 [110].

Of these analogues, mainly F$_2$Pmp 5 was repeatedly used for designing peptidic antagonists for Src [54, 110–112], Grb2 [110], p85α [113–115], and Abl [112] SH2 domains. F$_2$Pmp 5 is generally considered as an acceptable substitution for pTyr 1, maintaining remarkable affinity while imparting suffi-cient stability to intracellular phosphatases. F$_2$Pmp-containing peptidomimetic p85α SH2 domain antagonists were shown to have specific cellular effects upon microinjection, but proved to be cytotoxic in whole-cell assays [115]. In a related study, a fully protected diester of a F$_2$Pmp-containing dipeptide deri-vative 6 was used in a prodrug concept to block the Src and Abl SH2 domains [112].

6

The compound was readily taken up in Balbc3T3 cells and converted into the deprotected parent compound with acceptable reconversion rates.

In an alternative effort to improve the phosphonate ionization capability and hydrogen binding capacity, a Pmp analogue F$_4$Pmp 7 [2,3,5,6-tetrafluoro-4-(phosphonomethyl)phenylalanine] (Scheme 1) was designed and synthesized with a tetrafluoro phenyl ring [116]. Unexpectedly, the ionization properties of F$_4$Pmp 7 were even closer to Pmp 2 (pK$_{A2}$ (F$_4$Pmp 7): 6.9 relative to pK$_{A2}$ (Pmp 2): 7.1) than to pTyr 1 (pK$_{A2}$: 5.7). Consequently, F$_4$Pmp 7 incorporated into a standard peptide ligand for the Grb2 SH2 domain yielded even poorer affinity when compared to the non-fluorinated analogue Pmp 2, suggesting that the tetrafluorinated aromatic ring might sterically interfere with the pTyr-binding pocket, thus contributing to the reduced affinity [116].

Deletion of the phosphate ester oxygen of 1 results in 4-phosphono-phenylalanine, Phe(PO$_3$H$_2$) 8 (Scheme 1) which was incorporated in Abl and Src

SH2 domain-targeted peptidomimetic compounds [112] and displayed affinities comparable to those of F_2Pmp-, or pTyr-containing analogues.

Within the phosphonomethyl-series of pTyr analogues a pyridone-based mimic 9 (Scheme 1) was developed [117]. The design rationale is based on the assumption that the pyridone core structure carrying a methylene phosphonate moiety as phosphate mimic would result in a pK_A profile close to that of pTyr 1. The tripeptide derivative 10 was shown to inhibit CD19 phosphopeptide binding to the p85α N-terminal SH2 domain with an IC_{50} of only 50 µmol/l, while the pTyr-containing parent peptide is reported to inhibit this interaction with an IC_{50} of 77 nmol/l, respectively.

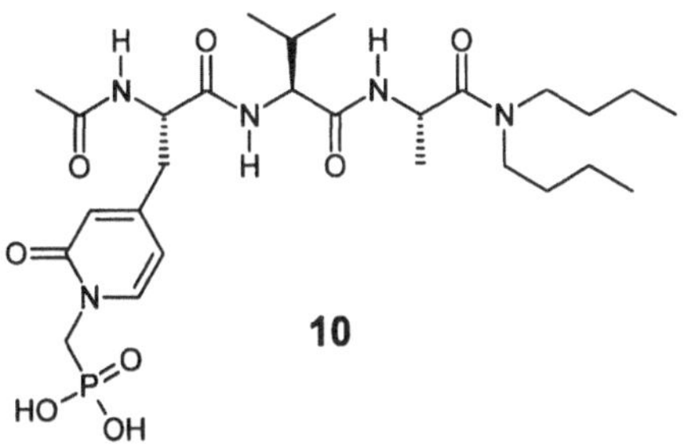

10

The dramatic loss in affinity when replacing pTyr 1 against the pyridone analogue 9 is attributed to an only moderate effect of the heterocycle on the pK_A profile of the attached methylene phosphonate, together with the disruption of π-stacking interactions normally seen between the pTyr aromatic ring and charged Arg sidechains aligning the binding pocket [117].

3.1.2
Monocarboxylic Acids

Non-phosphorus monocarboxylic acid-based pTyr mimics were developed in order to produce metabolically stable analogues that do not share the obvious disadvantage of two negative charges with the parent pTyr 1. The most simple analogue cmF 11 [4-(carboxymethyl)phenylalanine] (Scheme 2) incorporated in a Src SH2 domain-targeted peptide was shown to have only poor affinity [54]. Considerably higher affinity was retained in a recent study aimed to find leads against the Lck SH2 domain [118, 119], although when compared to the pTyr-

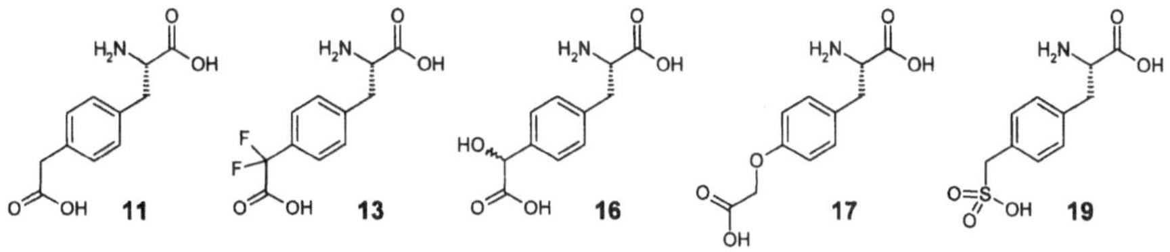

Scheme 2. Monocarboxylic acid-based pTyr mimics

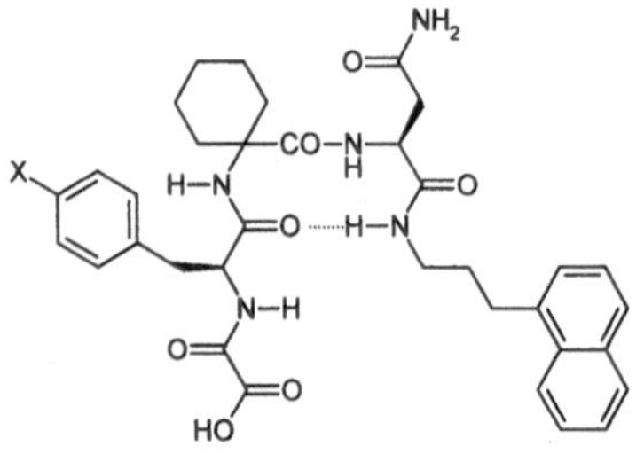

14: X = CH$_2$-CO$_2$H

15: X = CF$_2$-CO$_2$H

18: X = O-CH$_2$-CO$_2$H

Scheme 3. Grb2-targeted β turn peptide template used for evaluation of pTyr mimics 11, 13, and 17

containing parent peptides, the cmF-derived analogues were generally more than two orders of magnitude less active [118, 119]. Despite the low affinity of cmF peptides, researchers at Boehringer Ingelheim Pharmaceuticals succeeded in deriving a high-resolution structure of Ac-cmF-Glu-Glu-Ile-OH 12 docked into the binding canyon of the Lck SH2 domain (Fig. 8) [119].

Similar results were obtained for Grb2 SH2 domain-directed peptidomimetics containing cmF 11 as pTyr mimic [120, 121]. In order to enhance the pTyr-mimicking potential of cmF 11, a similar strategy as for the Pmp 2 → F$_2$Pmp 5 transition was chosen to generate F$_2$cmF 13 (Scheme 2). A Grb2 SH2 domain-targeted, β turn-stabilizing peptidomimetic developed at Novarits (see below) [68, 69, 122–124] has been used to assess the binding properties of cmF- and F2cmF-containing ligands (Scheme 3).

Contrary to initial expectations, introduction of fluorine into the bridging methylene group reduces affinity, as shown by 14 and 15 with IC$_{50}$ values of 0.6 µmol/l and 2.0 µmol/l, respectively (Scheme 3) [120, 121].

In the context of Lck SH2 domain ligands it was demonstrated that the introduction of an additional hydroxyl group onto the bridging methylene group, yielding the chmF 16 [4-(carboxyhydroxymethyl)phenylalanine] (Scheme 2), improved the potency threefold when compared to the corresponding cmF analogue [119].

A constitutional isomer of chmF 16 is cmT 17 [O-(carboxymethyl)tyrosine] (Scheme 2) which showed dramatically reduced affinity when incorporated in the Grb2-selective β turn peptidomimetic 18, causing an almost 30-fold drop in affinity when compared to the cmF-derived analogue 14 (Scheme 3). Obviously, the carboxylate group is displaced from the phenyl ring in 18 to such an extent that the corresponding registry with the native pTyr sidechain is disrupted. The application of cmT 17 was extended from SH2 domain antagonists to protein-tyrosine phosphatase inhibitors [125], thus demonstrating the versatility of the pTyr-pharmacophore for signal transduction therapy.

Apart from the monocarboxylic acid-based pTyr mimics, a sulfonic acid derivative 19 (Scheme 2) has been applied in Lck SH2 domain-targeting peptidomimetics, but displayed only moderate binding [119].

A non-classical, but still rational and structure-based, screening approach was applied by researchers at Abbott Laboratories, utilizing the SAR-by-NMR technique [126, 127] to identify biophysically pTyr mimics for the Lck SH2 domain [128]. The major advantage of this particular NMR-based detection principle is the sensitivity, since low-molecular weight compounds can be identified that bind in the micromolar to millimolar range. This allows one to screen for individual pTyr mimics prior to any synthetic efforts towards incorporation of these building blocks into any peptide or peptidomimetic scaffold. The binding affinity of Ac-pTyr-OEt **20** for the Lck SH2 domain was shown to be 0.3 mmol/l. Screening of over 3500 compounds revealed a number of molecules binding to the pTyr subsite on the SH2 domain.

Most interestingly, phthalamate analogues **21–23** were identified as potentially new pTyr mimics that may yield inhibitors equipotent to pTyr-containing compounds with the advantage of improved pharmacokinetic profiles.

3.1.3
Vicinal Dicarbonyls

The design of α-carbonyl-aldehydes, -ketones, or -carboxylic acids is based on the rationale that an increase in the "spatial concentration" of oxygen atoms resembles more closely the hydrogen bond acceptor properties of a phosphate group. Researchers at Glaxo described α-dicarbonyl moieties attached to the 4-position of the phenyl ring of Phe as non-charged phosphate mimics **24, 25** that turned out to be two to three orders of magnitude less active when compared to a corresponding pTyr analogue in the Src SH2 domain binding assay [129].

Oxamic acid **26** and *N*-hydroxyoxamic acid **27** derivatives were disclosed by the Boehringer Ingelheim group as moderately active Lck SH2 domain antagonists [119, 130].

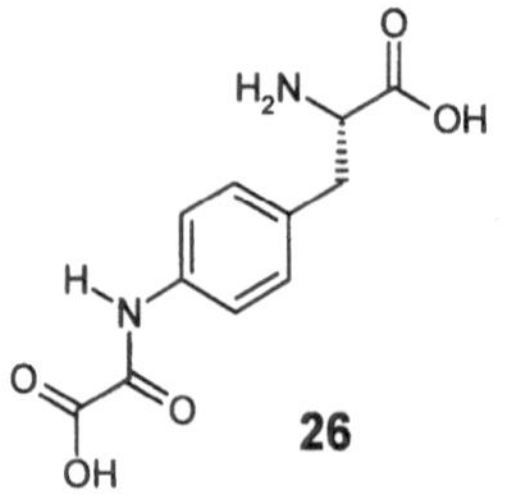

3.1.4
Dicarboxylic Acids

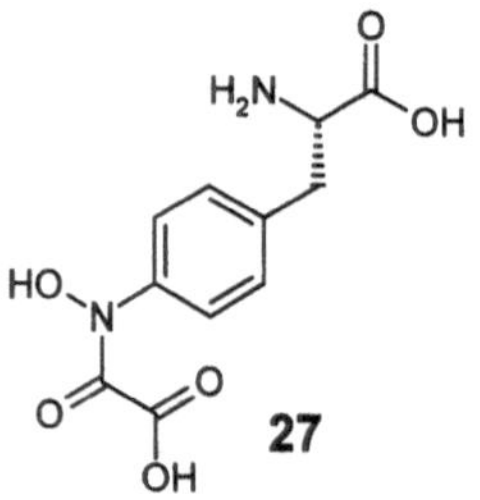

A further step towards increasing the spatial density of oxygen functionalities was made by introduction of various malonyl-derivatives such as OMT **28** [*O*-(2-malonyl)tyrosine], or FOMT **29** [*O*-(2-(2-fluoromalonly))tyrosine] (Scheme 4) [131–133].

Linear hexapeptide ligands of the C-terminal p85α SH2 domain containing both malonate derivatives displayed 100-fold decreased binding when compared to the pTyr peptide analogue [133]. Deletion of the phenolic oxygen atom in OMT **28** yields a new pTyr analogue, notably Pmf **30** [*p*-(2-malonyl)phenylalanine] (Scheme 4) [134]. Again, a β turn peptidomimetic inhibitor against the Grb2 SH2 domain developed by Novartis [135] was used to test the new Pmf building block. When changing from the OMT-containing peptidomimetic to the Pmf-containing derivative, a 15- to 20-fold increase in potency was observed. Finally, an N- and C-terminal capped tripeptide mimic was obtained with an inhibitory potential of $IC_{50} = 8$ nmol/l that showed an effective blockade of endogenous Grb2 binding to phosphorylated erbB-2 growth factor receptor and inhibition of downstream MAP kinase activation in a whole cell system [134].

A structurally rather dissimilar pTyr mimic was developed by the Boehringer Ingelheim group, namely a triazole-dicarboxylic acid derivative **31** [130], which was shown to yield highly active Lck SH2 domain antagonists.

3.1.5
Miscellaneous Derivatives

Researchers at ARIAD Pharmaceuticals described the incorporation of a 3,5-dinitrotyrosine **32** into peptide inhibitors of only double-digit micromolar affinity to the SH2 domains of Syk and Src [136].

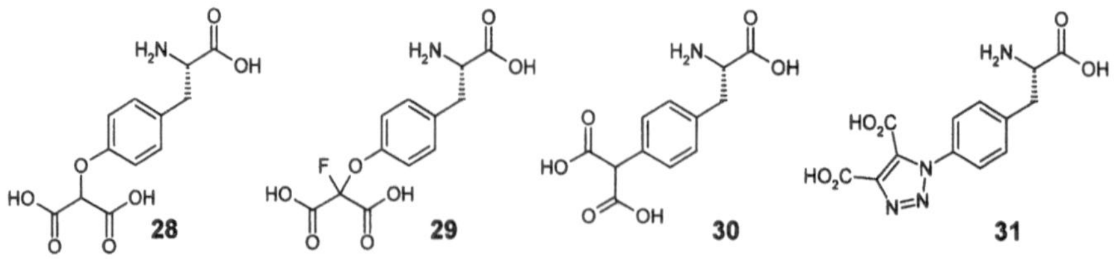

Scheme 4. Dicarboxylic acid-type pTyr mimics

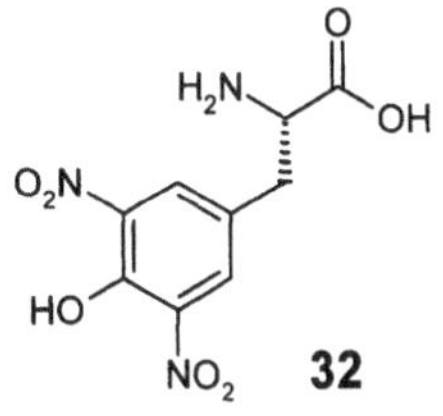

Due to the fact that the phosphate was replaced against entirely non-charged and phosphatase-stable analogues, the pTyr mimic is considered to have more favorable pharmacokinetic properties. However, aromatic nitro compounds bear an intrinsic potential of developing toxicological problems [136].

For designing covalent binding ligands, Glaxo employed the finding that within several X-ray and NMR structures of Src SH2 domain-peptide complexes the sidechain thiol group of Cys-188 is located in close proximity to the *meta*-position of the pTyr aromatic ring [137]. Since the proximity of that Cys residue to the ligand is unique for the Src SH2 domain, a thiol-capture strategy was chosen as a means to achieving selectivity for Src over other SH2 domains. Based upon the well-precedented inhibition principles of cysteine proteases [138], the aldehyde group was proposed to be a capturing agent by forming a thiohemi-acetal. Upon synthesis of the corresponding dipeptide derivative **33** (Scheme 5), its complex with the Src SH2 domain was investigated crystallographically [137].

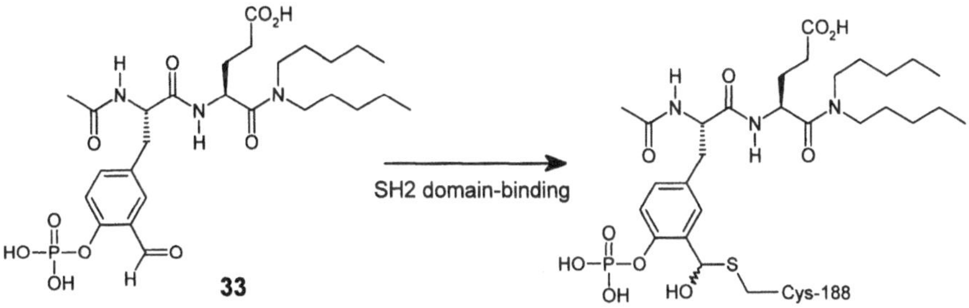

Scheme 5. Thiol-capture strategy for covalent target binding

As anticipated, **33** binds in the desired manner, capturing the thiol by the aldehyde group (Scheme 5). Unexpectedly, **33** binds with only a twofold improved affinity when compared to the parent dipeptide derivative containing the unmodified pTyr residue [137].

In conclusion, a variety of mutually different pTyr mimetics incorporated in peptide-type derivatives have been investigated. However, an overall satisfactory bioisosteric replacement has not yet been identified.

3.2
Peptidomimetic Scaffolding

After the discussion of several approaches aimed at substituting the metabolically unstable pTyr **1**, the following sections will introduce lead finding and optimization strategies with the ultimate goal of developing low-molecular weight, drug-like compounds. Due to the shortcomings related to solubility, stability, and bioavailability that generally arise with peptide-type analogues, most of the SH2 domain-targeted medicinal chemistry programs currently pursued apply a structure-based methodology towards an iterative design of small molecules with significantly reduced peptidic natures, rather than optimizing pTyr-containing oligopeptide sequences. In the following, the SH2 domain antagonist research programs will be presented according to the fragmentation of a peptide ligand depicted in Fig. 9. Rather than chronologically discussing the progress made by a certain research group for a special target, the compound classes will be introduced according to an increasing extent of de-peptidization.

3.2.1
Peptoids

The design of peptide sequence-derived peptoids is generally considered as a rather tiny step on the route from a peptide lead towards a non-peptide development candidate, since the polyamide backbone is retained without any significant conformational restriction [139, 140]. The Novartis group has synthesized a biased tetrapeptoid library for the discovery of monodentate ITAM mimics as potential inhibitors for the ZAP70 tandem SH2 domain [141]. The resulting tetrapeptoids were expected to show considerable affinity in vivo, due to the improved absorption characteristics of the peptoid skeleton [110]. From a 27 compound library with a constant 4-(phosphonomethyl)benzyl sidechain (Pmp (2) analogue), only a single ZAP70 antagonist 34 was identified that exhibited moderate binding (IC_{50}: 25 µmol/l).

34

For comparison, the ITAM-derived 19-mer bidentate peptide inhibits with an IC_{50} of 30 nmol/l [141]. Concluding, the chosen peptoid approach did not yield in the desired breakthrough for the design of useful SH2 domain inhibitors. Also in terms of stepwise reduction of the peptide character, no information for the subsequent design of next-generation analogues could be derived [141].

3.2.2
Non-Natural Amino Acids

The first step in a strict hierarchical procedure for transforming peptide lead sequences into promising non-peptide analogues generally utilizes the incorpo-

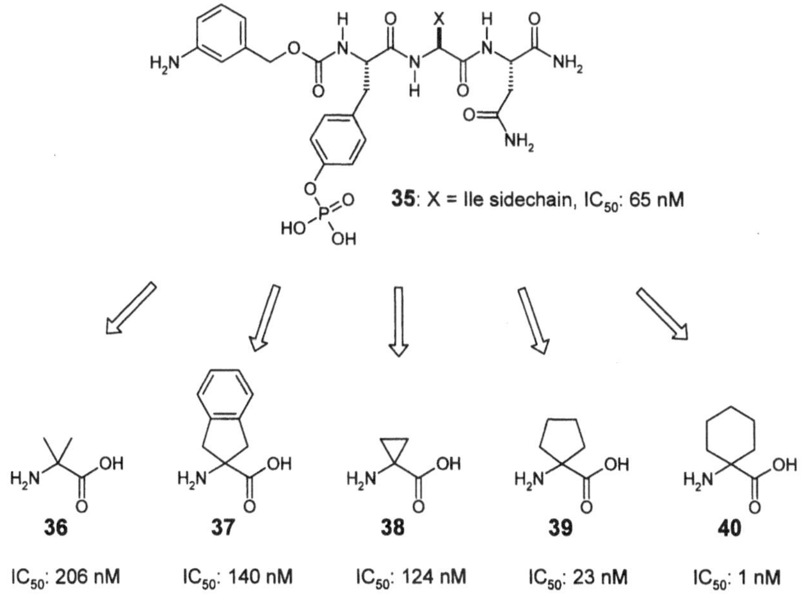

35: X = Ile sidechain, IC_{50}: 65 nM

36	**37**	**38**	**39**	**40**
IC_{50}: 206 nM	IC_{50}: 140 nM	IC_{50}: 124 nM	IC_{50}: 23 nM	IC_{50}: 1 nM

Scheme 6. Effects of α,α-disubstituted amino acids on Grb2 binding affinity

ration of non-natural amino acids [142]. Based on the submicromolar Grb2 SH2 domain-binding tripeptide derivative **35** (Scheme 6), the Novartis group systematically explored the pTyr+1 position following a conformational rationale [123, 143].

As shown in Fig. 10, the Grb2 SH2 domain is unique in that the peptide ligand is bound in a type I β turn conformation, reinforced by an intramolecular hydrogen bond.

To strengthen the turn-inducing potential of next-generation Grb2 SH2 domain ligands, α, α-dialkylated residues were probed in the i + 1 position of the reverse turn, since they are well-known to adopt kinked conformations as found in, e. g., 3_{10} helices [144–146]. From the X-ray structure of the Grb2 SH2 domain in complex with a peptide ligand (Fig. 10) [68] it was further shown that a hydrophobic moiety in pTyr+1 position is accommodated by a complementary binding pocket. Consequently, a series of hydrophobic α,α-dialkylated amino acids were introduced, resulting in a clear structure-activity relationship ranging from the least active Aib **36** to the most active Ac_6c **40** (Scheme 6). The 1-amino-cyclohexanecarboxylic acid Ac_6c **40** produced a compound with a 65-fold increased affinity when compared to the Ile-analogue **35**. The design rationales and underlying structural hypothesis that finally led to the proposed non-natural amino acids **36–40** (Scheme 6) have been totally confirmed retrospectively by X-ray crystallography [147]. The complex of a peptide derivative containing **40** in the pTyr+1 position reveals two functions for Ac_6c **40**, notably the stabilization of the βI turn conformation and the hydrophobic interaction with a binding pocket adjacent to the pTyr pocket [147]. These findings allowed

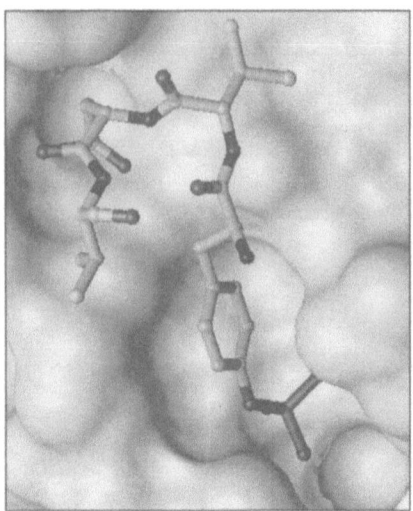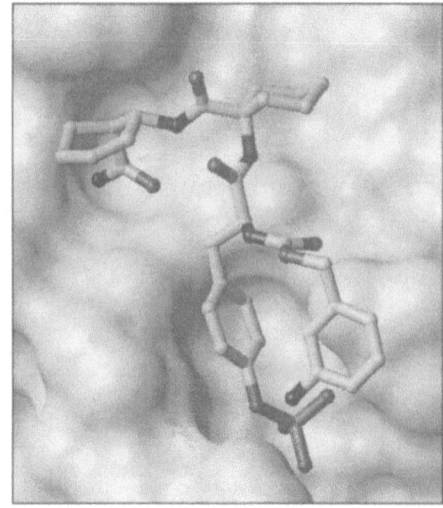

Fig. 10. Experimentally determined complexes of the Grb2 SH2 domain with a pTyr-containing oligopeptide (*left*, 1BMB.pdb [71]) and inhibitor **52** (*right*, 1CJ1.pdb [147]). Even though **52** lacks the pTyr+3 residue for forming the βI turn conformation, the residues pTyr, Ac₆c and Achec bind in the proposed mode to Grb2

the Novartis group to proceed further towards high-affinity Grb2 SH2 domain antagonists that were derived from that intermediate analogue and displayed even in vivo activity (see below).

3.2.3
Capping Groups as Amino Acid Analogues and Dipeptide Mimetics

A convenient strategy to shorten peptide lead sequences is the utilization of amino acid sidechain-derived capping groups. By attachment of simple amines onto the C-terminus, and carboxylic acids onto the N-terminus, the charged terminal functionalities can be omitted. Additionally, introduction of tailor-made substituents into capping groups allows one to explore further binding pockets on the target protein surface, thus allowing one to increase affinity and specificity of resulting analogues.

After successful replacement of a natural amino acid against Ac₆c **40** (Scheme 6) in the Grb2 SH2 domain-targeted lead, the Novartis group decided to probe further the Asn-following position by attaching different amines of the indolylpropylamine-type onto the Asn C-terminus as shown in Scheme 7 [135].

In the context of Pmp-containing tripeptide derivatives, the 3-indol-1-yl-propyl analogue turned out to exhibit high affinity. Substituting the indol 5-position with methyl or hydroxyl resulted in compounds **41** and **42** with subnanomolar IC₅₀ values (Scheme 7). These have been the first compounds of SH2 domain antagonists with subnanomolar affinity reported. These systematic studies enabled Novartis to push further analogues into in vivo studies [148,

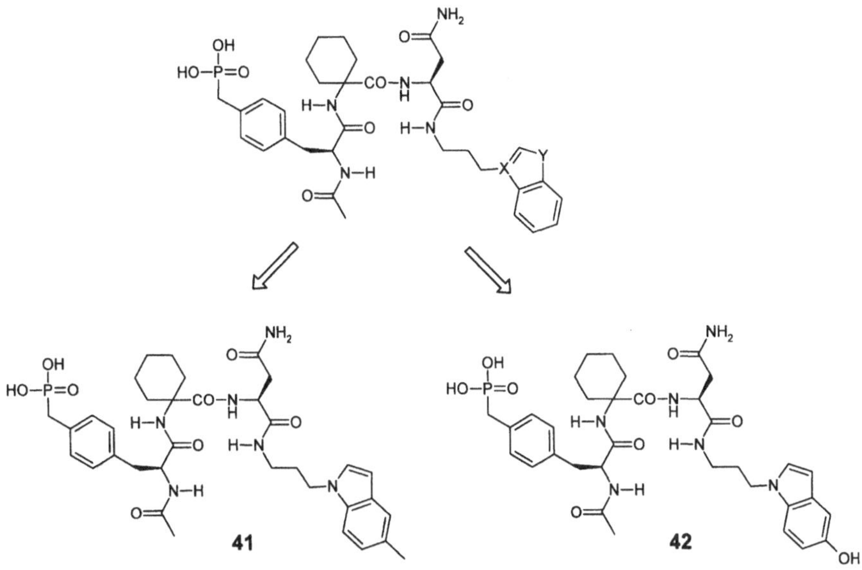

Scheme 7. Exploration of the pTyr+3 position in Grb2-targeted analogues

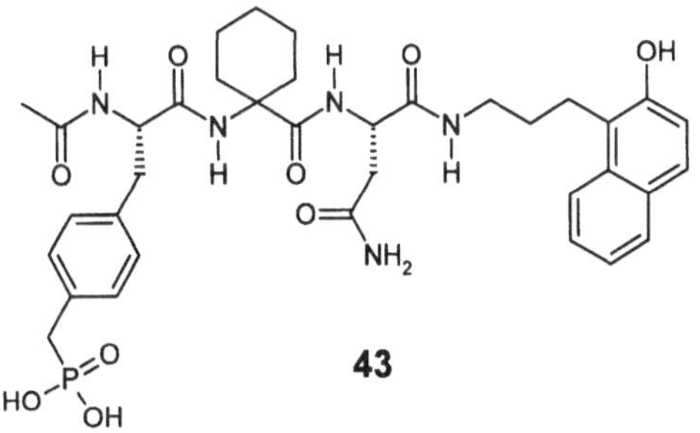

149]. The hydroxynaphthyl analogue CGP78850 **43** was shown to block the epidermal growth factor receptor (EGFR)-Grb2 interaction in living cells.

This analogue also inhibits the growth of cells transformed by receptor tyrosine kinases which transmit a proliferative signal through Grb2 to Ras, but not cells transformed by oncogenic Raf or cells containing activating Ras mutations [148], thus clearly supporting the Grb2 SH2 domain to be the molecular target of CGP78850 **43**. To achieve an improved penetration of **43** across cellular membranes, the phosphonamide prodrug CGP85793 **44** was synthesized. Upon hydrolysis by intracellular esterases, the active component CGP78850 **43** is liberated from its precursor **44**. The studies on **43** were extended mechanistically to interference into cell motility which generally correlates with oncogenic invasiveness and metastatic potential [150, 151]. It was demonstrated that inhibition of the Grb2 SH2 domain by **43** prevents HGF (hepatocytic growth factor)/SF (scatter factor)-induced A431 and MDCK (Madin-Darby canine kid-

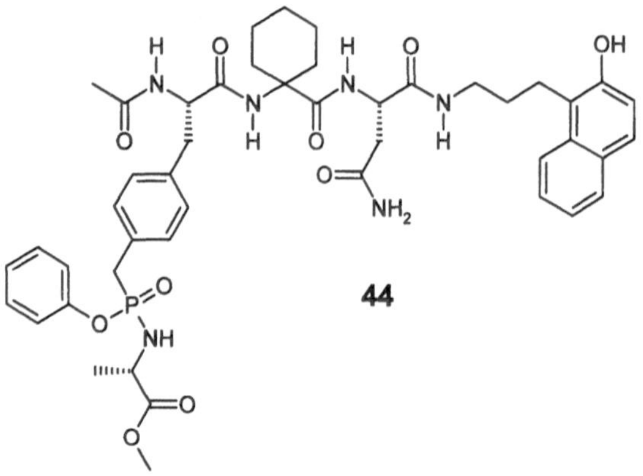

44

ney) cell motility and blocks the associated cytoskeletal rearrangement. CGP 78850 **43** is considered as a selective Grb2 SH2 domain antagonist that most likely opens an avenue for new cancer therapeutics, since it effectively controls the ability of tumor cells to move and traverse interstitial barriers.

While Novartis designed C-terminally blocked tripeptide analogues **41–44**, the Boehringer Ingelheim group succeeded in replacing a C-terminal dipeptide by a capping amine in their Lck SH2 domain antagonist program [152]. Based on the generic recognition motif -pTyr-Glu-Glu-Ile- the peptide portion downstream to pTyr was systematically replaced yielding a series of dipeptide derivatives possessing remarkable affinities. Lck SH2 domain ligands were developed that lack any carboxylic acid group. Interestingly, in the most active analogue **45** the Glu in pTyr+1 position was changed to Leu, and the C-terminal -Glu-Ile-fragment could be replaced by an (S)-1-(4-isopropylphenyl)ethylamine.

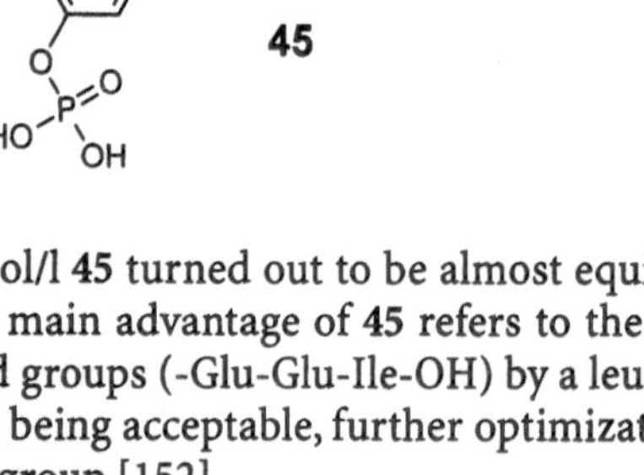

45

With a K_d of 0.2 µmol/l **45** turned out to be almost equipotent to a standard peptide analogue. The main advantage of **45** refers to the elimination of three distinct carboxylic acid groups (-Glu-Glu-Ile-OH) by a leucine amide. Since the affinity is still far from being acceptable, further optimization is ongoing in the Boehringer Ingelheim group [152].

A similar terminal-peptide-replacement strategy was used in the Src SH2 domain antagonist program pursued at the Parke-Davis Pharmaceutical Re-

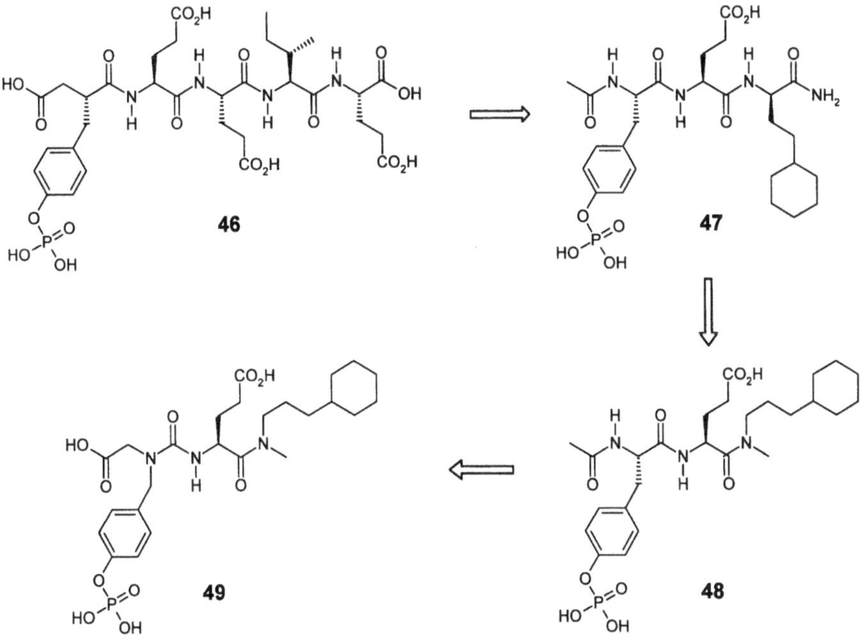

Scheme 8. Terminal-peptide-replacement strategy pursued by Parke-Davis

search Division [153] over several years. Systematic replacement of pTyr and downstream portions of the recognition motif -pTyr-Glu-Glu-Ile- led to a series of peptidomimetic analogues. For example, the Ac-pTyr backbone part was replaced by a substituted succinic acid yielding compound **46** with retained affinity (IC_{50}: 0.4 µmol/l) when compared to a standard lead peptide (Scheme 8). Replacement of the C-terminal dipeptide unit -Glu-Ile- against a D-homocyclohexylalanine amide (Hcy) resulted in the Ac-pTyr series in compounds of only moderately decreased affinity (**47**: IC_{50} = 1.8 µmol/l), while N-methylation of the Hcy-preceding amide bond together with the deletion of the terminal carboxamide brought back the original affinity (**48**: IC_{50} = 0.8 µmol/l) (Scheme 8).

With compound **48** the original pentapeptide sequence pTyr-Glu-Glu-Ile-Glu of the standard peptide was reduced to a dipeptide derivative with comparable binding characteristics. In an attempt to remove the chiral center from the pTyr building block, a urea analogue was synthesized based on an N-substituted glycine derivative which caused a tenfold drop in affinity (**49**: IC_{50} = 7 µmol/l). The crystal structure of the urea derivative **49** in its complex with the Src SH2 domain revealed an altered binding mode of the ligand with respect to the backbone conformation (Fig. 11). Most strikingly, the C-terminal amide bond adopts a *cis*-configuration, thereby displacing a highly-conserved protein-bound water molecule.

Due to the changed backbone conformation, several key interactions normally found in SH2 domain-peptide complexes are not formed, thus demanding further optimization in order to meet the spatial requirements of the target more precisely [153].

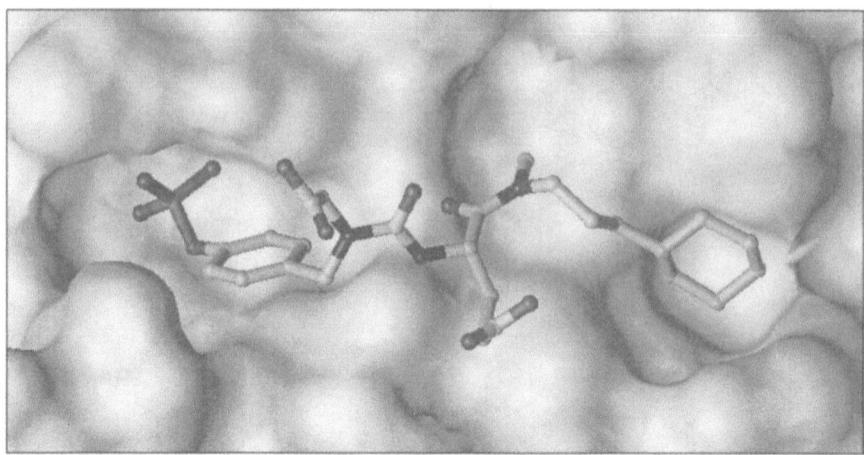

Fig. 11. Experimentally determined 3D structure of the Src SH2 domain in complex with compound **48** (1SKJ.pdb [153]). The cyclohexyl moiety binds into the pTyr+3 binding pocket (*right*)

Even though the Novartis group achieved promising in vitro as well as in vivo results with compounds of type **43** against the Grb2 SH2 domain, they made further attempts to reduce the peptide nature. Starting from the N-terminally 3-amino-Z-protected tripeptide amide **50** (IC$_{50}$: 1 nmol/l, Scheme 9), efforts have been made to replace the Asn in pTyr+2 position [147]. A computer-aided substructure search in the Novartis in-house database uncovered the *cis*-2-amino-cyclohex-3-enecarboxylic acid **51** (Achec, Scheme 9) to be a surrogate for the Asn residue in **50**.

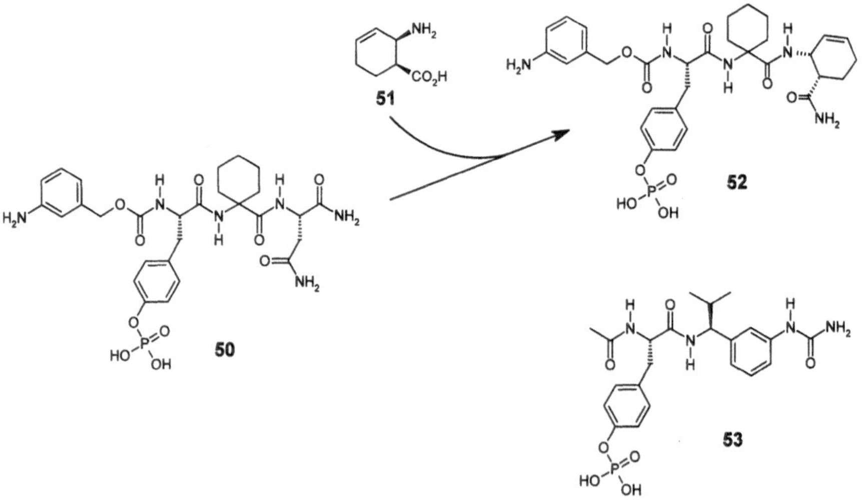

Scheme 9. Asn-replacement strategy utilizing the β-amino acid **51**

Upon synthesis of the dipeptide derivative 52, the (1S,2R)-isomer showed a comparable inhibitory potential with an IC_{50} of 1.6 nmol/l. Interestingly, the dipeptide mimic with the enantiomeric β amino acid (1R,2S)-Achec showed no binding at all. This study can be taken as a textbook example of a successful structure-based design approach utilizing a combination of molecular modeling, virtual screening of databases, medicinal chemistry, and experimental structure determination [147]. The binding mode of 52 to the Grb2 SH2 domain was shown by X-ray crystallography to be in perfect agreement with the predictions from molecular modeling (see Fig. 10) [147]. With 52, an analogue has been developed that reaches nanomolar binding without addressing the binding pocket of the pTyr+3 residue (see, for example, 41–44). While 52 is still a tripeptide analogue, alternative building blocks were evaluated that allowed for a simultaneous replacement of the C-terminal -Xaa-Asn- dipeptide fragment. By interactive design based on crystal structures of Grb2 SH2 domain-ligand complexes, the Novartis group replaced the terminal dipeptide against a 1,3-disubstituted phenyl ring (53 in Scheme 9) with the pTyr+1 sidechain attached to the aminomethyl substituent and a urea as Asn sidechain surrogate [154]. Unfortunately, the binding affinity of 53 was only in the micromolar range. However, this result proves it is possible to replace a dipeptide fragment with a single optimized non-peptide building block, yielding an SH2 domain ligand (53) that can be classified as a single amino acid derivative.

3.2.4
C-Terminal Tripeptide Mimetics

Apart from the Grb2, the Lck SH2 domain also served as target for peptidomimetic compounds that were designed along the line of replacing the entire C-terminal peptide sequence from the generic recognition motif -pTyr-Glu-Glu-Ile- by non-peptide structures [155, 156]. Applying a combinatorial chemistry approach, two libraries were designed and synthesized. An 84 compound library was used to identify the optimal -Glu-Glu- replacement, while a 900 compound array revealed a suitable Ile mimic. The pTyr derivative 54 was identified as one of the most active molecules of that series with an IC_{50} of 1.4 µmol/l [155, 156].

As in 53 (Scheme 9), the residual peptide sequence C-terminal to pTyr was exchanged against a non-peptide moiety. However, the low affinity of 54 can probably be attributed to an increased flexibility. Comparing 54 with the Grb2-targeted peptidomimetics 41, 42, or 43, it is reasonable to assume that the module C-terminally attached to pTyr in 54 serves as a tripeptide mimetic.

The Boehringer Ingelheim group chose a pyridone-based core structure to replace the -Glu-Glu-Ile- tripeptide of the generic SH2 domain recognition sequence [130]. From a series of more than 200 analogues, pyridones with mutually different decoration patterns were identified to result in pTyr derivatives with submicromolar affinities (e. g., 55, 56) for the Lck SH2 domain [130].

By focusing the entire compound collection on the pTyr-pyridone scaffold, the researcher made use of a "privileged" structural element, since the pyridone was repeatedly utilized to induce β strand conformations in serine and cysteine protease inhibitors [157, 158]. From this point of view, the potential of a modu-

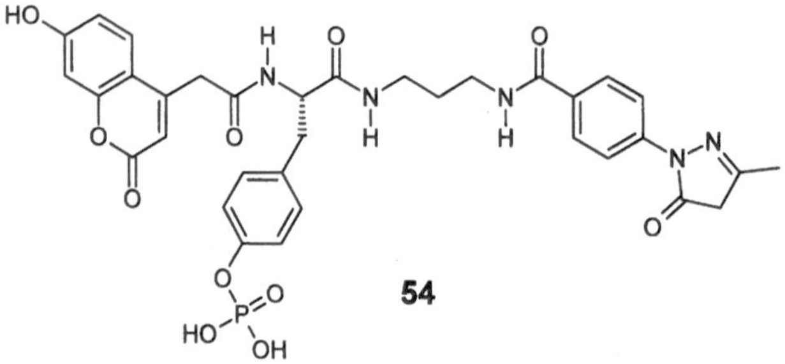

54

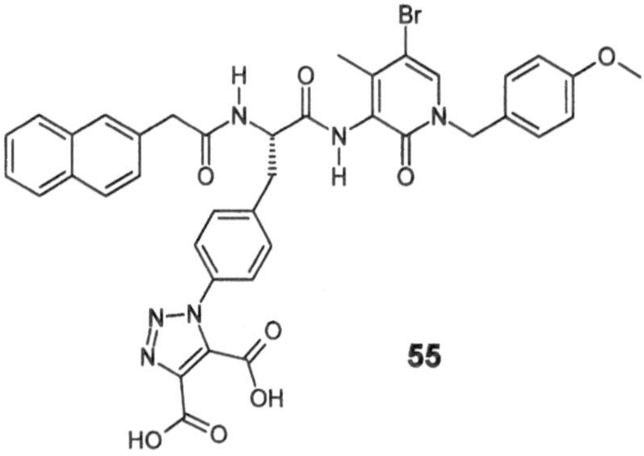

55

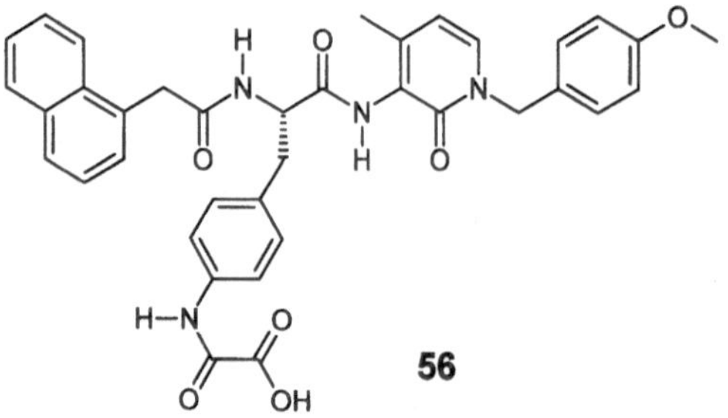

56

lar chemistry concept relying on versatile peptidomimetic building blocks for rapid and efficient lead finding becomes apparent.

Researchers at ARIAD Pharmaceuticals, interested over several years in developing Src SH2 domain inhibitors into therapeutic drugs, followed a strategy that would allow one to replace the C-terminal tripeptide of the generic recognition motif by a non-peptide module [159]. The design explicitly accounted for a hydrophobic protein surface patch separating the two "holes" of the "socket" that accommodate the "two-pronged plug-type" ligand. Consequently, a scaffold was required that accounted for that hydrophobic area, while appropriately delivering the two relevant ligand sidechains (pTyr and pTyr+3) into the corresponding pockets [159]. The initial design attempts resulted in a 1,3-disubstituted phenyl ring core that was decorated with the interaction-mediating groups as depicted in Scheme 10 (**57**: $IC_{50} = 769$ μmol/l; **58**: $IC_{50} = 306$ μmol/l (diastereomeric mixture)) [159].

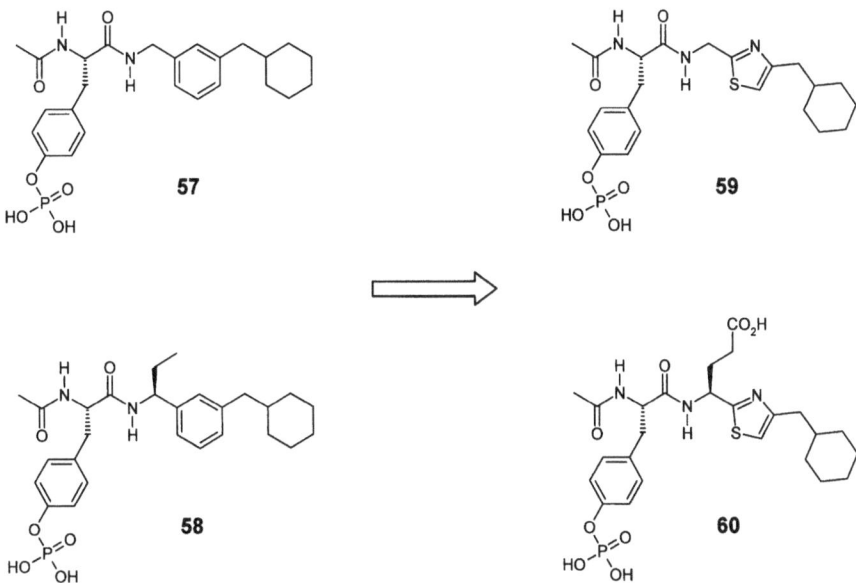

Scheme 10. Scaffolding the interaction-mediating sidechains on cyclic hydrophobic core templates

Due to synthetic reasons, the cyclic core was changed in favor of a heterocyclic scaffold, yielding the thiazole-based mimetics **59** (IC_{50}: 483 µmol/l) and **60** (IC_{50}: 26 µmol/l) [159]. Even though the binding affinities could not be improved with respect to the control peptide (IC_{50}: 6 µmol/l), a new scaffold replacing the C-terminal tripeptide was successfully designed (Scheme 10) and proven by X-ray crystallography to bind in the predicted manner [159]. These results served as an initial starting point for a series of follow-up studies at ARIAD Pharmaceuticals. Apart from the 2,4-substituted thiazoles as in **59** or **60** (Scheme 10), the study was extended to different heterocyclic core structures, such as 2,5-disubstituted thiazoles **61** (IC_{50}: 12 µmol/l), or 3,5-disubstituted 1,2,4-oxadiazoles [160]. A systematic comparison of different heterocyclic scaffolds suggested that the heterocycles were interchangeable without significant alterations in affinity, as long as a comparable decoration pattern was retained [160].

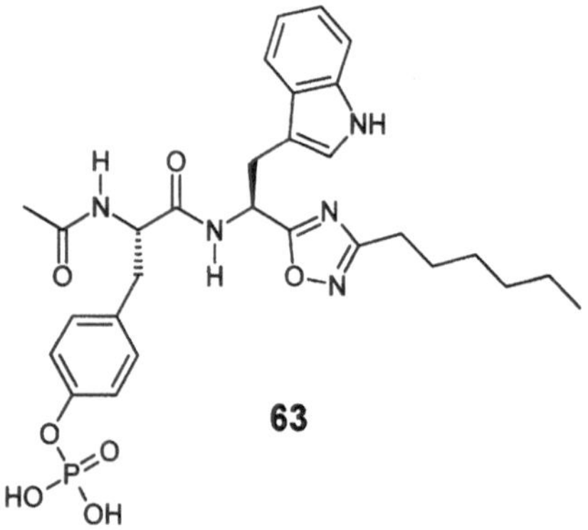

63

Since the oxadiazole series allowed an acceptable synthetic flexibility in terms of analogue preparation, this series was explored systematically [160]. Variations of the C-terminal lipophilic groups yielded analogues with affinities comparable to the standard peptide (IC$_{50}$: 6 μmol/l). Exchange of the Glu sidechain (e.g., **60**: IC$_{50}$ = 26 μmol/l; **62**: IC$_{50}$ = 7 μmol/l) against the Abu sidechain (**61**: IC$_{50}$ = 12 μmol/l), or the Trp sidechain (**63**: IC$_{50}$ = 16 μmol/l) resulted in a twofold loss in affinity, but allowed one to reduce the overall charge of the peptidomimetics, rendering these analogues as promising leads for further optimization [160].

Exactly along this line of modification, ARIAD Pharmaceuticals succeeded in generating an Src SH2 domain antagonist containing a bicyclic core element serving as tripeptide mimic, notably AP22161 **64** [161].

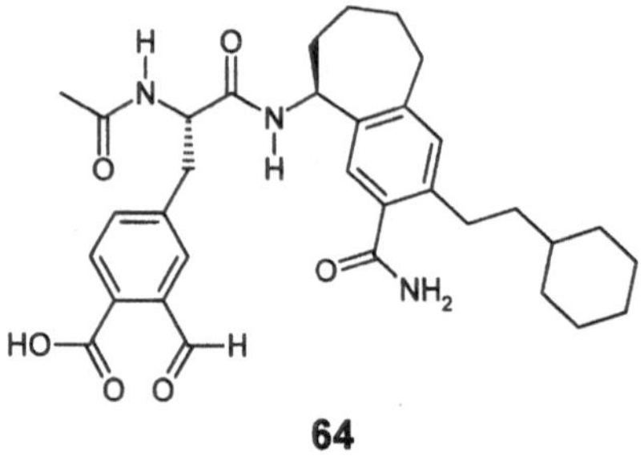

64

Reminiscent of the covalent inhibition principle employed by Glaxo with **33** (Scheme 5), AP22161 **64** covalently binds to the target SH2 domain. As well as **64** exhibiting a reasonable in vitro inhibition with an IC$_{50}$ of 0.24 μmol/l against the Src SH2 domain, it was also shown in vivo to diminish the osteoclast resorption activity in a mechanism-based cellular assay [161]. The concept of heterocyclic moieties bridging the distance between the two binding sidechains was extrapolated by ARIAD Pharmaceuticals from the Src SH2 domain onto the next representative of the target family, i.e., the ZAP70 SH2 domain [162]. Again the lead finding efficiency was significantly increased by developing privileged peptidomimetic building blocks that account for the target family-wide molecular recognition characteristics, followed by a repeated use of these modules in different medicinal chemistry programs. Variation of the C-terminal lipophilic group attached to C-3 of the oxadiazole core (e.g., **62, 63**) allowed the modulation of the SH2 domain specificity [162]. In a systematic iterative optimization procedure, detailed structure-activity relationships were obtained that finally resulted in a series of ZAP70-selective peptidomimetics, such as **65** (IC$_{50}$: 8 μmol/l) and **66** (IC$_{50}$: 1 μmol/l), respectively [162].

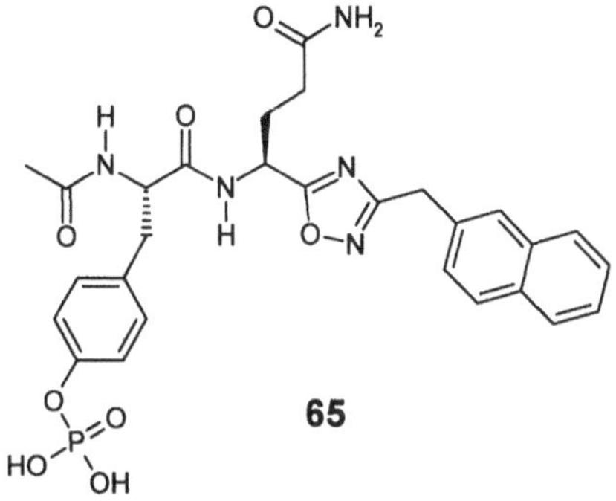

65

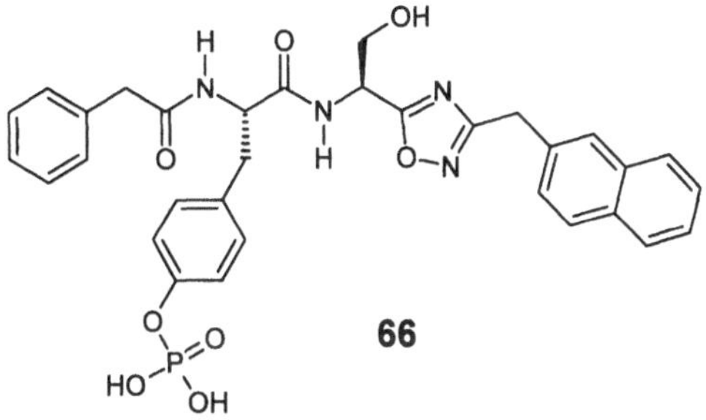

66

As in all cases discussed here, a tripeptide fragment was exchanged against a non-peptide module, resulting in considerably promising peptidomimetics.

3.2.5
Non-Peptide Peptidomimetics

The ultimate goal of any peptidomimetic approach is a non-peptide compound that more closely resembles the attributes of commonly used drug molecules, i.e., the peptidic origin is not apparent at first glance. Currently, only a few examples of SH2 domain-targeted non-peptide peptidomimetics have been reported, reflecting the novelty of this particular research area. From the main protagonists, Parke-Davis [163] and Novartis [164] published low-molecular weight analogues that can be classified into the group of non-peptide peptidomimetics.

Parke-Davis targeted a template structure with no formal charge replacing the Glu-Glu dipeptide bridging pTyr with Ile (pTyr+3), since the X-ray structure of SH2-peptide complexes suggested that the contribution of the two charged

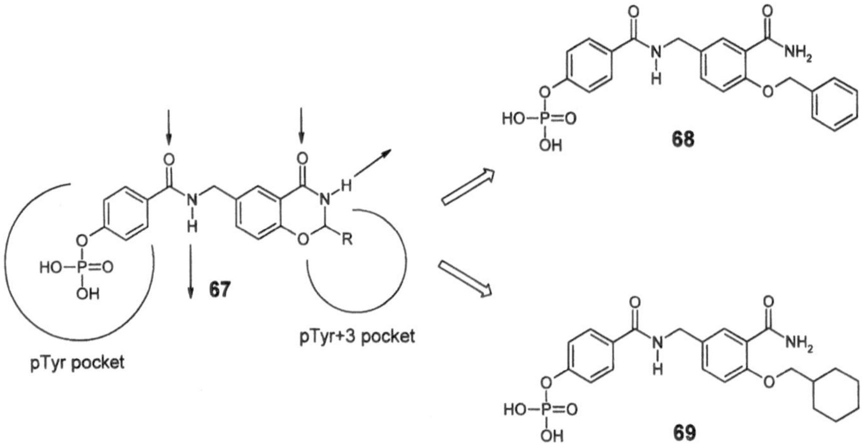

Scheme 11. Development of non-peptide Src SH2 domain antagonists emerging from a computer-aided molecular design strategy

carboxylates was nonessential for binding due to their solvent-exposed orientation [64]. The pTyr sidechain bound to its binding pocket was used as anchoring point for a protein-based de novo design of nonpeptide ligands. The design strategy aimed to incorporate a fairly rigid group as bridging moiety with defined hydrogen bonding capabilities. After definition of these structurally-derived geometric restraints, a database search of the Cambridge Crystallographic Database [165] revealed a benzoxazinone bicyclic core **67** as a high-ranking candidate (Scheme 11) [163].

The ring encoded *cis*-amide would further displace the aforementioned protein-bound water molecule, thus entropically contributing to binding affinity. Structurally fine-tuning and estimation on synthetic feasibility prompted the research group at Parke-Davis to follow a ring-opened form of the initial "virtual hit". The compound **68** is the first to be synthesized from that molecular modeling study and exhibited a remarkable affinity of 9.7 μmol/l (Scheme 11). Further optimization yielded a series of compounds with single-digit micromolar affinities for binding the Src SH2 domain, e. g., **69** (IC_{50}: 6.5 μmol/l), none of which reveals an obvious peptidic origin (Scheme 11) [163]. X-ray studies of these ligands in complex with the target SH2 domain revealed the excellent predictions made by molecular modeling. With these de novo designed compounds the first cycle in the iterative structure-based design application was completed and the information is further used to guide next-generation analogue design to improved levels of potency [163].

The Novartis group used the X-ray structure of a Grb2-peptide complex [68] as the structural basis for a design attempt that yielded entirely new non-peptide SH2 domain ligands [164]. As mentioned several times throughout this contribution, the interaction of the pTyr sidechain and the Asn sidechain in pTyr+2 position of the peptide ligand have been identified as key elements for molecular recognition (see Fig. 10). The obvious relevance of these two sidechain functionalities allowed the definition of a minimal pharmacophore pattern that

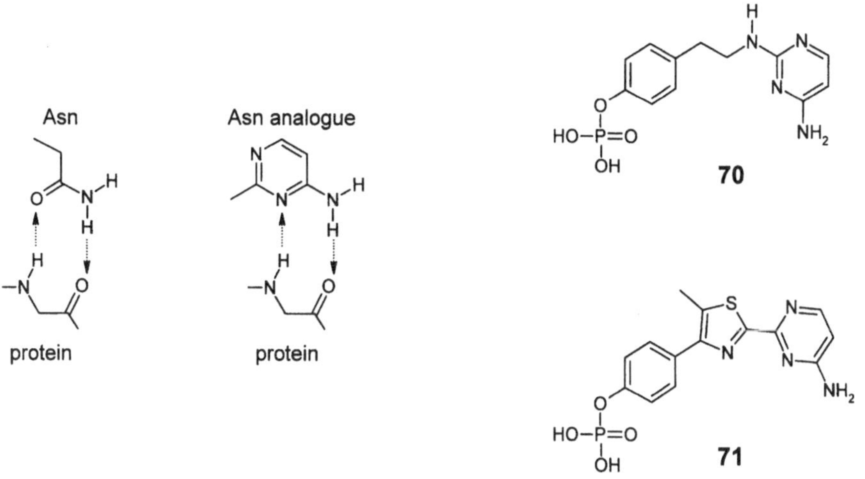

Scheme 12. Design of aminopyrimidines as Asn sidechain mimetics

served as structural basis for seeking non-peptide molecules containing these pharmacophoric elements. Based on thorough molecular modeling studies, the Asn sidechain was replaced by a 4-aminopyrimidine moiety, the latter being capable of retaining the hydrogen bond network formed in the complex structure (Scheme 12).

Further modeling studies led to the de novo design of entirely non-peptidic, "drug-like" compounds, **70** and **71** that were synthesized and biologically characterized [164]. It turned out that the open-chain compound **70** was inactive at concentrations up to 200 µmol/l, while the conformationally constrained thiazole analogue displayed binding affinity (IC_{50}: 25.9 µmol/l) which is within the range of the affinity of the reference tripeptide analogues (Scheme 12) [164].

These examples clearly prove the viability of a structure-based peptidomimetic design approach for developing non-peptide peptidomimetics for therapeutic interference into protein-protein interaction events.

4
Versatility of the Phosphotyrosine Pharmacophore

The recognition of pTyr-containing protein segments by downstream signaling adaptor proteins or enzymes is a central theme in intracellular signal transduction. As outlined in this contribution, selective inhibition of SH2 domains by low-molecular weight compounds has attracted considerable interest in recent years. A great deal of the synthetic work has been attributed to the development of useful pTyr mimics (Sect. 3.1). Future medicinal chemistry programs will undoubtedly benefit from the pioneering studies reviewed in this contribution, not only for potential SH2 domain antagonists but also for a relatively new class of pTyr-binding target proteins, i.e., the PI (phosphotyrosine-interaction) domains, more commonly termed PTB (phosphotyrosine-binding) domains

[26, 166]. Comparable to SH2 domains, PTB domains recruit signaling proteins to the vicinity of the intracellular domains of activated receptors. The first PTB domains were identified in two prominent RTK substrates, i. e., Shc and IRS-1 (insulin receptor substrate) [26, 27, 167, 168]. The association of PTB-containing proteins with signaling proteins involved in a variety of human diseases suggests that PTB domains might serve as novel promising targets for therapy [169]. For example, the PTB-containing multidomain proteins FE65, X11, and Dab bind to a cytoplasmic portion of the amyloid precursor protein (APP) [170–175] that is important for APP processing and internalization. At least FE65 and X11 antagonists are believed to block amyloid plaque formation [176] through inhibition of APP processing and may thus serve as useful therapeutics for Alzheimer's disease.

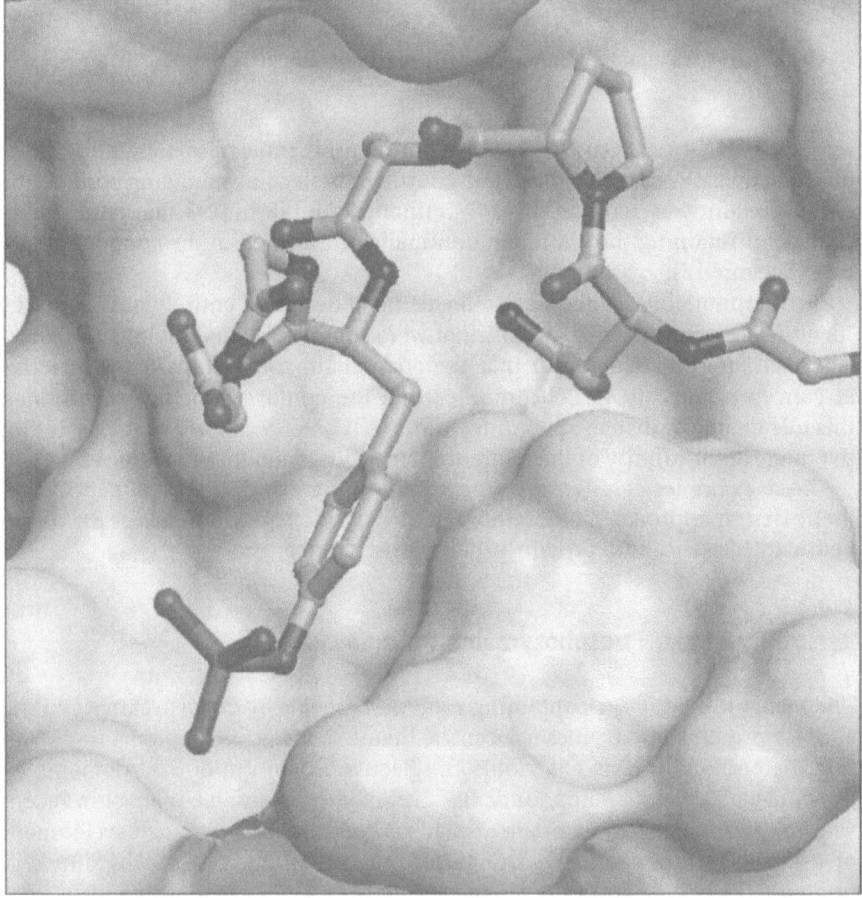

Fig. 12. 3D Structure of a pTyr-containing oligopeptide bound to the IRS-1 (insulin receptor substrate) PTB domain (1IRS.pdb). The Asn-Pro-Ala-pTyr tetrapeptide sequence adopts a regular βI turn conformation [181]

Although PTB domains and SH2 domains are both capable of phosphorylation-dependent binding [177], their ligand "read-out" and binding modes are substantially different. Specificity on PTB domain interaction is conferred by sequences N-terminal to pTyr (Fig. 12), rather than C-terminal as found for SH2 domains. PTB domains were identified based on their ability to recognize and bind -Asn-Pro-Any-pTyr- sequences [178].

Even though the strictness for a phosphorylated Tyr sidechain seems less pronounced for PTB domains [179], a modular peptidomimetics-based lead finding strategy could profit immediately by the knowledge and experiences gathered in SH2 domain-targeted medicinal chemistry projects.

Summarizing, pTyr emerged as a versatile pharmacophoric element over the last five years involved in the basic processes of intracellular signal transduction. Bioisosteric pTyr mimetics would not only allow one to interfere in the formation of activated multiprotein signaling complexes by blocking the adaptor protein recognition sites, but could also provide useful tools for PTP inhibitors. A certain degree of systematization of biomedical research projects from a target family point of view is required to permit cross-fertilization in terms of exchanging privileged building blocks and synthetic strategies, thus maximizing efficiency in lead finding and optimization.

5
Conclusion and Perspective

This contribution was intended to highlight pTyr-containing peptidomimetic compounds designed to interfere with signal transduction pathways by blocking protein-protein interactions mediated by SH2 domains. In addition, they bear a significant potential to be also applicable as PTB domain antagonists. Up to now, the majority of efforts aimed at developing specific SH2 domain antagonists have failed to reach an advanced stage of preclinical or even clinical development, due to the problems associated with generating nanomolar or subnanomolar binding entities. Most strikingly, a satisfactory structural replacement of pTyr has not yet been found. However, the design of therapeutically promising lead structures for SH2 domains presents novel challenges for medicinal chemistry. The excellent studies conducted by research groups at, among others, Novartis, Parke-Davis, ARIAD, and Boehringer Ingelheim clearly prove the viability of signal transduction therapy, which is undoubtedly still in its infancy. It remains difficult, if not impossible, to predict the outcome of a selective knockout of a distinct signal transduction element, such as an SH2 or a PTB domain. The applicability of adaptor protein-targeted inhibitors for therapeutic intervention in human diseases will not be proven until the discovery and subsequent clinical evaluation of potent and selective inhibitors of these pathways. The research projects discussed in this review will contribute to reaching this goal. Apart from the biomedical aspects addressed in this contribution, the area of SH2 domain antagonists turns out to be a textbook example of modern medicinal chemistry. According to a recent definition [180], this is still a chemistry-based discipline, being increasingly influenced by related areas such as biochemical, medical, pharmaceutical, and biophysical sciences. Instead of applying the traditional "hit-or-miss" method of finding new leads, a rational approach was preferred in all studies discussed, taking into account the available structural and functional information on the involved molecular components.

Acknowledgements. The author wishes to thank Dr. Andreas Schoop, Dr. Michael Schelhaas, and Dr. Marion Gurrath for carefully reading the manuscript, useful discussions, and valuable contributions to this review.

6
References

1. Levitzki A (1994) Eur J Biochem 226:1
2. Yu H, Schreiber SL (1994) Struct Biol 1:417
3. Hill CS, Treisman R (1995) Cell 80:199
4. Levitzki A (1996) Curr Opin Cell Biol 8:239
5. Saltiel AR, Sawyer TK (1996) Chem & Biol 3:887
6. Heimbrook DC, Oliff A (1998) Curr Opin Cell Biol 10:284
7. Botfield MC, Green J (1995) Annu Rep Med Chem 30:227
8. Bork P, Schultz J, Ponting CP (1997) Trends Biochem Sci 22:296
9. Schultz J, Milpetz F, Bork P, Ponting CP (1998) Proc Natl Acad Sci USA 95:5857
10. Schultz J, Copley RR, Doerks T, Ponting CP, Bork P (2000) Nucleic Acids Res 28:231
11. Burke TR Jr, Yao Z-J, Smyth MS, Ye B (1997) Curr Pharm Des 3:291
12. Hirschmann R (1991) Angew Chem, Int Ed Engl 30:1278
13. Horwell DC (1996) Bioorg Med Chem 4:1573
14. Adang AEP, Hermkens PHH, Linders JTM, Ottenheijm HCJ, van Staveren CJ (1994) Recl Trav Chim Pays-Bas 113:63
15. Olson GL, Bolin DR, Bonner MP, Bös M, Cook CM, Fry DC, Graves BJ, Hatada M, Hill DE, Kahn M, Madison VS, Rusiecki VK, Sarabu R, Sepinwall J, Vincent GP, Voss ME (1993) J Med Chem 36:3039
16. Hornby AS (2000) Oxford Advanced Learner's Dictionary. Oxford University Press, Oxford
17. Edwards DR (1994) Trends Pharmacol Sci 15:239
18. Pawson T, Scott JD (1997) Science 278:2075
19. Sadowski I, Stone JC, Pawson T (1986) Mol Cell Biol 6:4396
20. Russel RB, Breed J, Barton GJ (1992) FEBS Lett 304:15
21. Mayer BJ, Hamaguchi M, Hanafusa H (1988) Nature 332:272
22. Musacchio A, Gibson T, Lehto VP, Saraste M (1992) FEBS Lett 307:55
23. Pawson T, Schlessinger J (1993) Curr Biol 3:434
24. Mayer BJ, Baltimore D (1993) Trends Cell Biol 3:8
25. Pawson T (1995) Nature 373:573
26. Kavanaugh WM, Williams LT (1994) Science 266:1862
27. Blaikie P, Immanuel D, Wu J, Vajnik V, Margolis B (1994) J Biol Chem 269:32031
28. Bork P, Margolis B (1995) Cell 80:693
29. Fanning AS, Anderson JM (1996) Curr Biol 6:1385
30. Kim SK (1997) Curr Opin Cell Biol 9:853
31. Craven SE, Bredt DS (1998) Cell 93:495
32. Hughes RE, Fields S (1999) Nat Biotechnol 17:132
33. Cleveland J, Ihle JN (1995) Cell 81:479
34. Hofmann KO, Tschopp J (1995) FEBS Lett 371:321
35. Feinstein E, Kimchi A, Wallach D, Boldin M, Varfolomeev E (1995) Trends Biochem Sci 20:342
36. Golstein P, Marguet D, Depraetere V (1995) Cell 81:185

37. Bork P, Sudol M (1994) Trends Biochem Sci 19:531
38. Andre B, Springgael JY (1994) Biochem Biophys Res Commun 205:1201
39. Hofmann KO, Bucher P (1995) FEBS Lett 358:153
40. Sudol M, Chen HI, Bourgeret C, Einbond A, Bork P (1995) FEBS Lett 369:67
41. Chen HI, Sudol M (1995) Proc Natl Acad Sci USA 92:7819
42. Aitken A (1995) Trends Biochem Sci 20:95
43. Morrison D (1994) Science 266:56
44. Xiao B, Smerdon SJ, Jones DH, Dodson GG, Soneji Y, Aitken A, Gamblin SJ (1995) Nature 376:188
45. Marengere LEM, Pawson T (1994) J Cell Sci 18:97
46. Feng G-S, Pawson T (1994) Trends Genet 10:54
47. Cohen GB, Ren R, Baltimore D (1995) Cell 80:237
48. Pawson T (1988) Oncogene 3:491
49. Zhao Z, Shen S-H, Fischer EH (1995) Adv Prot Phosphatases 9:297
50. Schaffhausen B (1995) Biochim Biophys Acta 1242:61
51. Birge RB, Knudsen BS, Besser D, Hanafusa H (1996) Genes to Cell 1:595
52. Songyang Z, Shoelson SE, Chaudhuri M, Gish G, Pawson T, Haser WG, King F, Roberts T, Ratnofsky S, Lechleider RJ, Neel BG, Birge RB, Fajardo JE, Chou MM, Hanafusa H, Schaffhausen B, Cantley LC (1993) Cell 72:767
53. Marengere LE, Songyang Z, Gish CD, Schaffer MD, Stern JT, Cantley LC, Pawson T (1994) Nature 369:502
54. Gilmer T, Rodriguez M, Jordan S, Crosby R, Alligood K, Green M, Kimery M, Wagner C, Kinder D, Charifson P, Hassell AM, Willard D, Luther M, Rusnak D, Sternbach DD, Mehrotra M, Peel M, Shampine L, Davis R, Robbins J, Patel IR, Kassel D, Burkhart W, Moyer M, Bradshaw T, Berman J (1994) J Biol Chem 269:31711
55. Songyang Z, Carraway KL III, Eck MJ, Harrison SC, Feldman RA, Mohammadi M, Schlessinger J, Hubbard SR, Smith DP, Eng C, Lorenzo MJ, Ponder BAJ, Mayer BJ, Cantley LC (1995) Nature 373:536
56. Johnson TM, Perich JW, Bjorge JD, Fujita DJ, Cheng H-C (1997) J Peptide Res 50:365
57. Bazan JF (1990) Proc Natl Acad Sci USA 87:6934
58. Bazan JF (1990) Cell 61:753
59. Ihle JN, Kerr IM (1995) Trends Genet 11:69
60. Kuriyan J, Cowburn D (1997) Annu Rev Biophys Biomol Struct 26:259
61. Sicheri F, Kuriyan J (1997) Curr Opin Struct Biol 7:777
62. Sicheri F, Moarefi I, Kuriyan J (1997) Nature 385:602
63. Waksman G, Kominos D, Robertson SC, Pant N, Baltimore D, Birge RB, Cowburn D, Hanafusa H, Mayer BJ, Overduin M, Resh MD, Rios CB, Silverman L, Kuriyan J (1992) Nature 358:646
64. Waksman G, Shoelson SE, Pant N, Cowburn D, Kuriyan J (1993) Cell 72:779
65. Xu W, Harrison SC, Eck MJ (1997) Nature 385:595
66. Bradshaw JM, Mitaxov V, Waksman G (1999) J Mol Biol 293:971
67. Overduin M, Rios CB, Mayer BJ, Baltimore D, Cowburn D (1992) Cell 70:697
68. Rahuel J, Gay B, Erdmann D, Strauss A, García-Echeverría C, Furet P, Caravatti G, Fretz H, Schoepfer J, Grütter MG (1996) Nature Struct Biol 3:586
69. Rahuel J, García-Echeverría C, Furet P, Strauss A, Caravatti G, Fretz H, Schoepfer J, Gay B (1998) J Mol Biol 279:1013
70. Ogura K, Tsuchiya S, Terasawa H, Yuzawa S, Hatanaka H, Mandiyan V, Schlessinger J, Inagaki F (1999) J Mol Biol 289:439
71. Ettmayer P, France D, Gounarides J, Jarosinski M, Martin MS, Rondeau JM, Sabio M, Topiol S, Weidmann B, Zurini M, Bair KW (1999) J Med Chem 42:971
72. Lee C-H, Kominos D, Jacques S, Margolis B, Schlessinger J, Shoelson SE, Kuriyan J (1994) Structure 2:423
73. Pascal SM, Singer AU, Gish G, Yamazaki T, Shoelson SE, Pawson T, Kay LE, Forman-Kay JD (1994) Cell 77:461
74. Mulhern TD, Shaw GL, Morton CJ, Day AJ, Campbell ID (1997) Structure 5:1313

75. Poy F, Yaffe MB, Sayos J, Saxena K, Morra M, Sumegi J, Cantley LC, Terhorst C, Eck ME (1999) Mol Cell 4:555
76. Mikol V, Baumann G, Keller TH, Manning U, Zurini MGM (1995) J Mol Biol 246:344
77. Tong L, Warren TC, King J, Betageri R, Rose J, Jakes S (1996) J Mol Biol 256:601
78. Breeze AL, Kara BV, Barratt DG, Anderson M, Smith JC, Luke RW, Best JR, Cartlidge SA (1996) EMBO J 15:3579
79. Nolte RT, Eck MJ, Schlessinger J, Shoelson SE, Harrison SC (1996) Nature Struct Biol 3:364
80. Hoedemaeker FJ, Siegal G, Roe SM, Driscoll PC, Abrahams JP (1999) J Mol Biol 292:763
81. Narula SS, Yuan RW, Adams SE, Green OM, Green J, Philips TB, Zydowsky LD, Botfield MC, Hatada M, Laird ER, Zoller MJ, Karas JL, Dalgarno DC (1995) Structure 3:1061
82. Fütterer K, Wong J, Grucza RA, Chan AC, Waksman G (1998) J Mol Biol 281:523
83. Hatada MH, Lu X, Laird ER, Green J, Morgenstern JP, Lou M, Marr CS, Phillips TB, Ram MK, Theriault K, Zoller MJ, Karas JL (1995) Nature 377:32
84. Cowburn D (1996) Chem & Biol 3:79
85. Barford D, Neel BG (1998) Structure 6:249
86. Lemmon MA, Ladbury JE, Mandiyan V, Zhou M, Schlessinger J (1994) J Biol Chem 269:31653
87. Lemmon MA, Ladbury JE (1994) Biochemistry 33:5070
88. Ladbury JE, Lemmon MA, Zhou M, Green J, Botfield J, Schlessinger J (1995) Proc Natl Acad Sci USA 92:3199
89. Chen T, Repetto B, Chizzonite R, Pullar C, Burdhardt C, Dharm E, Zhao Z, Carroll R, Nunes P, Basu M, Danho W, Visnick M, Kochan J, Waugh D, Gilfillan AM (1996) J Biol Chem 271:25308
90. Ladbury JE, Hensmann M, Panayotou G, Campbell ID (1996) Biochemistry 35:11062
91. Charifson PS, Shewchuk LM, Rocque W, Hummel CW, Jordan SR, Mohr C, Pacofsky GJ, Peel MR, Rodriguez M, Sternbach DD, Consler TG (1997) Biochemistry 36:6283
92. Gay B, Furet P, García-Echeverría C, Rahuel J, Chêne P, Fretz H, Schoepfer J, Caravatti G (1997) Biochemistry 36:5712
93. McNemar C, Snow ME, Windsor WT, Prongay A, Mui P, Zhang R, Durkin J, Le HV, Weber PC (1997) Biochemistry 36:10006
94. Bradshaw JM, Grucza RA, Ladbury JE, Waksman G (1998) Biochemistry 37:9083
95. Bradshaw JM, Waksman G (1998) Biochemistry 37:15400
96. Bradshaw JM, Waksman G (1999) Biochemistry 38:5147
97. Grucza RA, Fütterer K, Chan AC, Waksman G (1999) Biochemistry 38:5024
98. Songyang Z, Gish G, Mbamalu G, Pawson T, Cantley LC (1995) J Biol Chem 270:26029
99. Birge RB, Hanafusa H (1993) Science 262:1522
100. Ladbury JE, Arold S (2000) Chem & Biol 7:R3
101. Sawyer TK (1998) Biopolymers 47:243
102. Blackburn GM (1981) Chem Ind 5:134
103. Blackburn GM, Perree TD, Rashid A, Bisbal C, Lebleu B (1986) Chem Scr 26:21
104. Engel R (1977) Chem Rev 77:349
105. Burke TR Jr, Russ P, Lim B (1991) Synthesis 11:1019
106. Shoelson SE, Chatterjee S, Chaudhuri M, Burke TR Jr (1991) Tetrahedron Lett 32:6061
107. Marseigne J, Roques BP (1988) J Org Chem 53:3621
108. Domchek SM, Auger KR, Chatterjee S, Burke TR Jr, Shoelson SE (1992) Biochemistry 31:9865
109. Eck MJ, Shoelson SE, Harrison SC (1993) Nature 362:87
110. Burke TR Jr, Smyth SM, Otaka A, Nomizu M, Roller PP, Wolf G, Case R, Shoelson SE (1994) Biochemistry 33:6490
111. Stancovic CJ, Plummer MS, Sawyer TK (1997) Peptidomimetic ligands for src homology-2 domains. In: Abell A (ed) Advances in amino acid mimetics and peptidomimetics. Jai Press, Greenwich, vol 1, p 127

112. Stancovic CJ, Surendran N, Lunney EA, Plummer MS, Para KS, Shahripour A, Fergus JH, Marks JS, Herrera R, Hubbell SE, Humblet C, Saltiel AR, Stewart BH, Sawyer TK (1997) Bioorg Med Chem Lett 7:1909
113. Roller PP, Otaka A, Nomizu M, Smyth MS, Barchi JJ, Burke TR Jr, Case RD, Wolf G, Shoelson SE (1994) Bioorg Med Chem Lett 4:1879
114. Burke TR Jr, Nomizu M, Otaka A, Smyth MS, Roller PP, Case RD, Wolf G, Shoelson SE (1994) Biochem Biophys Res Commun 201:1148
115. Eaton SR, Cody WL, Doherty AM, Holland DR, Panek RL, Lu GH, Dahring TK, Rose DR (1998) J Med Chem 41:4329
116. Liu W-Q, Roques BP, Garbay C (1997) Tetrahedron Lett 38:1389
117. Fu J-M, Castelhano AL (1998) Bioorg Med Chem Lett 8:2813
118. Tong L, Warren TC, Lukas S, Schembri-King J, Betageri R, Proudfoot JR, Jakes S (1998) J Biol Chem 273:20238
119. Beaulieu PL, Cameron DR, Ferland J-M, Gauthier J, Ghiro E, Gillard J, Gorys V, Poirier M, Rancourt J, Wernic D, Llinas-Brunet M (1999) J Med Chem 42:1757
120. Yao Z-J, Gao Y, Voigt JH, Ford H Jr, Burke TR Jr (1999) Tetrahedron 55:2865
121. Burke TR Jr, Luo J, Yao Z-J, Gao Y, Zhao H, Milne GWA, Guo R, Voigt JH, King CR, Yang D (1999) Bioorg Med Chem Lett 9:347
122. Furet P, Gay B, García-Echeverría C, Rahuel J, Fretz H, Schoepfer J, Caravatti G (1997) J Med Chem 40:3551
123. García-Echeverría C, Furet P, Gay B, Fretz H, Rahuel P, Schoepfer J, Caravatti G (1998) J Med Chem 41:1741
124. Furet P, Gay B, Caravatti G, García-Echeverría C, Rahuel J, Schoepfer J, Fretz H (1998) J Med Chem 41:3442
125. Burke TR Jr, Yao Z-J, Zhao H, Milne GWA, Wu L, Zhang Z-Y, Voigt JH (1998) Tetrahedron 54:9981
126. Shuker SB, Hajduk PJ, Meadows RP, Fesik SW (1996) Science 274:1531
127. Hajduk PJ, Meadows RP, Fesik SW (1997) Science 278:497
128. Hajduk PJ, Zhou M-M, Fesik SW (1999) Bioorg Med Chem Lett 9:2403
129. Mehrotra MM, Sternbach DD, Rodriguez M, Charifson P, Berman J (1996) Bioorg Med Chem Lett 6:1941
130. Betageri R, Beaulieu PL, Llinas-Brunet M, Ferland J-M, Cardozo M, Moss N, Patel U, Proudfoot JR (1999) WO 99/31066
131. Ye B, Burke TR Jr (1995) Tetrahedron Lett 36:4733
132. Ye B, Akamatsu M, Shoelson SE, Wolf G, Giorgetti-Perladi S, Yan XJ, Roller PP, Burke TR Jr (1995) J Med Chem 38:4270
133. Burke TR Jr, Ye B, Akamatsu M, Ford H, Yan XJ, Kole HK, Wolf G, Shoelson SE, Roller PP (1996) J Med Chem 39:1021
134. Gao Y, Luo J, Yao Z-J, Guo R, Zou H, Kelley J, Voigt JH, Yang D, Burke TR Jr (2000) J Med Chem 43:911
135. Schoepfer J, Fretz H, Gay B, Furet P, García-Echeverría C, End N, Caravatti G (1999) Bioorg Med Chem Lett 9:221
136. (1998) Exp Opin Ther Patents 8:333
137. Alligood KJ, Charifson PS, Crosby R, Consler TG, Feldman PL, Gampe RT Jr, Gilmer TM, Jordan SR, Milstead MW, Mohr C, Peel MR, Rocque W, Rodriguez M, Rusnak DW, Shewchuk LM, Sternbach DD (1998) Bioorg Med Chem Lett 8:1189
138. Otto H-H, Schirmeister T (1997) Chem Rev 97:133
139. Simon RJ, Kania RS, Zuckermann RN, Huebner VD, Jewell DA, Banville S, Ng S, Wang L, Rosenberg S, Marlowe CK, Spellmeyer DC, Tan R, Frankel AD, Santi DV, Cohen FE, Bartlett PA (1992) Proc Natl Acad Sci USA 89:9367
140. Kessler H (1993) Angew Chem, Int Ed Engl 32:543
141. Révész L, Bonne F, Manning U, Zuber J-F (1998) Bioorg Med Chem Lett 8:405
142. Horwell DC, Howson W, Rees DC (1994) Drug Design Discovery 12:63
143. García-Echeverría C, Gay B, Rahuel J, Furet P (1999) Bioorg Med Chem Lett 9:2915

144. Marshall GR, Hodgkiin EE, Lengs DA, Smith GD, Zabrocki J, Leplawy MT (1990) Proc Natl Acad Sci USA 87:487
145. Benedetti E, di Blasio B, Pavone V, Pedone C, Santini A, Crisma M, Valle G, Toniolo C (1989) Biopolymers 28:175
146. Toniolo C, Bonora GM, Bavoso A, Benedetti E, di Blasio B, Pavone V, Pedone C (1983) Biopolymers 22:205
147. Furet P, García-Echeverría C, Gay B, Schoepfer J, Zeller M, Rahuel J (1999) J Med Chem 13:2358
148. Gay B, Suarez S, Caravatti G, Furet P, Meyer T, Schoepfer J (1999) Int J Cancer 83:235
149. Gay B, Suarez S, Weber C, Rahuel J, Fabbro D, Furet P, Caravatti G, Schoepfer J (1999) J Biol Chem 274:23311
150. Thiery JP (1984) Cell Differ 15:1
151. Strauli P, Weiss L (1977) Eur J Cancer 13:81
152. Llinás-Brunet M, Beaulieu PL, Cameron DR, Ferland J-M, Gauthier J, Ghiro E, Gillard J, Gorys V, Poirier M, Rancourt J, Wernic D (1999) J Med Chem 42:722
153. Plummer MS, Holland DR, Shahripour A, Lunney EA, Fergus JH, Marks JS, McConnell P, Mueller WT, Sawyer TK (1997) J Med Chem 40:3719
154. Schoepfer J, Gay B, Caravatti G, García-Echeverría C, Fretz H, Rahuel J, Furet P (1998) Bioorg Med Chem Lett 8:2865
155. Lee TR, Lawrence DS (1999) J Med Chem 42:784
156. Lee TR, Lawrence DS (2000) J Med Chem 43:1173
157. Golec JMC, Mullican MD, Murcko MA, Wilson KP, Kay DP, Jones SD, Murdoch R, Bemis GW, Raybuck SY, Luong Y-P, Livingston DJ (1997) Bioorg Med Chem Lett 7:2181
158. Isaacs RCA, Cutrona KJ, Newton CL, Sanderson PEJ, Solinsky MG, Baskin EP, Chen I-W, Cooper CM, Cook JJ, Gardell SJ, Lewis SD, Lucas RJ Jr, Lyle EA, Lynch JJ Jr, Nayler-Olsen AM, Stranieri MT, Vastag K, Vacca JP (1998) Bioorg Med Chem Lett 8:1719
159. Buchanan JL, Bohacek RS, Luke GP, Hatada M, Lu X, Dalgarno DC, Narula SS, Yuan R, Holt DA (1999) Bioorg Med Chem Lett 9:2353
160. Buchanan JL, Vu CB, Merry TJ, Corpuz EG, Pradeepan SG, Mani UN, Yang M, Plake HP, Varkhedkar VM, Lynch BA, MacNeil IA, Loiacono KA, Tiong CL, Holt DA (1999) Bioorg Med Chem Lett 9:2359
161. Violette SM, Shakespeare WC, Bartlett C, Guan W, Smith JA, Rickles RJ, Bohacek RS, Holt DA, Baron R, Sawyer TK (2000) Chem & Biol 7:225
162. Vu CB, Corpuz EG, Merry TJ, Pradeepan SG, Bartlett C, Bohacek RS, Botfield MC, Eyermann CJ, Lynch BA, MacNeil IA, Ram MK, van Schravendijk MR, Violette S, Sawyer TA (1999) J Med Chem 42:4088
163. Lunney EA, Para KS, Rubin JR, Humblet C, Fergus JH, Marks JS, Sawyer TK (1997) J Am Chem Soc 119:12,471
164. Caravatti G, Rahuel J, Gay B, Furet P (1999) Biorg Med Chem Lett 9:1973
165. Allen FH, Bellard S, Brice MD, Cartwright BA, Doubleday A, Higgs H, Hummelink T, Hummelink-Peters BG, Kennard O, Motherwell WDS, Rodgers JR, Watson DG (1979) Acta Crystallogr B35:2331
166. Kavanaugh WM, Turck CW, Williams LT (1995) Science 268:1177
167. O'Neill TJ, Craparo A, Gustafson TA (1994) Mol Cell Biol 14:6433
168. Wolf G, Trub T, Ottinger E, Groninga L, Lynch A, White M, Miyazaki M, Lee J, Shoelson SE (1995) J Biol Chem 270:27407
169. O'Bryan JP (1999) Curr Opin Drug Discovery Development 2:505
170. Howell BW, Lanier LM, Frank R, Gertler FB, Cooper JA (1999) Mol Cell Biol 19:5179
171. Perez RG, Soriano S, Hayes JD, Ostaszewski B, Xia W, Selkoe DJ, Chen X, Stokin GB, Koo EH (1999) J Biol Chem 274:18851
172. Borg JP, Ooi J, Levy E, Margolis B (1996) Mol Cell Biol 16:6229
173. McLoughlin DM, Miller CC (1996) FEBS Lett 397:197
174. Sastre M, Turner RS, Levy E (1998) J Biol Chem 273:22351
175. Zambrano M, Buxbaum JD, Minopoli G, Fiore F, De Candia P, De Renzis S, Faraonio R, Sabo S, Cheetham J, Sudol M, Russo T (1997) J Biol Chem 272:6399

176. Sabo SL, Lanier LM, Ikin AF, Khorkova O, Sahasrabudhe S, Greengard P, Buxbaum JD (1999) J Biol Chem 274:7952
177. Zhou Y, Abagyan R (1998) Folding & Design 3:513
178. Shoelson SE (1997) Curr Opin Chem Biol 1:227
179. Li SC, Songyang Z, Vincent SJ, Zwahlen C, Wiley S, Cantley L, Kay LE, Forman-Kay J, Pawson T (1997) Proc Natl Acad Sci USA 94:7204
180. Wermuth C-G, Ganellin CR, Lindberg P, Mitscher LA (1998) Annu Rep Med Chem 33:385
181. Zhou MM, Huang B, Olijniczak ET, Meadows RP, Shuker SB, Miyazaki M, Trub T, Shoelson SE, Fesik SW (1996) Nat Struct Biol 3:388

Biophysical Characterization of the Ras Protein

Jürgen Kuhlmann · Christian Herrmann

Max Planck Institut für molekulare Physiologie, Otto-Hahn-Strasse 11, 44227 Dortmund, Germany
E-mail: juergen.kuhlmann@mpi-dortmund.mpg.de

Max Planck Institut für molekulare Physiologie, Otto-Hahn-Strasse 11, 44227 Dortmund, Germany
E-mail: christian.herrmann@mpi-dortmund.mpg.de

Ras is a central switch in regulation of cell proliferation and differentiation. It becomes activated by extracellular stimuli like growth factors and relays the signal into diverse cellular pathways. A balanced action of regulatory proteins is required to maintain the desired level of Ras activity. Oncogenic mutations in Ras lead to uncontrollable over-function, resulting in transformation of the cell. Therefore Ras became a target in many searches for anti-cancer agents. From the biochemical point of view Ras serves as the paradigm of the superfamily of GTP-binding proteins which fulfill cellular functions as diverse as transmission of vision and regulation of cell morphology. The common theme is switching between the GTP- and GDP-bound forms which correspond to the active and silent states, respectively. This article describes the function of Ras on the molecular level, leaving the enormous wealth of biological data in the Ras field largely aside. Rather, the focus is put on biochemical and biophysical methods which were used to analyze the interactions with the diverse partner molecules in a quantitative way to elucidate the mechanism of activation and regulation.

Keywords: Ras, Kinetics, Structure, Effectors, Regulators, GTPase, Fluorescence, Membrane

Topics in Current Chemistry, Vol. 211
© Springer-Verlag Berlin Heidelberg 2000

1
Introduction

1.1
Signal Transduction

All living organisms have two primary objectives: to maintain their vital functions and to relay their genetic information by reproduction. To exercise this task successfully a creature has to react flexibly upon its environment. It has to escape from enemies, it has to adapt to changes in temperatures like heat or coldness, drought or moistness. If food is available the metabolism has to be switched quickly for an optimized synthesis of an energy reserve and in the case of an injury damaged tissue has to be regenerated.

Multicellular organisms are especially constrained to allow cell division just at that time and place where it is necessary to build up new or enhance present functional structures during growth and differentiation.

Signal transduction now ensures that information is selectively addressed to single cell types, where it is registered and transformed in a common manner – e.g., in the case of a human being which is composed of some trillions of cells [1]. Hereby metabolic fluxes or cell division are controlled efficiently.

Animal cells are separated from each other by lipid membranes. During signal transduction this barrier has to be passed, which can be realized by permanently or temporarily opened channels or by an indirect mechanism without material flux between the extra- and intracellular lumen (Fig. 1).

In the latter case an extraneous messenger has to dock at the receptor's extracellular binding site on the cell surface. The information about the occupancy of the corresponding receptor is transmitted through the transmembrane part of the protein into its cytosolic domains by conformational changes. This structural response can be induced by an additional dimerization and results in a covalent modification of intracellular side chains. The new conformation is then recognized by cytosolic partner molecules. In this connection GTP binding pro-

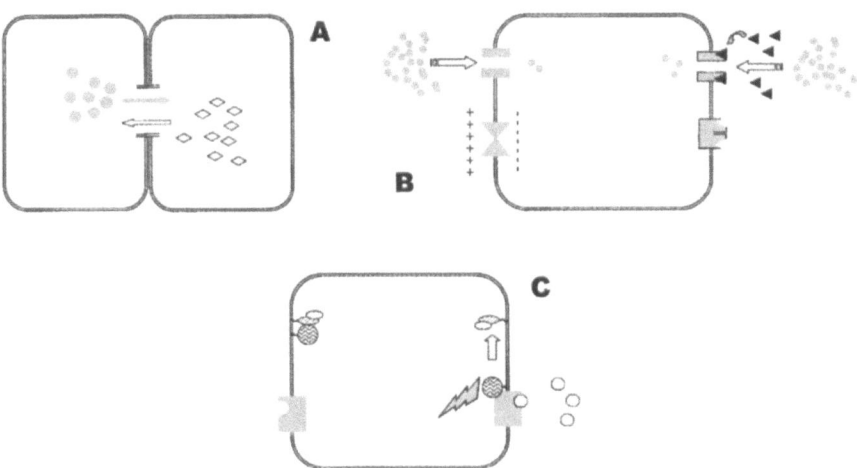

Fig. 1A–C. Communication between cells can be realized in three principal modes: A direct exchange of material via Gap Junctions is used to synchronize, e.g., heart muscle cells; B ligand and voltage-gated channels open for a short period of time to allow ion flux into or out of the target cell in response to ligand binding or a change in membrane potential, respectively; C indirect transmission of a signal. A transmembrane receptor is in its inactive state when no messenger is around. If a messenger molecule binds at the outer site of the receptor a change in conformation is induced. Now further proteins can bind to the intracellular site of the receptor and transmit the information

teins often work as an essential molecular switch [2]. GTP binding proteins are classified in five superfamilies:

- Translation factors in protein synthesis
- G_α subunits of heterotrimeric G-proteins
- Ras-like proteins of small GTPases
- Large GTP-binding proteins
- Subunits of signal recognition particles and their receptors

All GTP binding proteins in signal transduction share a common structural element – the "Ras-like" domain which is responsible for the specific complexation of guanosine diphosphate and -triphosphate and which contains catalytic residues that promote GTP-hydrolysis.

1.2
Heterotrimeric G-Proteins

G-Protein coupled receptors (GPCR) represent the start element in secondary messenger producing systems. They comprise a family of over 1000 structurally-related members. These membrane proteins are also called serpentine or seven-helix receptors due to their seven transmembrane domains with an α-helical conformation. Receptors belonging to this class respond to a variety of hormones and neurotransmitters, and they detect odorant molecules or light [3, 4].

The ligand complexed form of a GPCR allows binding of a heterotrimeric G-protein [5], which is anchored at the cytoplasmic side of the plasma membrane via hydrophobic modifications in two out of their three subunits. Binding to the receptor induces dissociation of the tightly bound GDP from the α subunit (G_α) of the G-protein, which is then replaced by cytosolic GTP. This step leads to dissociation of G_α from the $G_{\beta\gamma}$ dimer. $G_\alpha \cdot$ GTP now can interact with effector proteins, which themselves may generate second messenger molecules in the cell. Intrinsic α subunit GTPase activity catalyses GTP hydrolysis causing inactivation. GDP-bound α subunit re-associates with the $G_{\beta\gamma}$ subunit to form an inactive G protein heterotrimer which can again become activated by the receptor (Fig. 2).

So far four G protein subfamilies have been identified and classified according to the more than 20 known α subunits (Gq/11, Gi/0, Gs, and G12/13) [6]. There is a similar variety of beta and gamma subunits. G proteins can act either stimulatory or inhibitory.

1.3
Small GTP-Binding Proteins

The Ras superfamily of GTP-binding proteins is composed of several subfamilies [7] which all contain the "Ras-like domain" of approximately 160 amino acids and 5 consensus sequences. Two of these highly conserved motifs are responsible for specific recognition of the guanosine nucleotide, and three are necessary for binding of the phosphate groups and complexation of a Mg^{++} ion, which is found in all Ras-like proteins.

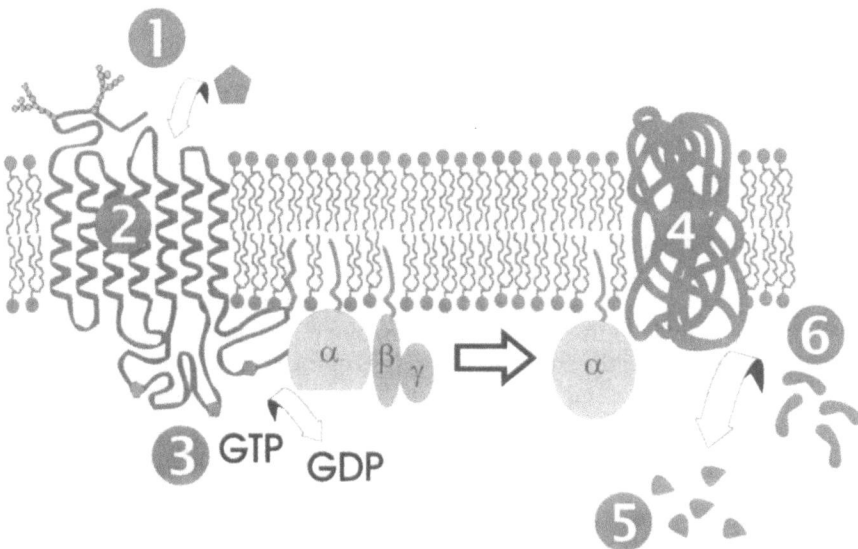

Fig. 2. An adrenaline molecule (1) binds to its binding site on the extracellular site of an adrenaline receptor (2). Thereby, the exchange of GDP by GTP in the G_α subunit of a hetero-trimeric G protein (3) is induced, followed by the dissociation of the G_α and $G_{\beta\gamma}$ subunits. G_α now binds and stimulates its effector adenylate cyclase (4), which produces cyclic AMP (5) from ATP (6). This second messenger starts a cascade of enzymatic reactions, which alter the behavior of the cell via several phosphorylation steps

The Ras and Rho proteins are involved in signal transduction; Ran, Rab and Arf proteins regulate events of the intracellular transport. The function of Rad-proteins is not yet clear.

All proteins of the Ras superfamily (Table 1) – with the exception of the Ran protein – undergo a posttranslational modification which introduces at least one hydrophobic modification. These lipid anchors qualify the members of the Ras family as peripheral membrane proteins, which stick into membrane structures from the cytoplasmic side only.

Table 1. Members of the Ras-superfamily

Family	Function
Ras	Cell growth and differentation
Rho	Organization of the cytoskeleton
Rab	Vesicular transport
Arf	Vesicular transport between endoplasmatic reticulum and cis-Golgi apparatus
Ran	Nucleo-cytoplasmic transport
Rad	–

1.4
Ras

Ras is the most prominent member of the small GTP-binding proteins. Homologs are found in all eukaryotic organisms and it is present in all types of cells. Four isoforms of Ras are known in human, namely H-Ras, N-Ras, and K-Ras4A and -4B. Their sequence is almost identical, with only the 20 C-terminal amino acids representing a hypervariable region. No clear evidence is available as to whether they are redundant in biological function or are responsible for different signal pathways. However, different effects of knock-outs of these genes in mice, which is lethal only when K-ras is disrupted [8], suggest a major importance for K-Ras. Recently, differential activities of the four isoforms with respect to cell transformation and stimulation of cell motility have been reported [9] (see [10] for review).

Ras plays a pivotal role in cellular signal transduction, as many different outer cellular inputs are relayed to Ras inside the cell and are transformed into specific biological responses by Ras activated signal pathways [11]. Dependent on the cellular context, this leads to cell differentiation or proliferation. Originally, Ras was discovered as an oncogene causing cancer (rat sarcomas [12]). Due to its potential when mutated to transform cells in the end to tumor cells, Ras is classified as a proto-oncogene. The molecular mechanism and aspects important for the biological function of Ras, like its cellular localization, its regulator and target proteins, are described in the following sections.

1.4.1
Membrane Anchorage

Ras is strictly localized to the inner side of the plasma membrane. A lipid anchor covalently attached to the C-terminus of Ras penetrates into the lipid bilayer. This membrane anchorage is essential for the biological activity of Ras. Hence, the inhibition of anchor attachment has become an attractive pharmacological target [13]. See: Waldmann H, Thutewohl M, Ras-Farnesyltransferase-inhibitors as promising anti-tumor drugs, this volume.

The primary target for postranslational modification is the so-called Caax box acting as a prenylation signal. These four letters represent the four C-terminal amino acids of almost all Ras-related proteins indicating a Cys and two aliphatic amino acids. Where x is a Ser or Met like in Ras the protein becomes farnesylated and where x represents Leu or Phe a geranylgeranyl group is attached. Ras becomes modified presumably immediately after synthesis. In the first postranslational modification step a farnesyl group is covalently linked to Cys-186 of Ras. This is catalyzed by farnesyl transferase localized in the cytoplasm and taking farnesylpyrophosphate as substrate [14]. Second, the three C-terminal amino acids aax are cleaved off by a prenylprotein specific endoprotease [15]. In the third step, the cystein is carboxymethylated by carboxymethyl transferase [15]. These two enzymes are located to the endomembrane system [16,17].

As soon as Ras sticks to the plasma membrane another lipid anchor is attached to it. A putative palmitoyl transferase which is assumed to reside in the plasma membrane catalyses the thio-esterification of another cysteine close to the C-terminus [18]. This process leads to trapping of Ras at the cell membrane [19]. K-Ras4B, which does not become further modified after carboxyme-thylation but which carries a set of six lysins in the hypervariable region, experiences enhanced binding to the negatively charged plasma membrane. Recently, a more complex model of Ras membrane targeting was suggested, involving intermediate localization of farnesylated Ras to the endomembrane system and vesicular transport of Ras to the plasma membrane [20]. Nevertheless, two signals are necessary for final docking of Ras to the cell membrane, i.e., farnesylation and palmitoylation or, alternatively, a polybasic stretch of amino acids.

1.4.2
GTPase Cycle

Ras is a GTPase with a weak intrinsic activity. It binds with similar, high affinity to both GTP and GDP, with a slow rate of dissociation. Due to its high molar excess in the cell it is prevailingly GTP which binds to Ras after dissociation of GDP. The biological function of Ras is based on the two different conformations it adopts in the GTP- and GDP-bound state, respectively. After GTP hydrolysis the release of the phosphate group triggers a conformational switch of two peptide stretches in Ras which were non-covalently bound to the γ-phosphate of GTP [21, 22]. As the change between the GTP- and GDP-bound form turns the biological activity on and off, respectively, Ras is often referred to as a molecular switch. In its resting state Ras is bound to GDP and only the GTP-bound form is able to bind strongly to effector molecules which become activated in order to transmit further the signal in their pathway (see Fig. 3). According to the following equation the relative amount of Ras bound to GTP and hence its biological activity, depends on the GTP hydrolysis and GDP dissociation rates which may be up-regulated as described in the two following sections:

$$\frac{[\text{Ras} \cdot \text{GTP}]}{[\text{Ras} \cdot \text{GTP}] + [\text{Ras} \cdot \text{GDP}]} = \frac{k_{\text{diss}}}{k_{\text{diss}} + k_{\text{hyd}}}$$

1.4.2.1
Exchange Factors

Ras activation, i.e., exchange of the bound GDP for GTP, is accelerated by guanine nucleotide exchange factors (GEF) (see Fig. 3). Sos is ubiquitously expressed whereas other GEFs are found in some tissues only [7]. Activation of Ras occurs after extracellular stimulation of the cell: for instance hormones or growth factors bind to their specific receptors at the outer cell membrane, leading to tyrosine kinase activity of the transmembranal receptor at the cytoplasmic side of the membrane. Autophosphorylation of receptor tyrosine residues leads to docking of the SH2 domain of Grb2 which, on its part, recruits Sos to

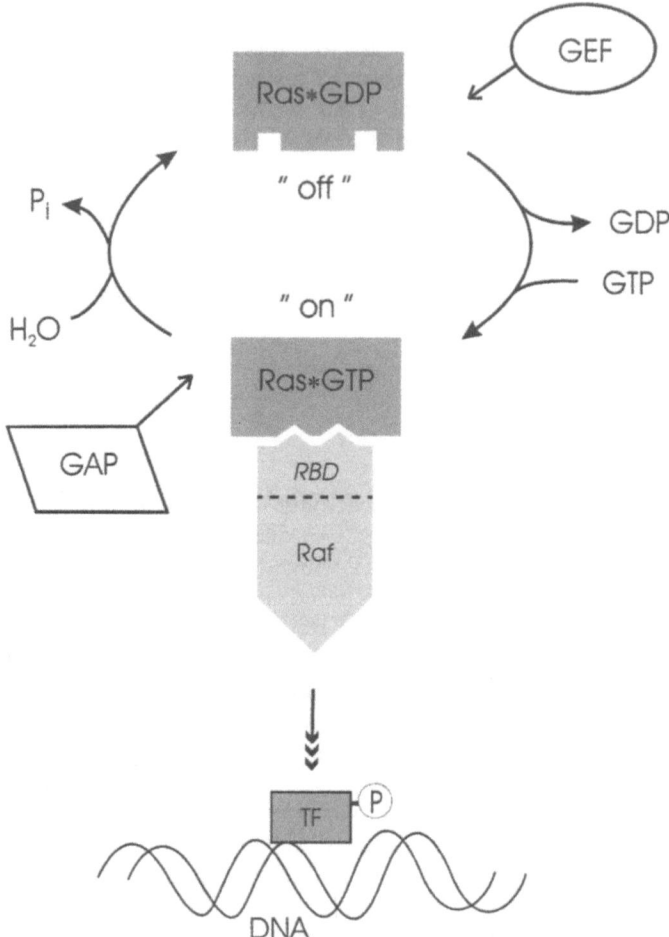

Fig. 3. Ras represents a typical GTPase, being active in the GTP-bound state and silent when GDP-bound. GTP loading can be accelerated by guanine nucleotide exchange factors (GEF), activating Ras, whereas GTP hydrolysis is upregulated by GTPase activating proteins (GAP), thereby turning Ras "off". In the "on"-state Ras can activate effector molecules like Raf through binding to the Ras binding domain (RBD) which seems to be common to (most) effector proteins. The cellular response in the case of Raf activation is – after a cascade of phosphorylation reactions – the activation of DNA transcription (TF = transcription factor in the nucleus) leading to cell division. Note that Ras is permanently located at the inner plasma membrane (not indicated here)

the membrane by binding with its SH3 domain to proline-rich regions in Sos. Three features common to many signaling pathways should be stressed here:

1. Phosphorylation leads to enhanced (or decreased) enzymatic activity or to altered binding affinities.

2. SH2 and SH3 domains refer to src homology 2 and 3, a non-receptor tyrosine kinase where these domains were identified originally. They bind to phosphorylated tyrosine and proline rich epitopes, respectively, with specificities each in respect to the neighboring amino acids. Grb2 is an adapter molecule par excellence having both SH2 and SH3 domains (see also the respective article in this volume).
3. The third feature is the recruitment of a molecule, in this case Sos, to the membrane in order to fulfill its function. Both increased local concentrations at the membrane and two-dimensional diffusion of the collision partners, Sos and Ras, lead to enhanced encounter of the two proteins and thus to accelerated nucleotide exchange on Ras.

Other activation systems have been identified lately, like activation of RasGRP (guanine nucleotide releasing protein) by diacyl glycerol/Ca^{2+} [23, 24]. Also $Cdc25^{Mm}$/RasGRF (guanine nucleotide releasing factor) is stimulated by Ca^{2+}-induced binding of calmodulin to its IQ-motif [25].

1.4.2.2
GTPase Activating Proteins

Ras returns to its resting state by hydrolysis of its cofactor GTP. As mentioned above, the GTPase activity of Ras is low but it can be enhanced up to 10^5-fold by the interaction with a GTPase activating protein (GAP) [26, 27] (see Fig. 3). The two most investigated GAPs in humans are p120^{GAP} and NF1. Whereas the former is thought to have an additional effector function [28, 29] the latter is classified as a tumor suppressor gene as disruption of its gene leads to neurofibromatosis [30]. This fact demonstrates the importance of having a regulatory machinery that not only activates Ras but which also allows it to be efficiently turned off. As with GEFs it is probably essential to have GAPs for all small GTPases for proper regulation and, in fact, specific GEFs and GAPs are known for many other Ras-related proteins.

The oncogenic potential of Ras is based on a gain of function, i.e., permanent activation of effectors, which actually is due to a loss of enzymatic activity [31]. Ras' Achilles heel is represented by residues Gly12 and Gln61. When either of these residues is mutated, Ras turns oncogenic which is based on the impairment of its GTP hydrolysis activity [31]. Moreover, GAPs cannot act on Ras any more [26]. This results in a permanently high concentration of Ras · GTP leading to permanent, unregulated activation of the downstream targets and hence to uncontrolled cell growth. Together with mutations of some other genes like p53 the cell becomes transformed which finally results in tumor growth. Although oncogenic Ras alone is not responsible for cell transformation but only in concert with a few other genetic disruptions [1, 32, 33], Ras is one of the most frequent oncogenes. It is found in 30% of all human tumors [34], playing a prominent role, e.g., in colon (50%) and pancreatic cancer (90%), apparently, being of little importance in, e.g., breast cancer. Designing compounds which re-establish GTPase activity of oncogenic Ras and its susceptibility to GAP represents another approach in the development of anti-cancer drugs directed against Ras [35, 235].

1.4.3
Effectors

After Ras has been activated, the signal is further transduced by interaction of Ras · GTP with downstream targets. Obviously, Ras may accomplish the diverse cellular functions which are known for it by activating various signal branches. Each of these pathways experiences additional regulation by inputs from other signal molecules so that the actual result of Ras activation depends on the physiological context of the cell. An intriguing finding was the partial loss of Ras function by distinct mutations in its effector region [36–39]. Some residues are essential for the interaction with one effector and others are more important for the interaction with another effector. Such mutants are useful to study the biological effect of suppressing one of the signal branches emerging from Ras.

Ras effectors are characterized by their nucleotide-dependent interaction with Ras, which is impaired by mutations in the effector region of Ras (e.g., D38A), and by their competition with GAP in binding to Ras suggesting that effectors and GAPs bind to the same or overlapping sites on Ras [26, 40]. In recent years many Ras effectors were identified like Raf, RalGDS, PI(3)K, PKCζ, MEKK, Byr2, AF-6, and Nore [41–54]. There is no sequence homology and no functional relationship among these effectors. There is no rule either for the location of the RBD in the effector reaching from the N-terminus in AF-6 to the C-terminus in RalGDS. So far only for three effectors, namely Raf, PI(3)K, and RalGDS, could the activation of their biological function by Ras be demonstrated. Figure 4 shows only five out of more than ten putative Ras effectors, also indicating the following pathways and the cellular response.

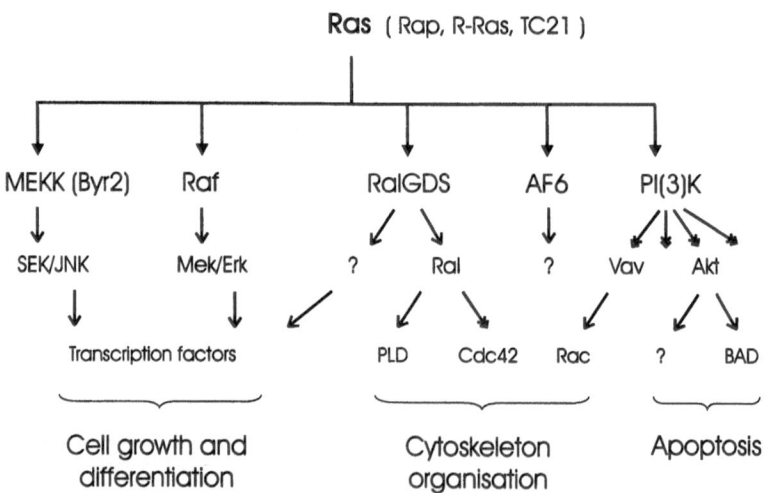

Fig. 4. Ras (in the GTP state) activates a variety of effectors each with different functions and cellular pathways. Only five out of more than ten putative effectors are shown. The signal cascades following effector triggering are schematically shown which lead to one of the possible cellular responses to activated Ras. Due to their high sequence homology, Rap, R-Ras, and TC21 may also interact with the Ras effectors. Scheme according to [55]

1.4.3.1
Raf

Adenylate cyclase was identified as the primary Ras target in yeast (*Saccharomyces cerevisiae*) [56] but it took a while before in 1993 several groups independently found Raf to be the effector of Ras in mammals [41–44]. Shortly afterwards it was realized that this is not the only target of Ras but up until now it appears to be the most prominent one. Raf is a Ser/Thr-specific protein kinase which phosphorylates and thereby activates Mek which in turn phosphorylates and activates Erk, leading to an amplification of the signal. Erk, also termed MAPK, has a plethora of phosphorylation targets, the most important of which are transcription factors such as Elk-1, leading to activation of the transcription machinery in the nucleus.

The catalytic domain of Raf is located in the C-terminal half whereas the N-terminal part is regarded as the regulatory domain [57]. Here two distinct domains adjacent to each other were identified, the Ras binding domain (RBD) comprising 80 amino acids and the zinc ion binding region or cysteine rich region (CRR). Much interest has been directed towards the Ras/Raf interaction and the mechanism of Raf activation, since this protein couple represents a crucial hinge in signal transduction and thereby a potential target for anti-cancer drugs. Through binding to Ras · GTP, Raf is translocated to the plasma membrane and there is still debate going on about further activation events on Raf. By fusing two membrane translocation signals to Raf, namely the C-terminus of K-Ras including the poly-lysine stretch and the CAAX box, the biological function of Raf could be fully activated, suggesting a mechanism of Raf activation simply by membrane recruitment [58, 59]. So far unknown factors also relating to the membrane were postulated to ensure sustained activation of Raf. Phosphorylation of critical Tyr residues in the middle part of the Raf molecule are discussed as additional activation steps [60, 61], although the isoform B-Raf does not have these sites. It has been suggested that 14-3-3 proteins play a role in activation which may convey allosteric or stability effects onto Raf [62–64] or which lead to dimerization and subsequent transphosphorylation of Raf in this complex [65, 66]. The role of RBD is obvious, i.e., docking the Raf molecule to Ras and thereby to the membrane, whereas the function of CRR in Raf activation is not so clear. Contradictory data are published about CRR involvement in Ras interaction [67–69], possibly binding to the farnesyl tail [70–72]. Furthermore, CRR might play a role, due to its hydrophobic character, in membrane attachment, thereby reinforcing membrane docking [73], or it might have a contact with the kinase domain which adopts a different structure when it is attached to the membrane. An alanine scanning study identified Raf activating and inhibitory mutations and defined distinct epitopes, demonstrating the involvement of CRR in both Ras interaction and lipid cofactor binding [74]. Observations with the isoform B-Raf, which can be activated by farnesylated Ras · GTP in a membrane-free assay, again in concert with 14-3-3 protein [75], are in favor of an allosteric model of Raf activation. Nevertheless, impairment of Ras/Raf interaction leads to abrogation of growth signaling, suggesting the most important signal pathway from Ras effectively to be cut off.

1.4.3.2
RalGDS

RalGDS is an exchange factor for a close relative of Ras, Ral, and it was identified as Ras effector, again independently by a number of research groups [45–47]. With Rlf and Rgl, RalGDS-related proteins were found which are also GEFs for Ral and bind to Ras via their C-terminal RBD [45, 76]. Elevated levels of Ras · GTP in the cell lead to increased loading of Ral with GTP [77]. Activated Ral in turn interacts with RalBP1 which is a GAP specific for Cdc42 and Rac, thereby linking Ras signaling to the Rho family of small GTPases [78] (see below). It was suggested before that oncogenic Ras achieves full cell transformation not by overactivation of Raf alone [79, 80]. Aberrant function in pathways regulating cell morphology is also required (see also Fig. 4). Mutants of Ras were found which block activation of one but not the other effector, the biological effects of which demonstrated the need for synergizing pathways [36]. A second function of RalGDS is transcriptional activation through c-fos [37, 77, 81] , the pathway of which is not quite understood yet. Rlf, another Ral-GEF, is also involved in transcriptional activation [76, 82].

In certain cell types RalGDS acts opposingly to other Ras effectors. It was shown to suppress neurite outgrowth in PC12 cells which is induced by Ras-activated Raf and PI(3)K [83]. This and other examples [84] demonstrate that the cellular response to Ras and the timing of biological events depend on the magnitude and the kinetics of activation of different effectors and their downstream targets which are also subject to the regulation by other signaling molecules. An example is given by the attenuated activation of Raf in contrast to RalGDS when these two effectors are phosphorylated by cAMP-dependent protein kinase A [85].

1.4.3.3
PI(3)K

PI(3)K is a well known enzyme involved in the production of second messenger molecules. α-, β-, and δ-isoforms in class I of this enzyme are heterodimers consisting of a regulatory subunit, p85, which contains SH2 and SH3 domains, and a catalytic subunit, p110. p110-γ does not bind to any p85 subunit and therefore cannot be activated by tyrosine kinase receptors but instead by α- and $\beta\gamma$-subunits of heterotrimeric G-proteins [86]. Additionally or alternatively, all four p110 isoforms bind to Ras · GTP [48, 49, 87, 88] and more or less are thereby activated [87, 89]. A mechanism for this could again be simply recruitment to the membrane where the major substrates for PI(3)K are located. p110 has not only the ability to phosphorylate phosphatidyl inositols but also proteins. PI(3)K is therefore involved in diverse cellular processes and Ras in particular seems to link it to the regulation of cytoskeletal rearrangements and cell division, the molecular pathways of which are not precisely traced so far. Targets for the second messengers produced by PI(3)K are parts from different signaling pathways like the Ser/Thr kinase Akt/PKB, Rac, and p70-S6 [90–95]. The biological consequences of PI(3)K activity appeared to be even more complex after finding that PI(3)K can also act upstream of Ras [96]. Finally, it is worthwhile to point out that the Ras/PI(3)K signal branch via Akt and Bad is involved in the control of cell survival, yet another factor important in cancer [97] (see also Fig. 4).

1.5
Relatives of Ras

Ras is the prototype of small GTPases but there are many other members in this protein superfamily specializing in distinct biological functions. The classification into subfamilies according to sequence homology coincides – not surprisingly – with functional relationships. While little is known about Rad proteins, the biological functions of members in the Rho, Rab, Arf, and Ran subfamilies shall be described briefly in the next sections but one. Yet before these cousins are reviewed, the brothers and sisters of Ras will be introduced in the following section.

1.5.1
Other Members of the Ras Subfamily

According to their high homology in primary sequence, Rap1/2 and R-Ras/TC21 are members of the Ras subfamily. In particular, they are characterized by their almost identical effector region (residues 32–40 [98]) where only Ral, also belonging to the Ras subfamily, shows significant deviations [11]. It is therefore not surprising that all these proteins except Ral can bind to the effectors described above, but the biological consequences of these interactions are very different. Concerning Ral, residues 36 and 37 in Ras-like proteins, respectively, were identified as tree-determinant in respect to effector specificity [99].

Rap was discovered as a gene that was able to revert the phenotype of cells transformed by K-Ras and was therefore originally termed K-rev [100]. Based on this observation and further evidence it was assumed that Rap acts as antagonist to Ras by sequestering Ras effectors in unproductive complexes. This view has changed in recent years to attributing a more active role to Rap. In fact, Rap can activate B-Raf [101, 102] and Rlf [103] demonstrating that Rap uses bona fide Ras effectors but leading to different biological effects (for review see [104, 105]). This could be explained by the different cellular localization of Ras and Rap or the occurrence of Rap in special cell types. In case the hydrophobic tail is involved in effector activation, it is interesting to note that Rap is geranyl-geranylated.

In contrast to the effectors which seem to be shared by the Ras subfamily members, specific GEFs and GAPs were identified for Rap. Whereas Ras is predominantly activated by a cell membrane-translocated GEF, many GEFs known for Rap are directly activated by second messenger molecules like cAMP [106], diacylglycerol and calcium ions [105, 107–109]. This may again have something to do with the localization of Rap at intracellular membranous compartments. Certainly the GAPs [110] of Rap play an important role in regulation as well since the intrinsic GTPase activity of Rap is ten times lower compared to Ras.

Ral has attracted much interest in recent years, not least because it was demonstrated to mediate part of Ras function as described above. In contrast to Rap, which rather inhibits Ras signaling, Ral is part of one of the essential Ras-activated pathways. Moreover, it has proved to be acting in parallel with the Raf pathway in cell transformation induced by oncogenic Ras [37, 77]. The case of Ral demonstrates the complexity – and the incomplete knowledge and understanding – of signal transduction. Ral can also be activated by Rap mediated by Rlf [103] and, alternatively, by binding of a calcium/calmodulin complex to the Ral C-terminus which obviously does not affect the nucleotide state of Ral [111].

Generally, the presence of several GEFs and GAPs for each GTPase introduces a lot of variability and complexity and it is not easy to recognize physiologically significant partnerships. R-Ras1 and 2, the latter also termed TC21, seem to share Ras effectors and to activate the corresponding pathways, not all of them and not to the same extent though. Both proteins were found to lead to cell transformation when their GTPase activity is deregulated, suggesting oncogenic potential at least in certain cell types [112–115]. The product of the proto-oncogene bcl-2, which is involved in control of apoptosis, binds to the C-terminus of R-Ras [116] but the mechanism of regulation is not known. For sake of completeness the other members of the Ras subfamily, Rheb, M-Ras, and Rit/Rin, should be mentioned here but little is known about their function [117–119].

1.5.2
Rho Subfamily

Of the eight subfamilies of small GTPases, the Rho family is closest related to Ras from the functional point of view and shall be described here briefly. For all three members in this subfamily, Rho, Rac, and Cdc42, particular functions in cytoskeletal organization are established [120]. Rho activation leads to adhesion and the formation of stress fibers, Rac is responsible for membrane ruffling, and Cdc42 is involved in filopodia formation. As described above for Ras, each Rho protein is activated and deactivated by specific GEFs and GAPs, respectively [121]. With the GDIs, a third class of regulators is found for Rho proteins. Recently, the structural basis was elucidated for their function in inhibition of both guanine nucleotide dissociation and GTPase activity [122]. Specific sets of effector proteins are known for all three family members. Through these and in part unknown proteins, the signals are mediated to F-actin or actin-binding proteins, which results in distinct processes of cytoskeletal rearrangement. In addition, and in analogy to Ras, MAPK modules are turned on by activated Rho proteins [123–125]. These kinase cascades lead to transcriptional activation which results here in morphological alterations of the cell.

A functional connection between the Rho proteins has been observed. Cdc42 may lead to the activation of Rac which for its part can activate Rho [126]. The communication between different pathways, often referred to as cross talk, is characteristic for signal transduction but for many pathways only vaguely described or postulated. Nevertheless, there are also links between Ras and Rho/Rac pathways and their synergism is essential for complete cell transformation [79, 80, 127].

1.5.3
Rab and Arf Subfamilies

There are more than 40 different Rab proteins known so far and the size of this subfamily is still growing [128]. They are involved in direction of vesicular transport between subcellular compartments like endosomes, endoplasmic reticulum, Golgi apparatus, etc. [129, 130]. In their GDP-bound form the prenylated Rab proteins stick to the donor membrane where GEF mediated exchange for GTP occurs. Bound to the detached vesicle Rab · GTP travels to the acceptor membrane and after fusion of the vesicle with this membrane GTP hydrolysis is enhanced by acceptor membrane-localized GAP. The way back to the donor membrane through the plasma is facilitated by company of a GDI protein which increases the solubility of Rab · GDP by masking the hydrophobic prenyl group. For a more detailed description of the role of Rab in target membrane recognition and the involvement of SNARE proteins in membrane fusion, one is referred to reviews in this field [130–132]. Members of the Arf subfamily play a role in vesicle transport as well [133]. A major difference in their function is that they recruit coating proteins onto the trafficking vesicle which dissociate after GTP hydrolysis.

1.5.4
Ran

Ran (the Ras-related nuclear protein) is the major regulator of nucleo-cytoplasmic transport [134] across the nuclear pore complex (NPC). Like other small Ras-like GTP-binding proteins, it switches between a GTP- and a GDP-bound form by GTP-hydrolysis and nucleotide exchange [135]. In contrast to its relatives, Ran does not undergo posttranslational modification.

The exclusive nuclear localization of the Ran exchange factor RCC1 [136] ensures that the generation of Ran · GTP is confined to the nucleus. The conversion of Ran · GTP into Ran · GDP is catalyzed by the GTPase activating protein RanGAP1 [137], which is present only in the cytoplasm or at the cytoplasmic site of the nuclear pore, efficiently depleting Ran · GTP from the cytoplasm (Fig. 5). This differential localization of the regulators of Ran's nucleotide-bound state should thus result in a Ran · GTP-gradient across the NPC which is believed to drive the import-export cycle [138].

Ran interacts with transport receptors which load and unload their cargo in the respective compartment depending on the nucleotide state of Ran [138]. An importin binds its cargo initially in the cytoplasm, gets translocated through the NPC, releases the cargo upon binding Ran · GTP in the nucleus [139–141], and returns to the cytoplasm as a Ran · GTP complex. The removal of Ran · GTP

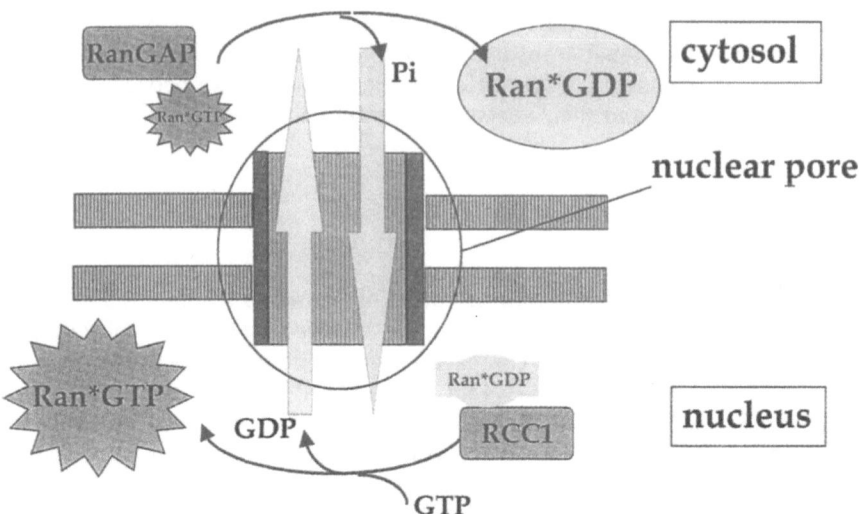

Fig. 5. The cellular compartment defines the nucleotide state of Ran. In the cytosol a Ran-specific GTPase activating protein (RanGAP) ensures that Ran exists almost exclusively in the GDP-complexed state. In the nucleus a Ran nucleotide exchange factor (RCC1) catalyzes the exchange reaction from Ran · GDP towards Ran · GTP. Import and export complexes recognize their actual cellular environment by interaction with the Ran protein

from the importin involves the hydrolysis of the Ran-bound GTP and allows the importin to bind and import the next cargo molecule.

Binding of substrates to exportins is regulated in a converse manner to importins. Exportins bind their cargoes preferentially in the nucleus, forming a trimeric cargo · exportin · Ran · GTP complex [142]. This trimeric complex is then transferred to the cytoplasm where Ran · GTP is converted to Ran · GDP. This results in Ran's dissociation from the complex, allowing the exportin to release its substrate, re-enter the nucleus, and to start the next export cycle.

The recycling process of Ran between its GDP and GTP complexed state is supported by another class of Ran-binding proteins, which support GTP-hydrolysis by RanGAP and are either soluble proteins [143] or part of the nuclear pore complex [144].

Sequences of Ran and Ran-interacting proteins are highly conserved between different species and allow interspecies exchange of GEFs and GAPs from, e.g., yeast to mammalian. Disturbing the Ran-system causes analogous effects on the organization of the nuclear skeleton, RNA-processing and transport, protein import, and cell-cycle control [145].

2
Biophysical Methods

Biophysical analysis of biomolecules like proteins, nucleic acids, or lipids utilizes intrinsic physical properties of the observed molecule itself or of an associated reporter molecule, which reflect information about structural characteristics, interactions, or reactions of the subject observed. In most cases the analysis (and the labels introduced) only interferes slightly with the interaction of interest and does not induce significant changes in the properties of the reactants.

2.1
Readout for Biological Macromolecules

Information about the state of an interaction can be achieved by a multitude of different readouts. Most of them are based on spectroscopical properties of the biomolecules or the labels introduced and allow real time analysis of the reactions. In some cases samples have to be retained and the analysis has to be performed subsequently.

2.1.1
Microscopic Properties and Macroscopic Signals

The main interest of biophysical analysis is to get information about function and mechanism of interactions between biomolecules. This is reflected by the typical way of presenting models for biological systems. Individual molecules bind to each other, catalyze an enzymatic reaction, or exhibit special structural features. The idea about the system is presented on a microscopic level.

Nevertheless most of the biophysical methods do not observe single molecules but a huge amount of players in a concentration range between 10^{14} and 10^{20} molecules/l (corresponding to 100 pmol/l to 1 mmol/l, which are reasonable concentrations in cellular systems). While the dynamics of the single reactants is stochastic the macroscopic readout of the measuring system normally can be described with ordinary differential equations.

2.1.2
Radioisotopes

Even nowadays the application of radioactive isotopes is the most sensitive method for the analysis of biomolecules or their reaction products. Besides the low detection limits, the replacement of a naturally overbalancing stable isotope by its radioactive analogue does not interfere with the physical or chemical properties of the enzyme (with some exceptions for hydrogens). Figure 6 lists some frequently used radioactive isotopes and their half-life periods.

$$^3_1H \rightarrow {^3_2}He + \beta^- + h\nu, \ t_{1/2} = 12.43 \ a$$

$$^{32}_{15}P \rightarrow {^{32}_{16}}S + \beta^- + h\nu, \ t_{1/2} = 14.3 \ d$$

$$^{35}_{16}S \rightarrow {^{35}_{17}}Cl + \beta^- + h\nu, \ t_{1/2} = 87.1 \ d$$

$$^{14}_{6}C \rightarrow {^{14}_{7}}C + \beta^- + h\nu, \ t_{1/2} = 5736 \ a$$

Fig. 6. Some frequently used Radioactive Nucleotides in Biochemistry. All nucleotides presented here are β-emitters with halflife periods from days (d) to thousands of years (a)

Concentrations of radiolabeled proteins, substrates, or products can be quantified by scintillation counters, which detect both emitters of weak (e.g., ^{3}H) and high energy (e.g., ^{32}P) by excitation of an organic solvent (e.g., toluene) which then emits fluorescence light. In commercial systems the primary fluorescence is transformed via one or two additional fluorescent dyes in the solution into a visible emission signal which can easily be detected by conventional photomultipliers.

A locally resolved detection of radioactive samples after chromatographic separation can be performed by imaging techniques which work either indirectly with Eu^{3+}- or P-doped sensor plates and laser activated emission or directly by a microchannel array detector which works like an open Geiger-Müller counter.

2.1.3
Staining Techniques

Direct staining of proteins (e.g., after electrophoretic separation in polyacrylamide gels) can be achieved by treatment with dyes like Coomassie Brilliant Blue R-250 [146] (Fig. 7), which binds positively charged proteins in an acidic fixation buffer, allowing detection down to 0.1 µg of protein.

An increase in sensitivity is realized by silver-staining, where residues containing sulfur (cysteine, methionine) or basic side chains (arginine, lysine, histidine) reduce Ag$^+$, leading to brown or black colored bands. Here, down to 0.1 ng of protein can be detected.

For quantitative analysis of protein concentration the colorimetric Bradford-assay [147] is most commonly used. Here another Coomassie dye, Brilliant Blue G-250, binds in acidic solutions to basic and aromatic side chains of proteins. Binding is detected via a shift in the absorption maximum of the dye from 465 nm to 595 nm. Mostly calibration is performed with standard proteins like bovine serum albumin (BSA). Due to the varying contents of basic and aromatic side chains in proteins, systematic errors in the quantification of proteins may occur.

2.1.4
Enzymatic Detection

Enzymatic reactions for the direct quantification of low molecular weight substrates and products are well established in clinical chemistry or nutrient analy-

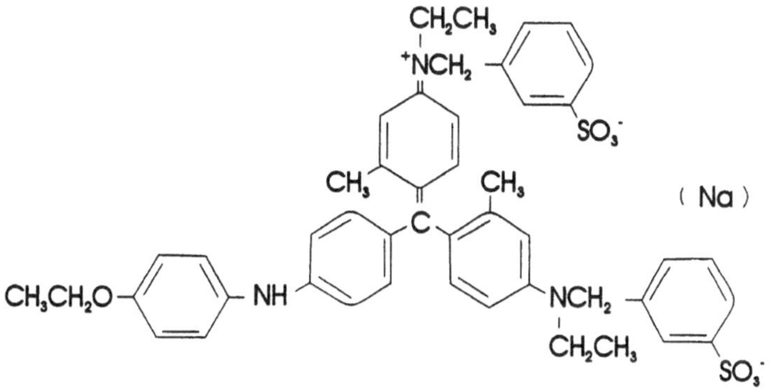

Fig. 7. Coomassie Brilliant Blue R-250 (I) is used to stain proteins, e.g., after gel-electrophoretic separation, its derivative G-250 (II) is applied in the Bradford assay for protein quantification

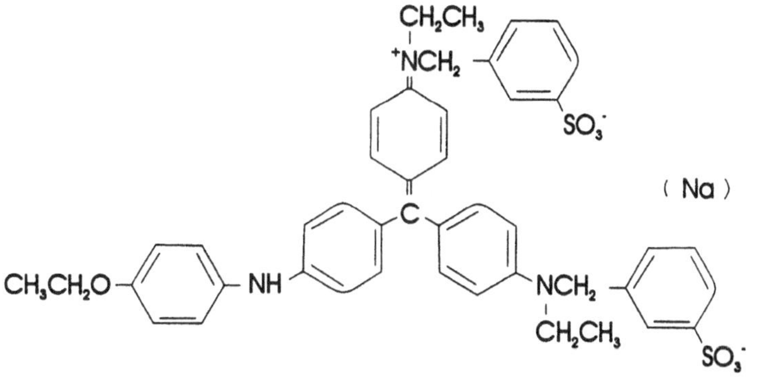

tics. For proteins enzymatic methods enhance sensitivity of detection and decrease the limits for quantification by orders of magnitude.

A classical approach is the enzyme linked immunosorbent assay (ELISA), where the antigen (e.g., the protein to be quantified) is immobilized on the surface of a well. A first antigen-specific antibody is applied to occupy all antigens, before a second antibody binds all primary antibodies on the well. The second antibody carries an enzyme, which now catalyzes a color reaction. If the substrate of the enzyme is given in high excess, the enzyme is saturated and the production of product is linear with time and concentration of second antibody and antigen (Fig. 8).

There exists a wide variety in the setup of ELISA assays (direct binding or competition setups) and the enzymatic reaction utilized [148]. A similar principle to enhance sensitivity by enzymatic coupling is realized after gel electrophoretic separation of proteins. Here proteins are transferred to nitrocellulose ("western blot") and detected by antibody-coupled enzymes.

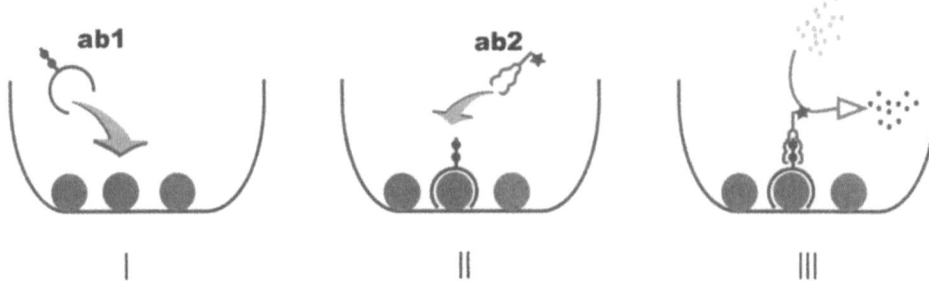

Fig. 8. Schematic presentation of a enzyme linked immunosorbent assay (ELISA). An antigen (●) is immobilized on the surface of a microtiter plate and incubated with its antibody (ab1). A second antibody (ab2) with a covalently linked enzyme (★, e.g., horseradish peroxidase) binds to the primary one and catalyzes a color reaction with its enzyme. All incubations are separated by washing steps

2.1.5
UV/Vis Spectroscopy

There are two principle ways for optical detection of protein concentrations: either the macromolecule or its label emits energy (after excitation by light) – then a fluorescence signal can be measured; or it absorbs energy from electromagnetic waves passing the sample – then the optical absorption of the sample can be measured by UV/Vis spectroscopy and concentrations can be calculated according to Lambert-Beers Law.

There is a strong limitation in the concentration range due to the logarithmic relationship between transmission and concentration (optical densities reasonably to measure range from 0.1 to 1.5). Nevertheless, protein quantification by direct UV-measurement or after staining with dyes in the visible range is a very robust method and can be found, e.g., as a common detection mode in HPLC or other chromatographic techniques.

2.1.6
Fluorescence-Based Assays

The most populated energy state of chemical species at room temperature is the ground state. Once a molecule has absorbed energy in the form of electromagnetic radiation, it returns to the ground state, which can occur via several routes, some of which are shown in the Jablonski diagram (Fig. 9).

Fluorescence is possible if the absorption of light (A) leads to a higher vibrational state for the excited molecule than in its ground state. Internal conversion (IC) by vibrational relaxation leads to a lower energy (or higher wavelength) of the emitted light (F) when the molecule returns to the electronic ground state. Lifetimes of fluorescent states are very short (10^{-5} s to 10^{-8} s). This process is in competition with vibrational relaxation, which usually occurs much faster and is enhanced by physical contact of an excited molecule with other particles. Therefore large portions of energy can be transferred through collisions.

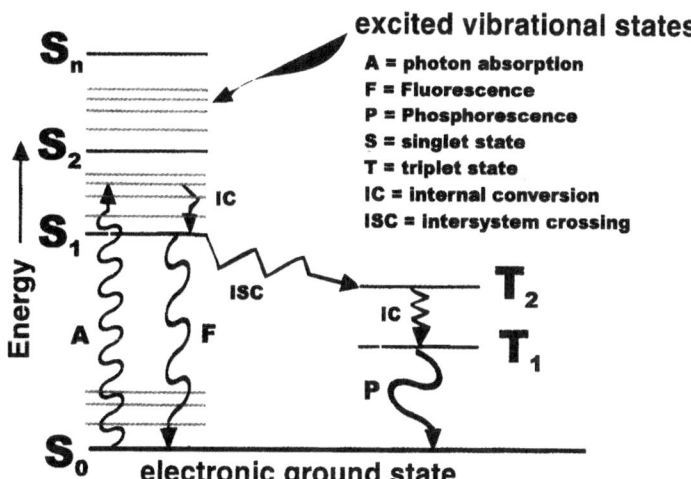

Fig. 9. Jablonsky Diagram for energy conversion pathways of an excited molecule. While fluorescence occurs between states of the same spin, an ISC (inter system crossing) leads to spin inversion and a delay in emission (phosphorescence halftimes from 10^{-4} s to minutes or even hours)

Due to the separation between excitation and emission in a fluorescence spectrometer, concentrations can be detected down to picomolar, with a wide linear range over up to five orders of magnitude. As a consequence of, e.g., vibrational relaxation, the amount of energy which is released as fluorescence (quantum yield) is strongly dependent on solvent and temperature.

A special option for measurements is realized in fluorescence polarization. Because a chromophore can absorb light energy only with high probability if the dipole moments of the transition state are parallel with the incident beam, only those molecules are excited by polarized light which have the corresponding orientation. During the lifetime of the fluorescence the molecule may change its orientation by rotation. The fluorescence light is again polarized with an axis parallel to the transition dipole moment of the chromophore.

If a sample is excited with polarized light and emission is measured through a second set of polarizers parallel and perpendicular to the first polarizer, the ratio of the two emission signals reflects the rotatory freedom of the chromophore. In practice, binding of a second molecule to the labeled one can be detected if the size of the chromophore complex increases considerably. The advantage of this method is that no changes in quantum yield are necessary for the observation of the binding reaction.

2.1.7
Surface Plasmon Resonance

The introduction of optical biosensors has made it possible to obtain data for a large number of macromolecular interactions without the necessity of additional labeling. Here several commercial instruments utilize the effect of Surface Plasmon Resonance (SPR) to detect accumulation of ligands in the sensor matrix.

In an SPR system a beam of polarized light is sent through a medium of higher refractive index towards a second medium with lower refractive index. If the incident light enters the interphase between the two phases beyond the critical angle, the beam is reflected totally, while an evanescent wave propagates from the interface into the lower refractive index medium, decaying within a distance of approximately one wavelength below the interphase.

The phenomenon of surface plasmon resonance is observed at thin metal films inserted between two phases of different refractive index. Now the free electron clouds (plasmon) of the metal are able to interact with the evanescent wave for specific angles of the incident beam. These angles are dependent on the ratio of the refractive indices of the two media and a reduced intensity of the reflected beam can be monitored. In the BIAcore system the high refractive index media is built by a glass prism coupled mechanically to the glass surface of the sensor chip (Fig. 10). The chip itself is covered with a 50-nm gold layer, which can be modified with dextran hydrogels or long-chain alkanethiol molecules in order to generate artificial surfaces [149].

In general the first reactant is immobilized on the surface, while the second one is applied in buffer solution. Enrichment of macromolecules (e. g., proteins) at the aqueous site of the sensor leads to an increase in refractive index of the corresponding phase and thereby to a change in resonance angle, which can be monitored directly in a time-resolved manner.

2.1.8
Circular Dichroism

Optically active chromophores show different absorption for left and right circular polarized light (where the orientation of the polarized light changes periodically). These substances modify a circular polarized beam in such a way that the light is elliptically polarized after leaving the sample, an effect called circular dichroism.

In proteins in particular the peptide bonds contribute to the CD-spectra of the macromolecule. Here, CD-spectra reflect the secondary structure of proteins, which are derived from CD-spectra of model macromolecules with only one defined secondary structure (like poly-L-lysine at given pH values) or based on spectra of proteins with known structures (e. g., from X-ray crystallography). The amount of α-helices or β-sheets in the unknown structure is calculated by linear combination of the reference spectra [150, 151].

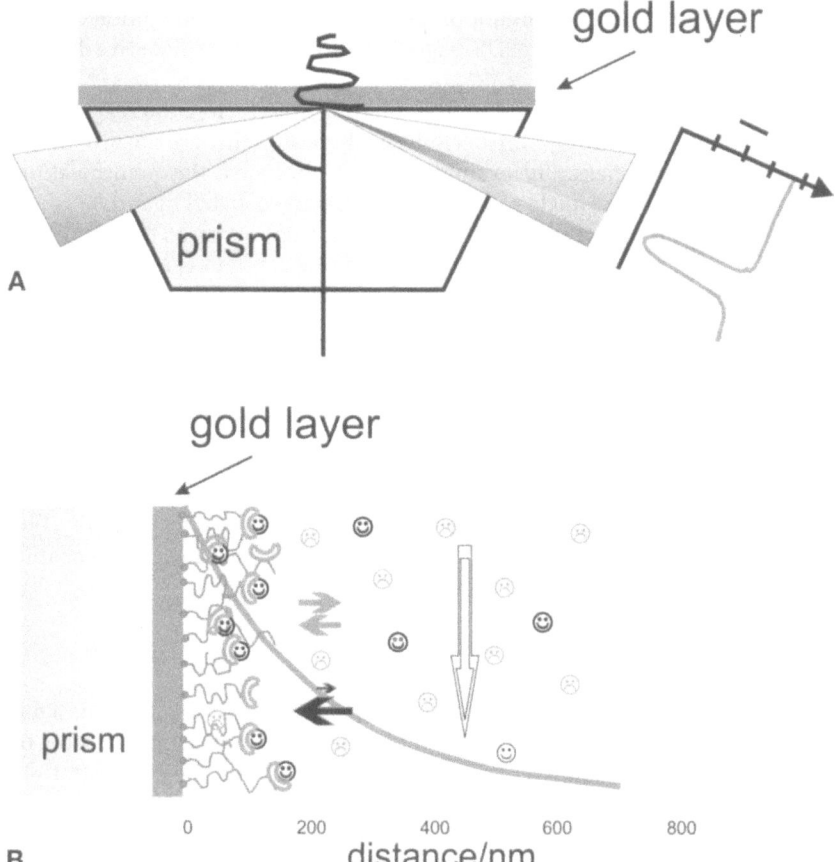

Fig. 10. A SPR Detection realized in a BIAcore system. A fan of polarized light passes a prism and is focused at the interface to an aqueous phase under conditions of total reflection. An evanescent wave enters the solvent phase. If the prism is coated with a thin gold layer at the interface the free electrons in the metal absorb energy from the evanescent wave for a distinct angle, depending on the refractive index of the solvent near the interface. **B** The gold layer can be modified with, e.g., a carboxydextrane matrix, where catcher molecules can be immobilized by standard chemistry. If a ligand is applied with the aqueous phase it may interact with the catcher and accumulate in the matrix, causing a shift in the resonance angle. If no specific binding occurs the refractive index in proximity of the sensor is less affected

2.2
Equilibria

2.2.1
Pull Down and Filter Binding Assays

A simple method to analyze binding equilibria with slow association and dissociation kinetics is the precipitation of one binding partner with an antibody or a specific affinity tag. The amount of free and complexed macromolecules can then be analyzed, e.g., after SDS-page separation with radioactive labels or staining techniques.

A related approach is realized in filter binding assays. Here the reaction solution is filtered, e.g., through nitrocellulose where proteins are absorbed, while small molecules can pass. One example of this technique is the quantification of protein bound and free nucleotides (with radioactive labeled ligands).

Because the precipitation (or affinity binding step) perturbs the equilibrium of interest it has to be ensured that removal of one reactant is much faster than the dissociation or association of the binding partners.

2.2.2
Fluorescence Titration

Measurements of binding curves without influencing the equilibria can be performed if the readout for complex formation is correlated with a change in a macroscopic signal. This can be either a change in fluorescence intensity, fluorescence polarization, optical absorption, or heat of association (see next chapter).

Assume an equilibirium

$$A + B \Leftrightarrow AB$$

If concentration of B is kept constant, then $[B]_{max}$ represents the maximum number of binding sites available. During the experiment increasing amounts of A are added, occupying more and more of B and increasing the concentration of AB.

In equilibrium there exist free A (concentration [A]), free B ([B]) and complex AB ([AB]), where the dissociation constant K_D describes the ratio:

$$K_D = \frac{[A] \times [B]}{[AB]}$$

After replacement of [B] by $[B]_{max} - [AB]$ this equation can be rearranged to

$$\frac{[AB]}{[A]} = \frac{B_{max}}{K_D} - \frac{1}{K_D} \times [AB]$$

A plot of [AB]/[A] vs [AB] (Scatchard plot, Fig. 11B) should therefore result in a straight line with the slope $1/K_D$ and the intercept $[B_{max}]/K_D$ [152].

Figure 11A shows a theoretical example of a titration curve A + B = AB, where the signal is proportional to the amount of complex. The solid lines represent conditions where B_{max} is equal to K_D. Here for both presentations of signal vs either $[A_{total}]$ (total concentration of A added to the preparation) or $[A_{free}]$ (concentration of non-complexed A in the solution, calculated as $[A_{total}] - ([AB])$) the plot is curved and allows discrimination between free and complexed binding partners. If $[B_{max}]$ is substantially higher than K_D the issue of active site

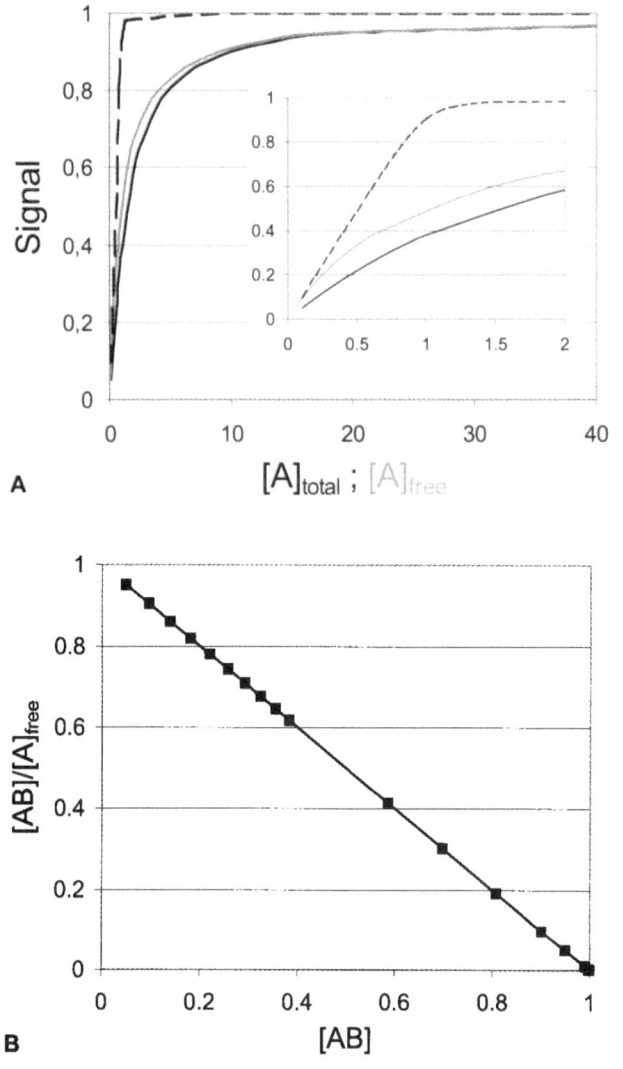

Fig. 11. A Theoretical binding curve for an A + B = AB interaction. *Black lines* indicate a plot of signal vs [A_{total}], *Grey lines* vs [A_{free}]. The *dashed line* shows an example of an active site titration. **B** Scatchard plot [154]

titration becomes relevant. In Fig. 11A the dashed line presents a plot of signal vs [A_{total}] for [B_{max}] = 100 × K_D. Under these conditions ligand A is complexed almost quantitatively by B, resulting in a straight line until almost all B is saturated by A. Although this titration has to follow the equations above, small errors in the determination of [A], [B], or [AB] make it impossible to quantify K_D reliably. On the other hand this kind of binding experiment is often used to determine the number of binding sites or, alternatively, the fraction of active protein. Alternative treatment of titration data is described elsewhere, e. g., [153].

For reasonable conditions [B] should be below the expected K_D value, and [A] should span a range from $0.1 - 10$ times K_D. Analysis of real binding curves has to consider additional sources of error such as equilibrium not being achieved, errors in the determination of the concentrations, or non-specific binding.

Complex binding reactions (more than one binding site for A on B, cooperative binding, etc.) can be described, e. g., by the Hill plot [154].

2.2.3
Isothermal Titration Calorimetry

Many of the methods described above may suffer from the modifications of the proteins which are necessary to obtain a detectable readout for the interaction under investigation. As a matter of fact, fluorescent labels are attached in close vicinity to the site of interaction, as otherwise binding to a partner is not likely to be detectable. It is conceivable that such labels invoke altered binding properties. Similar skepticism and arguments hold for immobilizing a protein. Non-invasive labeling is possible with radioactive isotopes but these methods bear other disadvantages.

Calorimetry takes advantage of an effect inherent to all interactions. Heat released or being consumed upon association is used to detect the interactions of proteins with their ligands or other proteins. This allows direct experimental access to the enthalpy of binding. A disadvantage is the relative high amount of material needed. Briefly, in this method one of the two reactants is added step by step from a syringe to a thermally insulated cell containing the other reactant, and the evolving heat is measured. A fit of a binding isotherm to the titration data yields the affinity constant and the stoichiometry of the complex formed, i. e., the number of binding sites [155]. A typical experiment is shown in Fig. 12. From these results the entropy of binding can be calculated. In addition, measuring the enthalpy in dependence of the temperature yields the heat capacity of binding. Altogether, isothermal titration calorimetry facilitates the investigation of biomolecule interactions in solution – without labeling or immobilization – leading to a detailed thermodynamic characterization.

2.3
Kinetic Analysis

2.3.1
Separation Techniques

As described for equilibria before, separation techniques like filter binding assays or pull down methods can be applied to separate educts and products of slow reactions too. Here modern HPLC systems with autosampler and time programmed injection offer a convenient approach for the analysis of reactions in the time scale of minutes and hours.

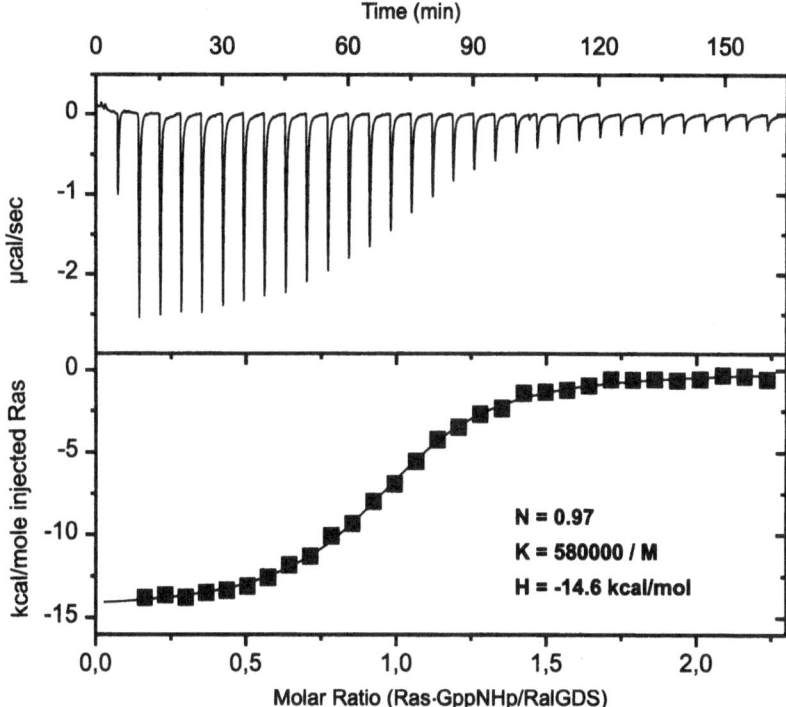

Fig. 12. Ras in complex with GppNHp at 600 µmol/l is injected into a solution of 45 µmol/l RalGDS-RBD. In the *upper panel* the (peakwise) change of the heating power is recorded which is necessary to keep the cell at constant temperature after each injection. The integrated peaks of the upper panel are plotted vs the molar ratio of Ras · GppNHp/RalGDS in the *lower panel*. The fitted curve yields the data in the box, where N indicates the stoichiometry, K the affinity constant and H the enthalpy of binding

2.3.2
Stopped Flow

If the reaction of interest is too fast to separate reaction products with classical techniques, or even to follow the change in signal in a standard cuvette by a UV/Vis or fluorescence spectrometer, stopped flow systems are a suitable alternative. Here, the reaction partners are filled in two syringes, which can be propelled by a pneumatic ram. The two reactants are then mixed in a small chamber, pass through a cuvette and end up in a stop syringe. As long as the piston of this syringe does not reach the trigger switch where it stops, there is a constant flow and the mixture in the cuvette will be exchanged continuously. After the stop the reaction proceeds in the cuvette and can be observed by spectroscopical (or other) readouts. Figure 13 shows the setup of a stopped flow system. Commercial instruments have dead-times of 1 – 2 ms and allow observations of reactions with half-times down to the ms range.

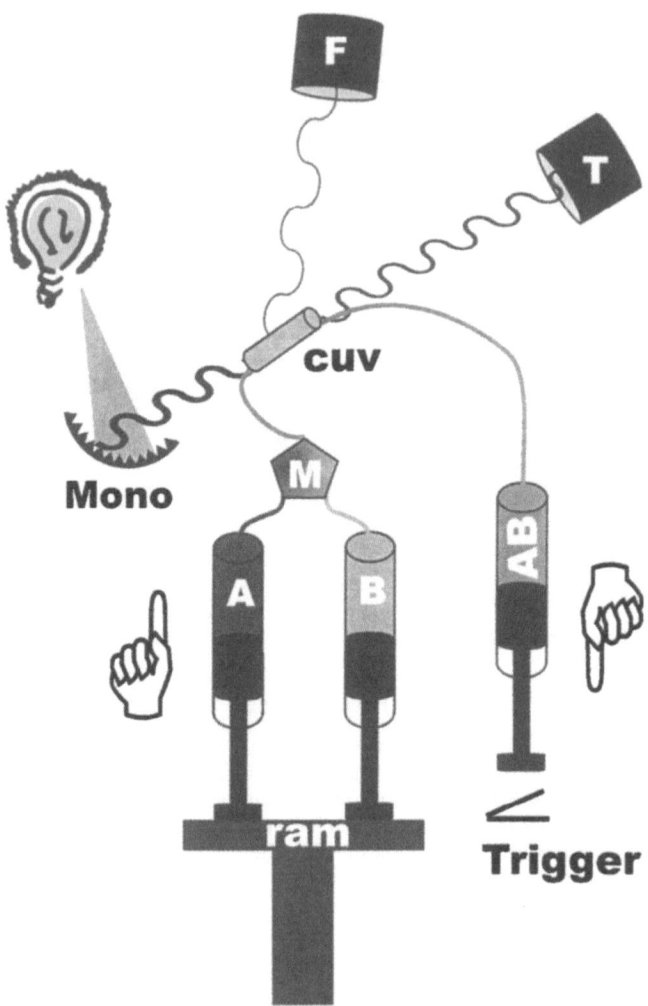

Fig. 13. Scheme of a stopped flow instrument. A pneumatic or electric driven ram pushes the contents of two driving syringes into a mixing chamber (*M*). The solution then passes a cuvette (*cuv*) and fills a stop syringe. When the piston of the stop syringe reaches the trigger switch the flow stops. Light of a selected wavelength passes through the cuvette. Reactions can be observed by changes in absorption (*T* = detector of transmitted light) or fluorescence signal (detector *F*)

To circumvent complex mathematical evaluations, stopped flow experiments are measured preferentially under pseudo first-order conditions. The rate of complex formation between protein A and protein or ligand B is described by

$$\frac{d\,[AB]}{dt} = k_{on} \times [A] \times [B] - k_{off} \times [AB]$$

[A] may be replaced by $[A_0] - [AB]$, and if $[B_0] \gg [A_0]$ the change in [B] may be neglected, i.e., it may be approximated as constant corresponding to pseudo first-order kinetics (where $[A_0]$ and $[B_0]$ are the concentrations of A and B at $t = 0$). Integration leads then to

$$[A]_t = [A]_0 \times \exp(-t \times (k_{on}[B]_0 + k_{off}))$$

Thus, the time traces can be fitted by an exponential equation and the linear concentration dependence of the observed rate constant yields the slope k_{on} and the ordinate k_{off}.

One advantage of (pseudo) first-order kinetics is the independence of the kinetic parameters in relation to the active concentration of ligand A. The only concentration which has to be determined precisely is that of the ligand in excess (B).

2.3.3
Surface Plasmon Resonance

As described for stopped flow experiments above, all commercially available SPR systems work under (pseudo) first-order conditions as well. This is realized either by a large excess of free ligand (in the large volume of the cuvette) compared with a nanoliter volume of the sensor layer [156] or by continuous replacement of free ligand in a flow injection system (e.g., BIAcore [157]).

The signal is measured in terms of resonance units (RU) and the mathematical treatment is as described above. An increase in RU signal is proportional to the concentration of complex AB. The maximum of the specific change in signal then corresponds to a complete saturation of the immobilized ligand.

Rinsing the sensor surface with buffer results in an irreversible dissociation, because all molecules which dissociate from immobilized protein are removed from the system by the buffer stream, allowing one to determine the rate constant of dissociation separately.

Again, care has to be taken for the non-ideal (or real) behavior of the measurement system. Applications are limited by non-specific absorption of molecules on the surface, mass transfer effects (under conditions of laminar flow a 1–5-μm layer between sensor surface and volume flow is not whirled and has to be passed by passive diffusion) or limited access for the immobilized molecules [158–160].

2.4
Structure

Determination of three-dimensional protein structures has become an important tool in protein research, indicated by the exponentially growing number of protein data bank entries during the last decade. They have contributed considerably to our understanding of function and mechanism of biomolecules. Structures of proteins alone or in complex with their substrates or binding partners like cofactors, DNA, or other proteins can visualize interactions at the atomic level. On the one hand mechanisms elucidated by biochemical methods may find further support or corrections. On the other hand the three-dimensional picture often proved to be the starting point for a detailed analysis of protein function and mechanism, particularly by mutating amino acid residues and studying the effect at key positions of the protein identified in its structure. Pharmaceutical companies began to direct their interest towards the structures of potential drug targets and the term *rational structure-based drug design* was able to supersede *drug screening* – at least for a short while. There are two major techniques for structure determination – X-ray analysis and NMR spectroscopy, both of which have their advantages and disadvantages, with the potential to complement each other.

2.4.1
X-Ray Analysis

For structure determination by X-ray analysis, single crystals of the protein are required. Depending on the X-ray source used, which can be a conventional X-ray rotating anode or a highly focussing synchrotron beam line, the minimum size of crystals may range from 0.3 mm down to 0.05 mm in diameter. The crystallization of the proteins is the key step towards their three-dimensional structure. Appropriate conditions are screened for by systematically changing the conditions in the solution like variation of temperature, pH, various salts and their concentrations, concentration of the protein, and, particularly, the type and concentration of precipitants (e.g., ammonium sulfate or diverse polyethylene glycols).

Once a suitable crystal is obtained and the X-ray diffraction data are collected, the calculation of the electron density map from the data has to overcome a hurdle inherent to X-ray analysis. The X-rays scattered by the electrons in the protein crystal are defined by their amplitudes and phases, but only the amplitude can be calculated from the intensity of the diffraction spot. Different methods have been developed in order to obtain the phase information. Two approaches, commonly applied in protein crystallography, should be mentioned here. In case the structure of a homologous protein or of a major component in a protein complex is already known, the phases can be obtained by molecular replacement. The other possibility requires further experimentation, since crystals and diffraction data of heavy atom derivatives of the native crystals are also needed. Heavy atoms may be introduced by covalent attachment to cystein residues of the protein prior to crystallization, by soaking of heavy metal salts into the crystal, or by incorporation of heavy atoms in amino acids (e.g., Se-methionine) prior to bacterial synthesis of the recombinant protein. Determination of the phases corresponding to the strongly scattering heavy atoms allows successive determination of all phases. This method is called isomorphous replacement.

The result is the electron density map of the protein crystal. The final task for the crystallographer is to build the appropriate protein model, i.e., putting amino acid for amino acid into the electron density. Routinely the theoretical amplitudes and phases are calculated from the model and compared to the experimental data in order to check the correctness of model building. The positions of the protein backbone and the amino acid side chains are well defined by X-ray structures at a resolution around 2 Å. Many examples of protein structures are known reaching down to a resolution of 1 Å, which is solely controlled by the quality of the crystal.

2.4.2
NMR Spectroscopy

Whereas structure determination by X-ray analysis, basically, is not restricted by the size of the molecule, there is an upper limit in NMR spectroscopy. The main reason for this is not the increasing complexity of the spectra with the size of the proteins but the increasing line widths of the resonance peaks. Due to a decreased tumbling rate with growing size of the molecule in solution, the spin/spin relaxation times become shorter, the inverse values of which define the width of the peaks. Depending on the strength of the magnetic field in the spectrometer, the biomolecule may be as large as 30–40 kDa. However, the newly introduced TROSY-spectroscopy which leads to strongly reduced line widths gives the promise that in future NMR structure determination will be possible even with much larger proteins [161, 162].

Two major advances have made NMR spectroscopy available for protein structure determination: the introduction of pulsed excitation combined with Fourier analysis and the still on-going development of multi-dimensional NMR techniques. These methods, also known as correlation spectroscopy, require labeling of the protein with isotopes like ^{15}N and ^{13}C, the nuclei of which have a spin quantum number of 1/2. This is achieved by feeding the bacteria exclusively with $^{15}NH_4Cl$ and ^{13}C-glucose and a cocktail of essential salts. The assignment of all peaks to the corresponding amino acids represents the bottleneck in protein structure determination by NMR spectroscopy. Basically, the information about the three-dimensional structure of the protein is extracted from NOESY spectra. The *nuclear Overhauser effect* underlying this two-dimensional technique generates cross-peaks in these spectra which originate from two spins each which are coupled through the space and not through covalent bonds. This allows for identification of neighbors other than those dictated by the sequence. Also the distances can be calculated and, based on this information (as a rule of thumb ≥ 10 NOE signals per residue), the three-dimensional structure can be constructed. Typically, short- and medium-range NOEs are used to build β-sheets and α-helices, which are also identified by their specific chemical shifts, whereas long-range NOEs are useful for arrangement of the tertiary structure.

The main advantage of NMR spectroscopy is its use with proteins in solution. In consequence, rather than obtaining a single three-dimensional structure of the protein, the final result for an NMR structure is a set of more or less overlying structures which fulfill the criteria and constraints given particularly by the NOEs. Typically, flexibly oriented protein loops appear as largely diverging structures in this part of the protein. Likewise, two distinct local conformations of the protein are represented by two differentiated populations of NMR structures. Conformational dynamics are observable on different time scales. The rates of equilibration of two (or more) substructures can be calculated from analysis of the line shape of the resonances and from spin relaxation times T_1 and T_2, respectively.

3
Analyzing Function and Properties of Ras

3.1
Properties of Ras

3.1.1
Nucleotide Binding

Ras proteins are the prototype for small GTP-binding proteins with the feature of high selectivity and affinity for guanosine di- and triphosphate. The first experiments to obtain information about the kinetics of nucleotide dissociation from Ras were performed with ^{3}H-labeled GDP or ^{32}P-labeled GTP in filter binding studies [163]. Here, Ras protein was loaded with radioactive nucleotides and incubated with an excess of non-labeled GDP for a given time. Separation of protein-bound and free nucleotide was achieved by filtration through nitrocellulose filters and resulted in a dissociation rate constant k_{diss} of 0.0079/min for GDP and 0.023/min for GTP with wild type protein at 37 °C. The dissociation reaction for GTP was superimposed by the intrinsic GTP-hydrolysis. Therefore k_{diss}(GTP) for the oncogenic RasG12 V protein (0.0047/min) may reflect the dissociation reaction more realistically.

Access to kinetic data for the association was made possible by the introduction of fluorescence labeled guanosine-nucleotides. Particularly helpful was the attachment of the methylanthraniloyl residue (mant) at the ribose, leading to an equilibrium mixture of the mant group at the 2'- and 3'-positions [164]. Non-hydrolyzable analogues of GTP commonly used are GppNHp or GTP-γS (see Fig. 14) [164].

The quantum yield of the mant fluorophore in mant-GDP or mant-GTP increases approximately by 100 % when the molecule changes from the aqueous environment into the nucleotide binding pocket of the Ras protein. Therefore the kinetics of complex formation between nucleotide free Ras and the mant analogues of GDP or GTP could be detected easily in a stopped flow system by an increase in fluorescence signal.

The association rate constant for mant-GDP with a value of $1.1 \times 10^6/(M \times s)$ at 25 °C together with the dissociation rate constant result in an affinity in the pmol/l range. In addition, the pseudo-first order kinetic experiments revealed a hyperbolic rather than a linear concentration dependence, indicating a two-step binding mechanism. Presumably, the initial formation of a weak complex is followed by a conformational change of the protein (induced fit) leading to tight binding, which is rate-limiting at high concentrations. All constants indicated in the following scheme could be obtained by stopped flow experiments where $K_D = K_1 \cdot k_{-2}/k_2$:

$$\text{Ras} + \text{mantGDP} \xleftrightarrow{K_1} \text{Ras} \cdot \text{mantGDP} \underset{k_{-2}}{\overset{k_2}{\rightleftharpoons}} \text{Ras*} \cdot \text{mantGDP}$$

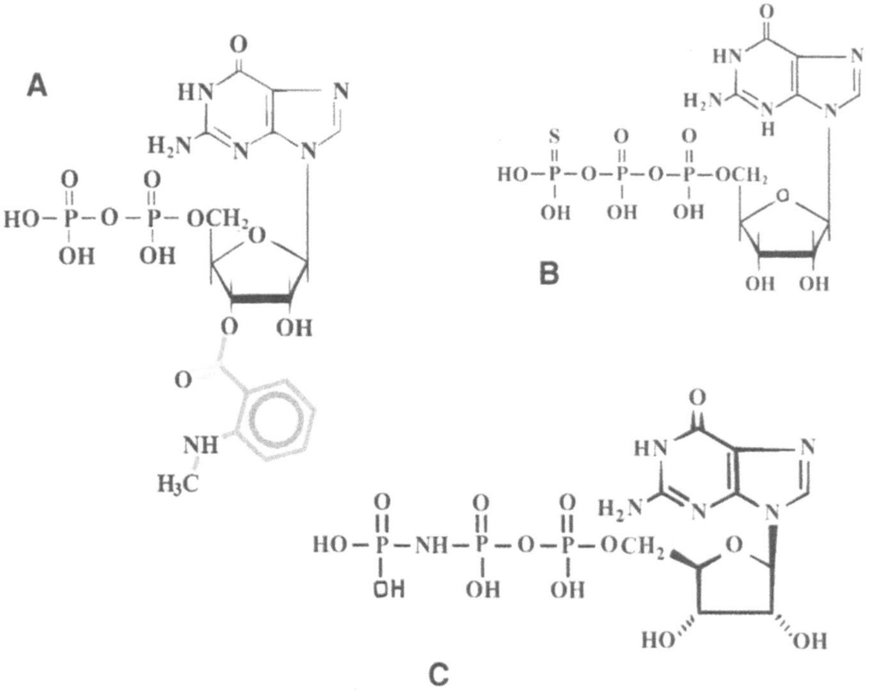

Fig. 14A–C. Guanosine-nucleotides for biophysical characterizations of Ras proteins: A guanosinediphosphate with a methylanthraniloylester (in *grey*) at the 3′-position of the ribose; B guanosine-5′-[β,γ-imido]-triphosphate (GppNHp); C guanosine-5′-O-(3-thiotriphosphate) (GTP-γS)

3.1.2
Intrinsic GTP Hydrolysis

The Ras isoforms are rather slow GTPases. In the absence of catalytic partners the intrinsic rate of GTP hydrolysis is 0.01/min for H-Ras [163] and 0.02/min for N-Ras [165] at 37 °C. Oncogenic mutations in Ras (which in most cases concern codons 12 for glycine or 61 for glutamine) result in an approximately tenfold reduction of the intrinsic GTPase rate [166–168], and complete loss of activation by GAPs (see Sect. 3.2.1). Therefore the ratio of Ras · GTP and Ras · GDP in the cell becomes unbalanced in favor of the GTP state leading to permanent (and uncontrolled) signaling for proliferation.

GTP hydrolysis by Ras, which is considerably slow, has been analyzed, usually by single turn-over measurements. Ras is loaded with GTP by incubation of Ras (being in the GDP form after preparation) with a large excess of GTP in the presence of EDTA, which leads to faster dissociation of the nucleotide. After change of buffer and removal of excess nucleotide, aliquots are taken at certain time intervals and the nucleotide composition is analyzed with the help of filter binding assays or HPLC. Series of experiments with mutations in the neighborhood of the phosphates and the attacking water (see below) led to a correlation between the resulting pK_a value of the γ-phosphate (titrated by NMR) and the rate of GTP hydrolysis [169]. The authors concluded that the γ-phosphate is the general base, rather than any amino acid residue, which activates the water molecule for nucleophilic attack.

3.2
Regulators and Effectors

3.2.1
GTPase Activating Proteins

The slow intrinsic GTP-hydrolysis by Ras can be accelerated by orders of magnitudes upon interaction with its GTPase activating proteins (GAPs). The cytosolic RasGAPs p120[RasGAP] and NF1 (neurofibromin) are the main factors, which ensure that cellular Ras exists predominantly in its inactive GDP-complexed state [26]. The mechanism of GTP-hydrolysis and its stimulation has been the object of controversial disputes for over a decade.

The hydrolysis of mant-GTP bound to Ras can be monitored by a slight decrease in fluorescence. Binding experiments of N-Ras with the non-hydrolyzable GTP-analogue mant-GppNHp showed a biphasic increase in fluorescence. The slow phase had the same amplitude as the decrease observed for the hydrolysis of Ras · mant-GTP which led to the hypothesis that a conformational change in the Ras protein proceeds GTP-hydrolysis [170] and represents the rate limiting step:

$$\text{Ras} \cdot \text{mant} - \text{GTP} \Rightarrow \text{Ras}^* \cdot \text{mant} - \text{GTP} \Rightarrow \text{Ras} \cdot \text{mant} - \text{GDP} + P_i$$

In this model the function of RasGAP was to accelerate the isomerization step and to bring Ras into a conformation competent for GTP cleavage without further support of GAP residues. A different analysis of the macroscopic readout interpreted the small change in amplitude as the hydrolysis reaction or the release of inorganic phosphate, and rejected the model of a GAP-induced conformational change in Ras [171], but this was again disputed by supporters of the conformational model [172].

Introduction of fluorescence anisotropy methods allowed direct measurement of Ras binding to p120[GAP] and NF1, respectively. The affinities for Ras · mant-GppNHp and the catalytic domains of the two Ras-GAPs were rather low, with a K_D around 20 µmol/l with p120[GAP] and around 1 µmol/l with NF1 [173]. It was demonstrated by this technique that a p120[GAP] mutant, which was deficient in catalyzing the GTP-hydrolysis in Ras, can still bind Ras · mant-GppNHp (Fig. 15). This observation is in support of the hypothesis of direct involvement of GAP residues in GTP hydrolysis.

The weak binding of p120[GAP] to Ras in the GTP-state was enhanced considerably when GTP (or GppNHp) was replaced by GDP and fluoroaluminate. In the α subunits of heterotrimeric G-proteins AlF_4^- and GDP were shown to mimic a kind of transition state of GTP hydrolysis. For Ras this state could only be shown

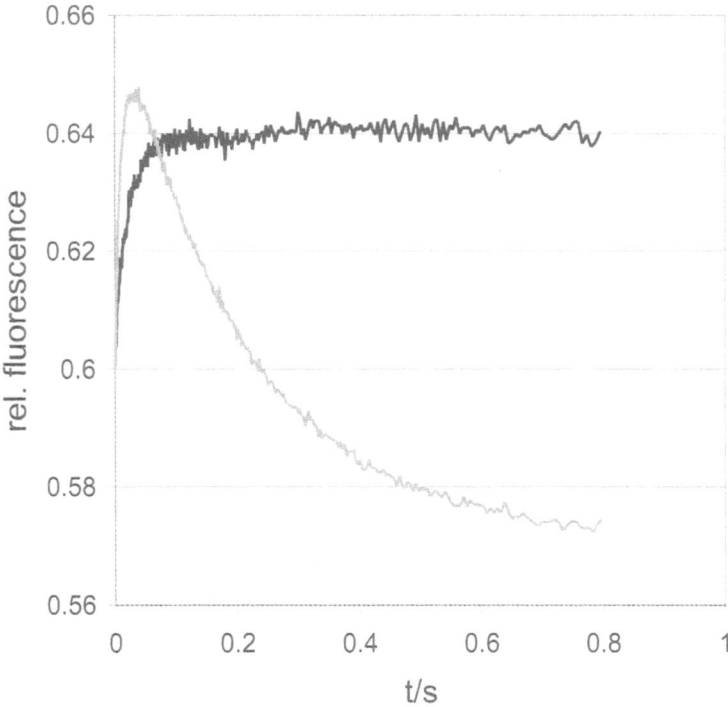

Fig. 15. H-Ras · mGTP shows a transient increase in fluorescence signal when mixed with the catalytic domain of the Ras-GAP neurofibromin (NF1). When wildtype NF1 is applied (*gray trace*) a decrease in signal follows, indicating the hydrolysis of mant-GTP. A NF1 mutant bearing an alanine at the position of the catalytic arginine residue (*black trace*) can only bind to Ras · mant-GTP (increase) but cannot induce mant-GTP hydrolysis (no decrease). Figure kindly committed by Reza Ahmadian

in the presence of additional Ras-GAP, indicating an additional stabilization of the transition state by p120^{GAP} and NF1 [174].

A further understanding of the GAP-mechanism was achieved by the arginine-finger hypothesis, where the role of a highly conserved arginine residue in Ras-GAPs of several species in completing the catalytic machinery of Ras was demonstrated by a combination of biophysical and structural (see Sect. 3.3.2) experiments [175].

In actual research, setups developed for the analysis of Ras interaction are adapted for the screening of new therapeuticals, e. g., chemicals which overcome the lack in GAP-stimulated GTP-hydrolysis of oncogenic Ras mutants [35].

3.2.2
Exchange Factors

Guanine nucleotide exchange factors (GEFs) are the immediate activators of Ras. Upon interaction of a GEF with Ras the binding to the nucleotide is weakened and its dissociation is accelerated. According to the large molar excess of GTP over GDP in the cell, it is predominantly GTP which rebinds to Ras and transforms it into its active state. The catalytic domain responsible for GEF activity is present in different exchange factors, like Sos, RasGRF, Scd25, and Cdc25, and this protein fragment has been used for many studies.

A detailed biochemical analysis by Cool and coworkers has revealed the mechanism of nucleotide exchange promoted by GEF. By use of mant-labeled guanine nucleotides and SPR they have studied the kinetics and equilibria of nucleotide and exchange factor binding to Ras [176]. The equilibrium dissociation constants (K_D) for the binary Ras/nucleotide and Ras/exchange factor complexes lie in the picomolar and nanomolar range, respectively. The ternary complex may dissociate either by nucleotide release, the K_D value of which lies in the micromolar range, or by exchange factor dissociation with a K_D value in the millimolar range [177]. This means that exchange factor binding to Ras lowers its nucleotide affinity by six orders of magnitude. The microscopic process responsible for this drastic change was proposed to be the conformational switch from a tight to a loose binding state which was identified to be rate limiting for nucleotide dissociation and which is accelerated by more than 10^5-fold in the ternary complex [177]. Notably, the exchange factor is able to displace the nucleotide from Ras but, vice versa, rebinding nucleotide does induce dissociation of the exchange factor from the ternary complex. In the same year the underlying structural details were published which are discussed below.

Dominant-inhibitory mutants have been very helpful in elucidating the biological effects of Ras in diverse cellular systems [178]. The most widely used mutant is RasS17N bearing two biochemical features which lead to the suppression of Ras function. This mutant fails to bind to downstream effectors but it binds to GEFs, resulting in sequestering the activator for endogenous Ras in an unproductive complex. It is not so much the (fivefold) tighter binding to the exchange factor but more the (1000-fold) weaker binding to the nucleotide which favors S17N over wild type Ras in binding to the exchange factor. According to basic thermodynamic considerations this leads to a 5000-fold stronger binding of exchange factor to S17N than to wild type Ras in the presence of nucleotide, i.e., in the cell. The mutant D119N has a similar effect as S17N, leading to suppression of endogenous Ras function. In contrast to S17N this mutant shows transforming activity at higher concentrations because it has not lost the ability to bind to the effectors. At low concentrations D119N acts as an inhibitor on wild type Ras present in the cell by sequestering the exchange factor [179]. At high concentrations it activates the effectors because it gets into the GTP form in a GEF-independent manner due to its high intrinsic nucleotide dissociation rates [180].

Mutants corresponding to S17N in Ras-related proteins like Rho or Ran have created dominant-inhibitory effects likewise, despite the fact that completely different exchange factors are involved. This may be based on the fact that 17N impairs Mg ion binding and therefore generally leads to weaker nucleotide binding, resulting in the effect described above. For Rap, closely related to Ras, contradictory data about the S17N mutant are reported but no efficient dominant-negative mutant could be identified so far, demonstrating that weaker nucleotide binding does not necessarily lead to the inhibitory effect. Despite high structural similarity between Rap and Ras and between the respective exchange factors C3G and Sos/Cdc25, van den Berghe et al. could point out some subtle differences in the importance of distinct residues for the two systems which might explain the different behavior [181]. Nevertheless, an efficient dominant-negative mutant of Rap is still sought, as it is so crucial for studying biological effects of Rap.

3.2.3
Multiple Effectors

Many different biological effects are known for Ras which depend on the type and the state of the cell. It is therefore not surprising that so many potential Ras effectors were identified. The degree of activation of each pathway, triggered by the interaction of Ras with the respective effector, is modulated by the input of other signals in parallel or counteracting pathways. Obviously, synergism is the keyword to understand the complex processes in regulation of cellular functions. Rap may be an example that even a closely related Ras protein, possibly by competition, can contribute to suppression of Ras function [100]. Characterization of Ras effector interaction has been extensively reviewed [55, 153]; therefore only some aspects which have to do with Ras itself will be discussed here.

The methods used for identification of new Ras targets, e.g., two hybrid screening in yeast, are of a qualitative nature and were able to demonstrate interactions of the effectors with Ras relatives as well. Further characterization of all possible interactions between the Ras-related proteins on the one side and the different effectors on the other was necessary in order to differentiate true partnerships. Many of the quantitative methods resulting in estimates of the binding affinity require immobilization of Ras or the effector. Another possibility to quantify Ras/effector interaction is competing GAP binding to Ras taking the rate of GTP hydrolysis as the readout [42]. For the equilibrium dissociation constant (K_D) values are reported in the range 5–500 nmol/l, the big variance of which is possibly caused by different experimental conditions. Another method based on the inhibition of guanine nucleotide dissociation (GDI assay) allowed systematic determination of the K_D values for all proteins in solution [182, 183]. A high degree of specificity for the members of the Ras subfamily became evident in respect of their interactions with the different effectors [184, 185]. The K_D values cover three orders of magnitude and the values for Ras are in the nanomolar range with some effectors and in the micromolar range with others. Intriguingly, Rap shows both high and low affinities but to other effectors as Ras. For instance, Rap binds strongly to RalGDS and weakly to Raf, and, in contrast, Ras binds tightly to Raf but weakly to RalGDS. The molecular basis for this pronounced specificity was discovered to be residue Lys31 in Rap and Glu31 in Ras, respectively [186]. The partner residue on the Raf side is Lys84, being attracted by oppositely charged Glu31 and repelled by Lys31. In the case of RalGDS a patch of acidic residues is suggested to be the counterpart of residue 31 in the Ras protein [187].

It is worthwhile to note that abrogation of interaction as detected in biological assays, often reported for mutants or other modifications, does not really mean complete loss of binding. Ras · GDP binds Raf-RBD 1000 times less strongly than Ras in the GTP state, i.e., with a K_D value still in the micromolar range [182]. Other differences are far more subtle. The above-mentioned *partial loss of function mutants*, which refers to the observation that certain mutations in Ras lead to the loss of binding to distinct effectors, shows this clearly. For example E37G can still activate RalGDS but not Raf whereas T35S acts in the opposite way [36, 37]. Other examples are known, including selective activation of PI(3)K (D38E, Y40C) [127]. When measuring the increase of the K_D values for these mutants, it turns out that there are no dramatic differences with the effectors. Apparently a fivefold difference between the values of Raf and RalGDS with Ras T35S (Herrmann, unpublished) is sufficient to result in a plainly perceptible alteration of the biological readout. These observations also suggest that effectors or Ras proteins have to pass concentration thresholds, which are in relation to the respective K_D values, in order to provoke a biological effect. This again is explainable by the high degree of synergism and cooperation in cellular signal transduction. Therefore, when detection limits in cellular assays are turned up by heavily overexpressing one of the interaction partners, the results should be interpreted with caution. The issue discussed here may also have implications for drug development. Only slight inhibition of interaction can be enough to achieve the desired effect – the main requirement for the drug remains the specific binding to one target. Both modest inhibition and high selectivity may lead to little side effects.

The dissociation of Ras and Raf was shown to be very fast, in contrast to the intuitive belief that a stable complex should be long-lived [188, 189]. This highly dynamic equilibrium of Ras/Raf interaction ensures that GAP, which does not bind to Ras simultaneously with Raf, can turn off Ras in between. The mechanism of binding was addressed by stopped flow kinetics [189]. For the rate constant of Ras/Raf association saturation behavior was observed with increasing concentration, indicating a two-step binding mechanism. Presumably, formation of a weak complex is followed by a conformational change of one of the proteins (induced fit), being rate limiting at high protein concentrations.

3.3
Structure

One may say that most structures relevant for understanding Ras regulation and function have been solved, with some limitations. Ras is known in its active and inactive state, but with the 20 C-terminal amino acids including the hydrophobic modifications missing. For structure determination of the regulators in complex with Ras, the catalytic fragments of the proteins were used. The reason for this is the difficulty in obtaining large proteins in a soluble and stable form in reasonable quantities and at high concentrations. For the same reason the structures of the Ras binding domains of the effectors in complex with Ras only are known. Recently the structure of PI(3)Kγ was solved, revealing the same topology of the RBD as found for other effectors (see below) but the binding to Ras could only be modeled in this case [190]. The following paragraphs describe structural aspects of Ras activation and deactivation and of the interaction with downstream targets.

3.3.1
Paradigm of GTPase

Although the structure of Ras, complexed to the non-hydrolyzable GTP analog GppNHp and solved by X-ray analysis [21, 191], was not the first in the superfamily of GTP-binding proteins, being second to ribosomal EF-Tu [192], Ras is established as the prototype for this large family. Ras represents the minimal G-domain, which was found later for many other Ras-like proteins such as Ran, Rab, and Rho proteins [193–196], in larger GTPases like in Gα [197] and, most recently, in hGBP1, a member of the subfamily of large GTPases [198]. The center of the G-domain is constituted by a six-stranded β-sheet with all strands running in parallel except β2 at the edge of the protein, representing the effector region in part (see Fig. 16). Again, with one exception, between β2 and β3, an α-helix is inserted each between the β-strands, kind of positioned around the central β-sheet. The structure of Rap, a member from the same subfamily, is almost identical to that of Ras [199]. Members from other subfamilies of small GTPases have a similar structure, each with specific features, like an additional α-helix in Ran embracing effector-like proteins [200]. The larger relatives of Ras also show the Ras G-domain core, according to their size with a few insertions like the heli-

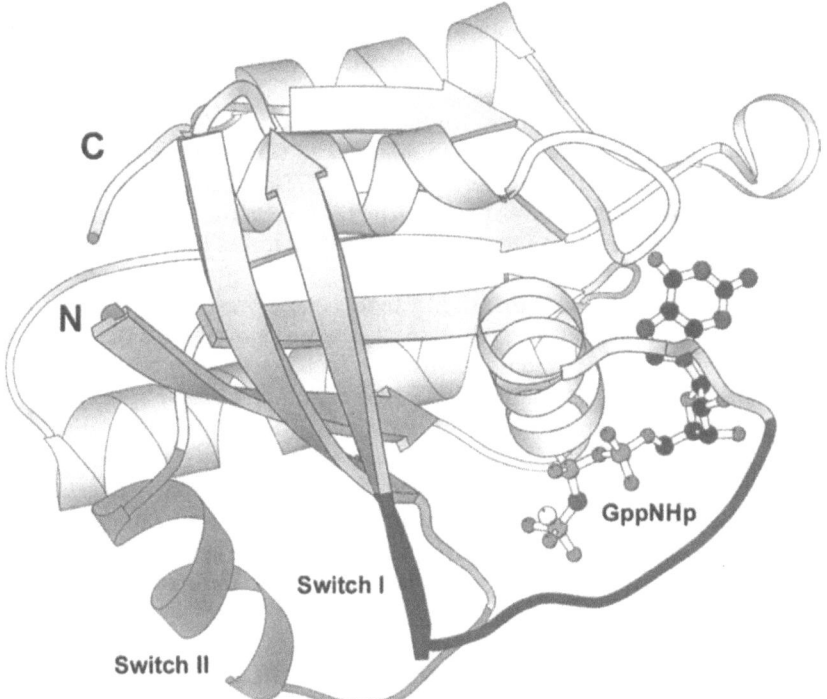

Fig. 16. X-ray structure of Ras in complex with GppNHp and a Mg^{2+} ion. Protein in ribbon presentation and the nucleotide in ball and sticks. The secondary structure elements are connected by short peptide loops, where loop 2 and 4 (see text) are in *dark grey*

cal GAP domain in G_α (see below) or the effector domain in EF-Tu [201]. Even in the structure of hGBP1 from the family of large (dynamin-related) GTPases, the Ras G-domain is found, also with a few insertions with presumably specific functions and two large helical domains extended at the C-terminus [198].

The most important property and common to all Ras-like proteins is the ability to bind GTP and GDP and to adopt different conformations, which is the basis of their switch function. As all the GTP-binding proteins have different regulators and effector proteins it is not surprising that all conserved amino acid motifs are concerned with nucleotide binding. They are almost exclusively located in the peptide loops connecting the secondary structural elements described above. Residues in loops 8 and 10, carrying NKx(D/T) and SAK motifs, respectively, interact with the guanine base of the nucleotide. Asp119 in loop 8 contributes most strongly to recognition of the nucleotide due to the formation of two specific hydrogen bonds to the guanine base. Loop 1, also known as P-loop and representing the GxxxxGK(S/T) motif, and loops 2 and 4 build the catalytic center in Ras by coordinating the phosphate groups with main chain and side chain atoms. Particularly amino acids Lys16 and Gln61 from loops 1 and 4, respectively, are believed to be of immediate catalytic importance, with a set of further residues being important for their proper positioning.

When comparing to the structure of GDP-bound Ras, the largest structural change is observed in loop 2, and to a lesser extent in loop 4 [22]. The former is also known as effector loop (followed by $\beta2$, see above) whereas the latter carries catalytic residues like Gln61 and, due to their conformational switching between the GDP and GTP states, they are also referred to as switch 1 and 2, respectively. In the GTP state these two loops are held in position by main chain contacts of Thr35 (invariant in all sequences) and Gly60 (from DxxG motif) to the γ-phosphate, respectively. After hydrolysis of GTP and subsequent loss of the phosphate the two loops move apart which may be regarded as a relaxation to a lower energy conformation [202]. Apparently, GTP forces Ras into a conformation amenable to effector binding.

The nucleotide is bound more tightly to Ras in the presence of Mg^{2+}, and this ion supports catalysis of GTP hydrolysis. It is bound through 6 oxygens near to ideal octahedral coordination. One oxygen each is contributed by β- and γ-phosphates, and the side chains of S17 (see exchange factors) and Thr35. The two apical positions are occupied by water molecules which are stabilized by Asp33 (from the effector loop) and Asp57 (DxxG motif), respectively. The contact to the Thr35 side chain and, obviously, to the γ-phosphate oxygen is lost after GTP hydrolysis but Mg^{2+} remains bound to Ras · GDP. Altogether many features have been identified in the Ras structure which are responsible for the catalytic activity, e.g., by stabilizing negative charges on the γ-phosphate by conveying leaving group characteristics to GDP and, alternatively, by positioning and activating a water molecule for in-line attack of the γ-phosphate.

Time-resolved X-ray crystallography has brought further insight into the mechanism of GTP hydrolysis and has confirmed former conclusions. With this method it was possible to obtain the structure of Ras bound to GTP, rather than non-hydrolyzable analogs like GppNHp or GppCH$_2$p, and, moreover, to follow the structural changes in Ras due to GTP hydrolysis [203]. Initially, Ras is bound to a photolabile and non-hydrolyzable compound, caged GTP, and set up for crystallization. The crystal mounted in an X-ray beam is flashed with UV-light which leads to cleavage of the protecting 2-nitrophenylethyl group leaving Ras · GTP behind. A white X-ray beam is now irradiated which allows data collection within a few minutes according to the Laue method. As GTP hydrolysis in the crystal is as slow as in solution (half-life of 40 min at room temperature) the structure of Ras · GTP and snapshots during the time course of hydrolysis can be obtained. GppNHp and GppCH$_2$p seem to be suitable analogs since the structure of Ras with GTP is very similar to those with the analogs showing the catalytic features described above. Comparison to the Ras · GDP structure, obtained by monochromatic X-ray diffraction after complete hydrolysis, shows nicely the structural changes in switch 1, here particularly residues 32–36 (note around Thr35), and in switch 2. Usage of pure diastereoisomers of caged GTP and rapid freezing technique led to higher resolution of the Ras · GTP structure [204]. Comparison of the high resolution structures of GTP- and GppNHp-bound Ras showed that they are almost identical, implying that there is no rate limiting conformational change between two GTP states (one being represented by the equilibrated GppNHp structure) prior to hydrolysis. This GTP structure also contributed to the discussion about the importance of distinct water molecules and catalytic residues which shall not be further detailed here.

Ras in complex with caged GTP was also used in time resolved infrared spectroscopy. After vibration band assignment with help of isotope labeling the phosphate groups of Ras-bound GTP could be studied, revealing lower frequency for the $\beta(PO_2^-)$ vibration compared to free GTP [205]. This was taken as an indication of extensive interaction of the β-phosphate with the protein environment, leading to electron withdrawal from the β/γ-bridging oxygen, and hence to bond weakening between this oxygen and γ-phosphorus atom. The authors see the splitting of GTP as "preprogrammed" by the β-phosphate interactions with Ras.

3.3.2
Regulators and Effectors

3.3.2.1
Exchange Factors

Kuriyan and coworkers used a fragment of the exchange factor Sos comprising the catalytic domain and in addition, as it turned out after structure determination, a region which stabilizes the catalytic region [206]. Both are purely helical and the catalytic domain is described as a bowl in the center of which Ras is bound. Interactions of Ras with Sos are observed for the P-loop, switch 1 and 2, as anticipated, and for helix α3. Hydrophobic residues from switch 2 act as a binding anchor buried into the hydrophobic core of Sos. Most further contacts are hydrophilic in nature, opening the way to fast displacement of Sos by competing water molecules after nucleotide binding. An obvious catalytic feature is an α-helical hairpin protruding from Sos into the nucleotide binding pocket of Ras which is nucleotide-free in this structure. It shifts away switch 1, leaving big structural changes behind in this functional element of Ras (compared to the structure of nucleotide-bound Ras). At the same time it inserts a hydrophobic residue, thereby blocking magnesium binding, and positions an acidic side chain at the site normally occupied by the α-phosphate of the nucleotide. Also Sos-induced changes of switch 2 are made responsible for impaired nucleotide binding. In consideration of the need for a non-durable competition of exchange factor and nucleotide for binding to Ras, and in view of the mechanism described above, the authors point out an intriguing observation in the Ras/Sos structure. Whereas the binding sites for the phosphate part of the nucleotide and the associated magnesium ion are disturbed by Sos, the binding pockets for the ribose and the base of the nucleotide are not fundamentally altered. This allows gradual rebinding of the nucleotide to Ras by entering the guanine base and the ribose moiety and then expelling of Sos by the phosphate groups. Regardless of which of the conformational changes is the rate limiting event for overall nucleotide exchange, both, biochemical and structural studies have shown that more than disruption of magnesium ion binding is responsible for efficient enhancement of nucleotide exchange rates.

3.3.2.2
GTPase Activating Proteins

The breakthrough in crystallization of GAP in complex with Ras was the discovery that GDP and AlF_4^-, together, are bound by Ras in the presence of stoichiometric amounts of GAP, thereby mimicking a transition state of GTP hydrolysis [174]. The historic background is that G_α, bound to GDP, was known to be activated by AlF_4^-, which occupies the position and imitates the role of γ-phosphate of GTP. The reason for this is that G_α carries a helical domain, inserted into its canonical Ras-like G-domain, which is responsible for the higher GTPase activity in comparison to Ras by contributing additional catalytic amino acid side chains. These residues are interpreted to stabilize the transition state of GTP hydrolysis as evidenced in the three-dimensional structure [207]. Now, Ras · GDP alone is not able to bind AlF_4^- except in complex with GAP which brings the helical domain with the stabilizing residues into the game. Also Rho in complex with Rho-GAP could be crystallized by use of GDP and AlF_4^- as GTP- or transition state-analog [208].

At first glance the structure of the catalytic GAP fragment resembles that of the exchange factor as it is purely helical, forming a "shallow groove" [209] rather than an "oblong bowl" [206]. Ras seems to dive in the same orientation into this groove/bowl and the contacts are again described to be via the P-loop, switch 1 and 2, and helix $\alpha3$. A closer look reveals many important differences of course, explaining many biochemical findings and the high degree of conservation of distinct residues in the family of GAP proteins. Most interesting is a loop exposed from the GAP fragment towards the nucleotide binding site of Ras and allowing an arginine to penetrate into the phosphate region. This "arginine finger" is interpreted, from biochemical experiments hypothesized, as the GAP residue which contributes catalytic assistance to GTP hydrolysis on Ras. It appears that a negative charge developing on the γ-phosphate during GTP hydrolysis is stabilized by the arginine side chain, leading to faster hydrolysis.

This is taken as further evidence for an associative (S_N2-like) mechanism of GTP hydrolysis. More details about this debate are reviewed by [210].

A further catalytic effect of GAP binding is the stabilization of switch 2, thereby fixing Ras residue Gln61 in a catalysis-competent position close to the attacking water molecule. For Gln61, one of the two most frequently mutated amino acids found in oncogenic Ras, the involvement in GTP hydrolysis was apparent from the Ras structure alone. Yet its exact role in the mechanism is not clear. Now the oncogenic effect of mutated Gly12 could also be rationalized by the Ras/GAP structure. Any amino acid side chain at this position leads to a clash with the main chain of the arginine finger, presumably leading to deranged conformations in the catalytic center of the complex.

It seems difficult to derive strategies for the development of anti-cancer drugs from the Ras/Sos structure because oncogenic Ras can fully dispense with exchange factors. In contrast, the Ras/GAP structure may help to create new ideas on how hydrolysis activity of mutated Ras might be restored by addition of a GAC (= GTPase activating compound). A first approach is given in the investigation of the mechanism of DABA-GTP hydrolysis [35]. See also [235] for review.

3.3.2.3
Ras Binding Domains of Effectors

In one of the first papers identifying Raf as Ras effector, the Ras binding domain (RBD) was defined as well [41]. By comparison of Ras-binding fragments of Raf obtained in a two-hybrid screen, the minimal binding region was mapped to residues 51–131 in c-Raf-1. Two years later the NMR structure of this fragment in solution and the X-ray structure in complex with the Ras homolog Rap were solved [199, 211]. The Raf-RBD shows the ubiquitin superfold consisting of a five-stranded β-sheet, a three-turn, and a one-turn α-helix (see Fig. 17). The interaction with the Ras protein is mediated by residues of the N-terminal β-hairpin, thereby forming an apparent antiparallel β-sheet with $\beta2$ and $\beta3$ from the Ras side. In addition, Raf residues from the C-terminal end of the α-helix contribute further contacts to the effector region of Ras. Apart from a small hydrophobic patch the interaction is mediated predominantly by oppositely charged or hydrophilic side chains forming specific salt bridges and hydrogen bonds, respectively. The different specificities of Ras and Rap in respect to Raf binding, which are caused by residue 31 each as described above, can be comprehended in the structures. Whereas Raf residue Lys84 is pushed out by Lys31 in Rap, it forms a stable salt bridge with Glu31 and Asp33 in the Ras-like complex [186]. NMR experiments in solution suggest that the salt bridge partner of Lys84 is Asp33 rather than Glu31 [212].

Again two years later and another two years later the structures of the RBDs of RalGDS and PI(3)K were published, respectively, the former as NMR and X-ray structure [213, 214], and the latter as part of PI(3)Kγ in full length [190]. Intriguingly, all RBD structures of effectors known so far show the same topology despite their lack of sequence homology. The mode of docking to the Ras protein is similar as well, as described above for Raf-RBD, and also observed in the Ras/RalGDS complex [187, 215] and as modeled for Ras/PI(3)Kγ [190]. When

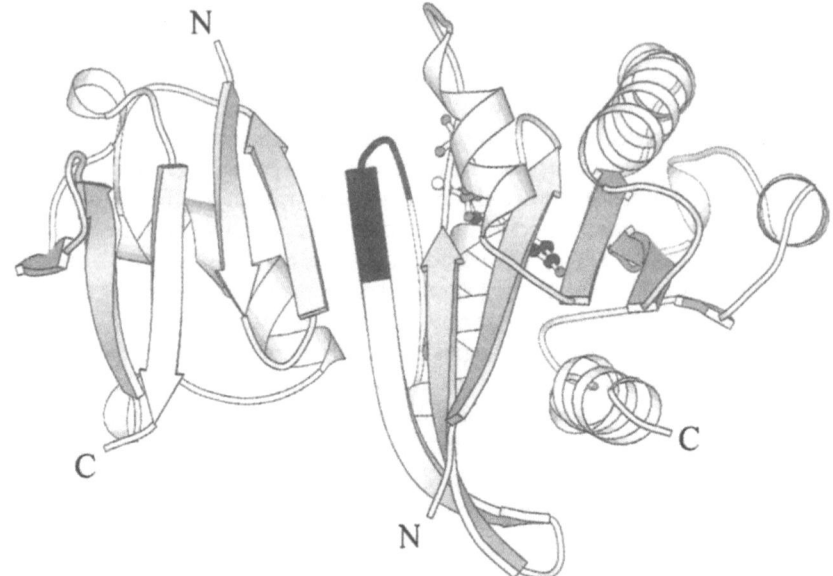

Fig. 17. The X-ray structure of Rap, highly homologous to Ras, in complex with Raf-RBD. The structure of Rap is almost identical to the one of Ras (see Fig. 16, Rap shown here in a different orientation). Raf-RBD is shown on the *left*

comparing uncomplexed RBDs in complex with Rap or Ras, no drastic structural changes are observed, which suggests that Ras binding does not induce significant structural changes of the RBD, thus not favoring an allosteric model of effector activation.

Ras, at the interface to the RBD, exposes predominantly negatively charged amino acid residues. Accordingly, arginines and lysines are found on the RBD side of Raf and RalGDS, each giving rise to specific recognition of Ras and effector proteins. Despite all these structural similarities, an important difference in the thermodynamics is observed for Raf and RalGDS interaction with Ras. Whereas Raf binding is favored by a positive entropy change, RalGDS binding is accompanied by a huge decrease in enthalpy in order to compensate for a negative entropy change (Herrmann, unpublished). In order to design drugs based on the structure of their protein targets it was desirable to understand the quantitative relation between structure and energy of interaction, but as yet we are quite a bit away from this.

3.3.3
NMR Analysis

The NMR solution structure of Ras · GDP [216] is very similar to the X-ray structure [22], except the position of helix 2, which matches neither with the corresponding helix in the GDP nor in the GppNHp X-ray structure, respectively. The authors argue that such "energetically finely balanced conformational states" detectable in a solution structure may be masked by crystal packing forces [216]. Nevertheless, the same regions, namely loops 2 and 4, are found to be mobile which demonstrates that ill-defined electron density in the X-ray structure is not due to crystallization artifacts but corresponds to loop dynamics, at least in this case. From measurements of the heteronuclear NOEs and of T_1 and T_2 in ^{15}N-labeled Ras, the dynamics are calculated to lie in the nanosecond time range.

Another type of NMR experiment revealed equilibration of two or more stable conformations of the Ras protein. ^{31}P-NMR spectra of GppNHp bound to Ras recorded at 5 °C show the resonances of the α and β phosphorus atoms to be split into two lines which coalesce at higher temperatures (30 °C) [217]. The two different chemical shift values observed were each interpreted to result from the positioning of Tyr32, which may be oriented towards the phosphates in one conformation and pointing away in the other. The dynamics of this flip were calculated by line shape analysis to be in the ms time range. Consistent with the implied mobility of the effector loop, one of these two states could be stabilized by the addition of effector protein. One of the two states is preferred by the interacting effector, resulting in only one resonance line for each phosphorus atom. Similar conclusions were made after analyzing line widths of fully isotope labeled Ras [218]. Extreme broadening was observed for residues of loop 2 and 4 (switch 1 and 2), and part of loop 1, taken as indication for "regional polysterism" of the Ras molecule in the GTP-bound form. The authors discuss the possibility that Ras, with many different interaction partners which bind to loop 2 and 4, presents different conformations one of which is selected by a distinct partner. Again, binding of Raf-RBD resulted in fixation of one of the polysteric forms.

The solution structures of Raf-, Rlf-, and RalGDS-RBD were solved and, in addition, NMR spectroscopy was used to probe the interaction with Ras [211, 213, 219]. Only the RBD was labeled with ^{15}N and the shifting of distinct cross peaks (HSQC spectra) due to binding of Ras allowed one to identify the binding surface on the RBD. Consistent with X-ray structures of the complexes (see above) in both Raf and RalGDS, amino acid residues located in $\beta1$ and $\beta2$ and the C-terminal part of $\alpha1$ are involved in the interaction with Ras.

3.4
Posttranslational Modification

Ras proteins and almost all other GTP binding proteins involved in signal transduction, regulation, and intracellular transport are anchored in membranes via hydrophobic modifications such as isoprenylation and palmitoylation [220]. Ha- and N-Ras as well as K-Ras4A undergo four post-translational modification steps starting with the prenylation of the last C-terminal cysteine in the CaaX box of the polypeptide chain by a protein farnesyltransferase. Subsequent removal of the last three C-terminal amino acids by a prenyl protein specific endoprotease and donation of a methyl group to the S-farnesylated cysteine by a prenyl protein specific methyltransferase occur in the endoplasmatic reticulum. Modification is completed by palmitoylation of one or two cysteines by a puta-

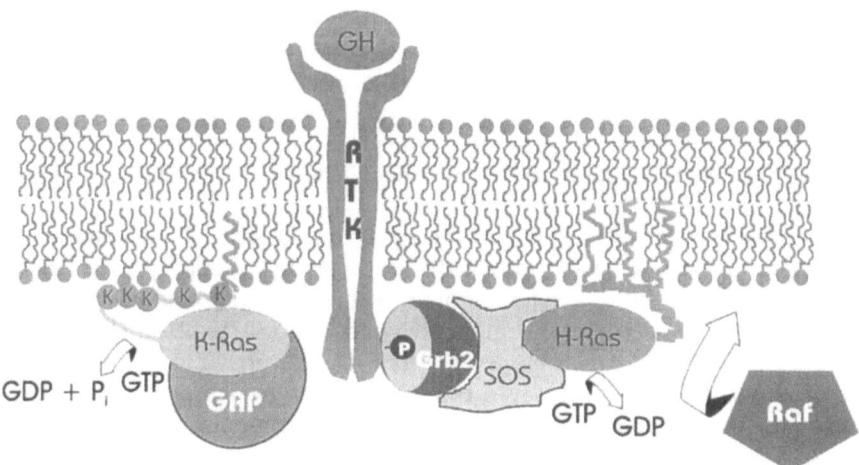

Fig. 18. Interaction between Ras and its regulators and effectors at the plasma membrane. Binding of a growth hormone (GH) at the extracellular binding site of its receptor (RTK, receptor tyrosine kinase) leads to autophosphorylation of intracellular tyrosines of the RTK, which allow binding of the cytosolic adapter protein Grb2 and thereby brings the Ras guanosine exchange effector Sos to the plasma membrane. Sos now catalyzes the exchange of GDP complexed with Ras by GTP. Thereby, activated Ras · GTP now recruits Raf-kinase to the plasma membrane, where Raf initiates a cascade of phosphorylation reactions leading to cell division or differentiation. GTPase activating proteins like RasGAP stimulate the intrinsic GTP-hydrolysis in Ras · GTP up to 10^5-fold and bring Ras back to its resting state. While Ha-Ras is inserted in the plasma membrane with its farnesyl group and up to two palmitoylthioesters at its C-terminus (shown on the *right*), the isoprenoyl membrane anchoring of K-Ras4B is supported by electrostatic interaction between a positively charged poly-lysine stretch and the negative head groups at the inner side of the plasma membrane (shown on the *left*)

tive prenyl protein specific palmitoyltransferase presumably localized at the plasma membrane. K-Ras4B undergoes the same modifications as the other Ras isoforms, but is not palmitoylated and achieves its membrane association by a combination of hydrophobic (farnesyl group and carboxymethylation) and electrostatic interaction of its basic C-terminus with the negatively charged plasma membrane [221]. All interactions between Ras proteins and their effectors or regulators have to occur at the inner side of the plasma membrane (Fig. 18). If the membrane localization of Ras is prevented, e.g., by inhibitors of the farnesylation reaction, even oncogenic Ras mutants cannot transform a cell – an idea which is strongly investigated for cancer treatment [222].

3.4.1
In Vitro vs In Vivo

All steps of the Ras pathway from ligand binding to receptor tyrosine kinases, down to activation of effectors like Raf kinase, occur at the plasma membrane. However, most biophysical studies on protein/protein interactions involved in this scenario have been carried out with bacterially synthesized proteins lacking

C-terminal modification. This was caused by big problems in the synthesis and isolation of completely modified proteins [223] and good accessibility to the non-modified proteins expressed in E. coli.

Of course, insertion of proteins into intracellular membranes may have effects upon the kinetic and thermodynamic properties of the corresponding biological interactions. Direct contributions of hydrophobic modifications of the Ras proteins to their interaction with regulators and effectors have been demonstrated or are the subject of discussion. Catalytic activity of the exchange factor hSos (human Sos) with H-Ras is stimulated by the isoprenylation of Ras [224] and a similar effect is described for smgGDS (small guanosine dissociation stimulators) and K-Ras [225]. The G-protein responsive phosphoinositide 3-kinase p110γ shows markedly increased binding of Ras when Ha-Ras or K-Ras4B are farnesylated [226]. Additionally, the membrane itself can contribute to further modifications of the protein-protein interactions. It can provide additional electrostatic and hydrophobic interactions distinct from the lipid anchorage and thereby affect conformation and/or activity of membrane associated proteins.

3.4.2
Artificial Membranes

A first approach to study the interaction of posttranslational modified Ras proteins with membranes was the analysis of binding and exchange of isoprenylated peptides with and between lipid vesicles utilizing a fluorescent bimanyl label. Studies with K-Ras peptides revealed that a single isoprenyl group is sufficient for membrane association only if supported by carboxymethylation of the C-terminal cysteine [227, 228].

Stable membrane binding for K-Ras peptides was achieved when hydrophobic and electrostatic contributions act in concert. The effect of a combination of positively charged side groups and an isoprenoid modification upon membrane binding of the C-termini of K-Ras4B was synergistic [229]. Due to the long stretch of basic amino acids the electrostatic interaction of the K-Ras4B peptide with negatively charged vesicles results in an approximately 10^3-fold increase in binding compared with a neutral membrane.

Further investigations with bimanyl-labeled K-Ras4B peptides demonstrated that relatively small differences in membrane charging (approximately 10 mol %) are sufficient for an electrostatic enrichment in the more negative environment [230]. With the farnesyl group as a hydrophobic anchor, the peptide is still mobile and can swap between vesicles but may find its target membrane with the sensitive surface potential-sensing function of its lysine residues.

The H- and N-isoforms of Ras support the first (isoprenoid) hydrophobic modification by additional thioester formation with palmitoylic acids [18]. At physiological temperature (37 °C) the dissociation of doubly modified lipopeptides with an isoprenyl thioether and a palmitoyl thioester is very slow and characterized by half-times in the order of 50 h. Here, the relative effect of the carboxymethylation is significantly reduced. Palmitoyl groups with their C_{16} alkane chain length contribute more efficiently to membrane anchoring than the farnesyl modification.

These findings led to the conclusion that the regulation of membrane anchored proteins has to be achieved by mechanisms other than spontaneous dissociation. In principal, binding to an "escort protein" or de-S-acylation may induce dissociation of the lipoproteins out of the membrane structure.

As well as fluorescence-based assays, artificial membranes on the surface of biosensors offered new tools for the study of lipopeptides. In a commercial BIA-core system [231] a hydrophobic SPR sensor with an alkane thiol surface was incubated with vesicles of defined size distribution generating a hybrid membrane by fusion of the lipid vesicles with the alkane thiol layer [232]. If the vesicles contain biotinylated lipopeptides their membrane anchoring can be analyzed by incubation with streptavidine. Accordingly, experiments with lipopeptides representing the C-terminal sequence of N-Ras show clear differences between single and double hydrophobic modified peptides in their ability to persist in the lipid layer [233].

New insights into the analysis of hydrophobically post-translational modified proteins could be achieved by the construction of lipidated proteins in a combination of bioorganic synthesis of activated lipopeptides and bacterial expression of the protein backbone (Fig. 19). The physico-chemical properties of such artificial lipoproteins differ substantially from those of the corresponding lipopeptides. The pronounced dominance of the hydrophilic protein moiety (e.g., for the Ras protein 181 amino acids) over a short lipopeptide with one or two hydrophobic modifications provides solubility up to 10^{-4} mol/l, while the biotinylated or fluorescence labeled lipopeptides exhibit low solubility in aqueous solutions and can be applied in the biophysical experiments only in vesicle integrated form or dissolved in organic solvent.

Thus lipoproteins could be injected over the surface of a lipid covered SPR sensor in a detergent free buffer solution, and showed spontaneous insertion into the artificial membrane [234]. Again, two hydrophobic modifications are

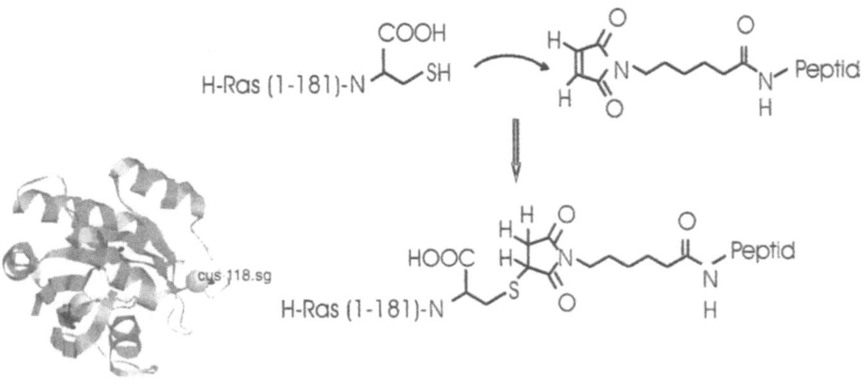

Fig. 19. Lipopeptides with the aminoacid sequence of the Ras C-terminus and the natural or artificial lipid-modifications can be coupled with C-terminally truncated Ras via a maleimidocaproyl linker. This electrophile reacts with free thiol groups (here, a C-terminal cysteine at the Ras moiety)

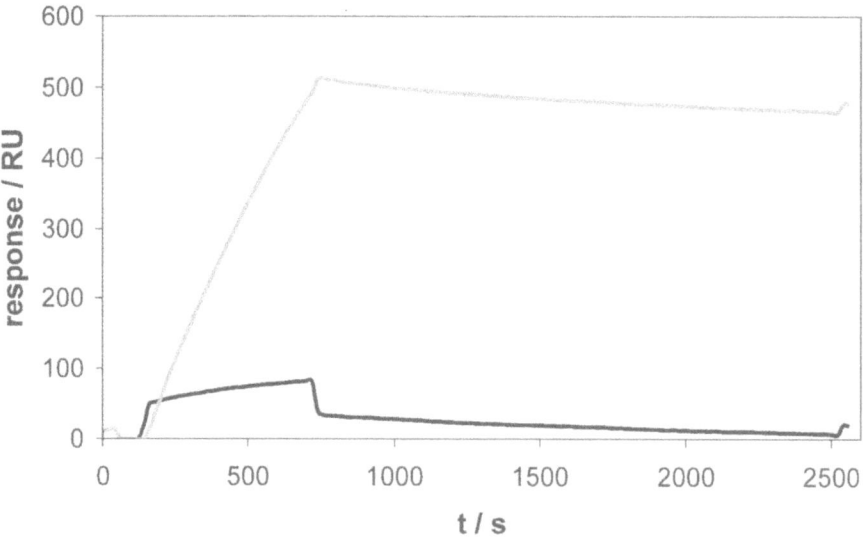

Fig. 20. The surface of a hydrophobic SPR-sensor (covered with long chain alkanethiols) was enlarged to an artificial membrane by application of lipid vesicles with a defined size distribution. Application of a Ras-lipopeptide construct with both a farnesyl- and a palmitoyl-modification leads to increase in resonance signal assumed to indicate membrane insertion (*grey trace*). Washing with buffer induces the slow decrease in signal. A Ras protein without hydrophobic modification (*black trace*) does not lead to signal increase

necessary for stable insertion into the lipid layer, whereas lipoproteins with a farnesyl group only dissociate significantly faster out of the membrane (Fig. 20). Therefore the isoprenylation of a protein is sufficient to allow interaction with membranous structures, while trapping of the molecule at a particular location requires a second hydrophobic anchor [19].

Interaction between the Ras protein and its effector Raf-kinase depends on complex formation of Ras with GTP (instead of the Ras · GDP complex, present in the resting cell).

The SPR setup can now be applied to the study of interactions between membrane associated proteins and their effectors and regulators in a membrane environment mimicking the situation in the living cell.

3.5
Outlook

Ras and its relatives are subjects of intensive investigations by biological, biochemical, biophysical, and medical studies. Within just one decade more than 17,000 articles (Medline, 1966–2000) deal with function and properties of this protein. Structural and functional data, based on Ras as a prototype, have provided insight into the basic principles of GTP-binding proteins, their activation, de-activation, and signal transmission.

Here biophysical methods contributed a reductionistic approach. By analyzing a limited number of actors under well defined conditions, the mechanisms of nucleotide exchange, intrinsic and stimulated GTP-hydrolysis, effector binding, and membrane attachment have been elaborated to present a comprehensive model.

Despite the data reviewed here, the work is not finished at all. Physicochemical data have to be correlated with biological activity and the complexity of the living cell has to be reflected in the biophysical setups. This covers the use of full-length proteins instead of only functional domains, processed proteins instead of the straightforward bacterial expression, and the introduction of a membranous environment vs simple experiments in solution.

Novel techniques as single molecule detection will focus on molecular localization and dynamics and may open insights into the micro-compartmentalization of cellular systems.

In future, two developments may be expected. On the one hand, elaborated biophysical methods will measure interactions in the cellular context and dissect complex biological patterns. On the other hand, established physico-chemical techniques mainly based on fluorescence will be utilized in high throughput screens to investigate new substances upon their interference with Ras function.

For general literature on biophysical analysis, see [236–241].

4
References

1. Varmus H, Weinberg RA (1994) Gene und Krebs – Biologische Wurzeln der Tumorentstehung. Spektrum – Akademischer Heidelberg
2. Bourne HR, Sanders DA, McCormick F (1991) Nature 349:117
3. Gilman A (1989). Annual Review of Biochemistry 56:615
4. Stryer L, Bourne HR (1986) Annual Review of Cell Biology 2:391
5. Bowler WB, Gallagher JA, Bilbe G (1998) Frontiers in Bioscience 3:769
6. Gudermann T, Kalkbrenner F, Schultz G (1996) Annual Review of Pharmacology & Toxicology 36:429
7. Boguski MS, McCormick F (1993) Nature 366:643
8. Koera K, Nakamura K, Nakao K, Miyoshi J, Toyoshima K, Hatta T, Otani H, Aiba A, Katsuki M (1997) Oncogene 15:1151
9. Voice JK, Klemke RL, Le A, Jackson JH (1999) Journal of Biological Chemistry 274:17,164
10. Reuther GW, Der CJ (2000) Current Opinion in Cell Biology 12:157
11. Wittinghofer A, Herrmann C (1995) FEBS Letters 369:52
12. Scolnick EM, Papageorge AG, Shih TY (1979) Proc Natl Acad Sci USA 76:5355
13. Cox AD, Der CJ (1997) Biochimica et Biophysica Acta-Reviews on Cancer 1333:F51
14. Cox AD, Der CJ (1992) Current Opinion in Cell Biology 4:1008
15. Zhang FL, Casey PJ (1996) Annual Review of Biochemistry 65:241
16. Schmidt WK, Tam A, Fujimura-Kamada K, Michaelis S (1998) Proc Natl Acad Sci USA 95:11,175
17. Dai Q, Choy E, Chiu V, Romano J, Slivka SR, Steitz SA, Michaelis S, Philips MR (1998) Journal of Biological Chemistry 273:15,030
18. Dunphy JT, Linder,ME (1998) BBA-Mol 1436:245
19. Shahinian S, Silvius JR (1995) Biochemistry 34:3813
20. Choy E, Chiu VK, Silletti J, Feoktistov M, Morimoto T, Michaelson D, Ivanov IE, Philips MR (1999) Cell 98:69

21. Pai EF, Kabsch W, Krengel U, Holmes KC, John J, Wittinghofer A (1989) Nature 341:209
22. Milburn MV, Tong L, DeVos AM, Brünger A, Yamaizumi Z, Nishimura S, Kim S-H (1990) Science 247:939
23. Ebinu JO, Bottorff DA, Chan EYW, Stang SL, Dunn RJ, Stone JC (1998) Science 280:1082
24. Tognon CE, Kirk HE, Passmore LA, Whitehead IP, Der CJ, Kay RJ (1998) Molecular & Cellular Biology 18:6995
25. Freshney NW, Goonesekera SD, Feig LA (1997) FEBS Letters 407:111
26. Trahey M, McCormick F (1987) Science 238:542
27. Gideon P, John J, Frech M, Lautwein A, Clark R, Scheffler JE, Wittinghofer A (1992) Molecular & Cellular Biology 12:2050
28. McCormick F (1992) Philosophical Transactions of the Royal Society of London – Series B: Biological Sciences 336:43
29. Leblanc V, Delumeau I, Tocque B (1999) Oncogene 18:4884
30. Feldkamp MM, Gutmann DH, Guha A (1998) Canadian Journal of Neurological Sciences 25:181
31. Marshall C (1984) Nature 310:448
32. Weinberg RA (1982) Cell 30:3
33. Weinberg RA (1982) Advances in Cancer Research 36:149
34. Barbacid M (1987) Ann Rev Biochem 56:779
35. Ahmadian MR, Zor T, Vogt D, Kabsch W, Selinger Z, Wittinghofer A, Scheffzek K (1999) Proceedings of the National Academy of Sciences of the United States of America 96:7065
36. White MA, Nicolette C, Minden A, Polverino A, Vanaelst L, Karin M, Wigler M (1995) Cell 80:533
37. White MA, Vale T, Camonis JH, Schaefer E, Wigler MH, Joneson T, Barsagi D (1996) Journal of Biological Chemistry 271:16,439
38. Joneson T, White MA, Wigler MH, Barsagi D (1996) Science 271:810
39. Rodriguezviciana P, Warne PH, Khwaja A, Marte BM, Pappin D, Das P, Waterfield MD, Ridley A, Downward J (1997) Cell 89:457
40. Schaber MD, Gibbs, JB (1995) Methods in Enzymology 255:171
41. Vojtek AB, Hollenberg SM, Cooper JA (1993) Cell 74:205
42. Zhang X, Settleman J, Kyriakis JM, Takeuchi-Suzuki E, Elledge SJ, Marshall MS, Bruder JT, Rapp UR, Avruch J (1993) Nature 364:308
43. Warne PH, Viciana PR, Downward J (1993) Nature 364:352
44. Van Aelst L, Barr M, Marcus S, Polverino A, Wigler M (1993) Proc Natl Acad Sci USA 90:6213
45. Kikuchi A, Demo SD, Ye Z-H, Chen Y-W, Williams LT (1994) Mol Cell Biol 14:7483
46. Hofer F, Fields S, Schneider C, Martin GS (1994) Proc Natl Acad Sci USA 91:11,089
47. Spaargaren M, Bischoff JR (1994) Proc Natl Acad Sci USA 91:12,609
48. Kodaki T, Woscholski R, Hallberg B, Rodriguez-Viciana P, Downward J, Parker PJ (1994) Current Biology 4:198
49. Rodriguez-Viciana P, Warne PH, Dhand R, Vanhaesebroeck B, Gout I, Fry MJ, Waterfield MD, Downward J (1994) Nature 370:527
50. Russell M, Lange-Carter CA, Johnson GL (1995) J Biol Chem 270:11,757
51. Diazmeco MT, Lozano J, Municio MM, Berra E, Frutos S, Sanz L, Moscat J (1994) Journal of Biological Chemistry 269:31,706
52. Vanaelst L, White MA, Wigler MH (1994) Cold Spring Harbor Symposia on Quantitative Biology 59:181
53. Kuriyama M, Harada N, Kuroda S, Yamamoto T, Nakafuku M, Iwamatsu A, Yamamoto D, Prasad R, Croce C, Canaani E, Kaibuchi K (1996) Journal of Biological Chemistry 271:607
54. Vavvas D, Li X, Avruch J, Zhang XF (1998) Journal of Biological Chemistry 273:5439
55. Vojtek AB, Der CJ (1998) Journal of Biological Chemistry 273:19,925
56. Gibbs JB, Marshall MS (1989) Microbiological Reviews 53:171
57. Daum G, Eisenmann-Tappe I, Fries HW, Troppmair J, Rapp UR (1994) Trends in Biochemical Sciences 19:474

58. Stokoe D, Macdonald SG, Cadwallader K, Symons M, Hancock JF (1994) Science 264: 1463
59. Leevers SJ, Paterson HF, Marshall CJ (1994) Nature 369:411
60. Marais R, Light Y, Paterson HF, Marshall CJ (1995) EMBO Journal 14:3136
61. Dent P, Reardon DB, Morrison DK, Sturgill TW (1995) Molecular & Cellular Biology 15:4125
62. Freed E, Symons M, Macdonald SG, McCormick F, Ruggieri R (1994) Science 265:1713
63. Irie K, Gotoh Y, Yashar BM, Errede B, Nishida E, Matsumoto K (1994) Science 265:1716
64. Fu H, Xia K, Pallas DC, Cui C, Conroy K, Narsimhan RP, Mamon H, Collier RJ, Roberts TM (1994) Science 266:126
65. Farrar MA, Alberolaila J, Perlmutter RM (1996) Nature 383:178
66. Luo ZJ, Tzivion G, Belshaw PJ, Vavvas D, Marshall M, Avruch J (1996) Nature 383:181
67. Ghosh S, Bell,RM (1994) Journal of Biological Chemistry 269:30,785
68. Chuang E, Barnard D, Hettich L, Zhang X, Avruch J, Marshall MS (1994) Mol Cell Biol 14:5318
69. Brtva TR, Drugan JK, Ghosh S, Terrell RS, Campbellburk S, Bell RM, Der CJ (1995) Journal of Biological Chemistry 270:9809
70. Hu CD, Kariya K, Tamada M, Akasaka K, Shirouzu M, Yokoyama S, Kataoka T (1995) Journal of Biological Chemistry 270:30,274
71. Pumiglia K, Chow YH, Fabian J, Morrison D, Decker S, Jove R (1995) Molecular, Cellular Biology 15:398
72. Luo ZJ, Diaz B, Marshall MS, Avruch J (1997) Molecular & Cellular Biology 17:46
73. Ghosh S, Xie WQ, Quest AF, Mabrouk GM, Strum JC, Bell RM (1994) Journal of Biological Chemistry 269:10,000
74. Daub M, Jockel J, Quack T, Weber CK, Schmitz F, Rapp UR, Wittinghofer A, Block C (1998) Molecular & Cellular Biology 18:6698
75. Yamamori B, Kuroda S, Shimizu K, Fukui K, Ohtsuka T, Takai Y (1995) Journal of Biological Chemistry 270:11,723
76. Wolthuis RMF, Bauer B, Vantveer LJ, Devriessmits AMM, Cool RH, Spaargaren M, Wittinghofer A, Burgering BMT, Bos JL (1996) Oncogene 13:353
77. Urano T, Emkey R, Feig LA (1996) EMBO Journal 15:810
78. Feig LA, Urano T, Cantor S (1996) Trends in Biochemical Sciences 21:438
79. Qiu R-G, Chen J, Kirn D, McCormick F, Symons M (1995) Nature 374:457
80. Qiu RG, Chen J, McCormick F, Symons M (1995) Proceedings of the National Academy of Sciences of the United States of America 92:11,781
81. Okazaki M, Kishida S, Hinoi T, Hasegawa T, Tamada M, Kataoka T, Kikuchi A (1997) Oncogene 14:515
82. Wolthuis RMF, Deruiter ND, Cool RH, Bos JL (1997) EMBO Journal 16:6748
83. Goi T, Rusanescu G, Urano T, Feig LA (1999) Molecular & Cellular Biology 19:1731
84. Marshall CJ (1995) Cell 80:179
85. Kikuchi A, Williams,LT (1996) Journal of Biological Chemistry 271:588
86. Stoyanov B, Volinia S, Hanck T, Rubio I, Loubtchenkov M, Malek D, Stoyanova S, Vanhaesebroeck B, Dhand R, Nurnberg B, Gierschik P, Seedorf K, Hsuan JJ, Waterfield MD, Wetzker R (1995) Science 269:690
87. Rubio I, Rodriguezviciana P, Downward J, Wetzker R (1997) Biochemical Journal 326:891
88. Vanhaesebroeck B, Leevers SJ, Panayotou G, Waterfield MD (1997) Trends in Biochemical Sciences 22:267
89. Rodriguezviciana P, Warne PH, Vanhaesebroeck B, Waterfield MD, Downward J (1996) EMBO Journal 15:2442
90. Franke TF, Yang S-I, Chan TO, Datta K, Kazlauskas A, Morrison DK, Kaplan DR, Tsichlis PN (1995) Cell 81:727
91. Franke TF, Kaplan DR, Cantley LC (1997) Cell 88:435
92. Burgering BMT, Coffer PJ (1995) Nature 376:599
93. Didichenko SA, Tilton B, Hemmings BA, Ballmerhofer K, Thelen M (1996) Current Biology 6:1271

94. Kauffmanzeh A, Rodriguezviciana P, Ulrich E, Gilbert C, Coffer P, Downward J, Evan G (1997) Nature 385:544
95. Chung J, Grammer TC, Lemon KP, Kazlauskas A, Blenis J (1994) Nature 370:71
96. Hu Q, Klippel A, Muslin AJ, Fantl WJ, Williams LT (1995) Science 268:100
97. Datta SR, Dudek H, Tao X, Masters S, Fu HA, Gotoh Y, Greenberg ME (1997) Cell 91:231
98. Marshall MS (1993) TIBS 18:250
99. Bauer B, Mirey G, Vetter IR, Garcia-Ranea JA, Valencia A, Wittinghofer A, Camonis JH, Cool RH (1999) Journal of Biological Chemistry 274:17,763
100. Kitayama H, Sugimoto Y, Matsuzaki T, Ikawa Y, Noda M (1989) Cell 56:77
101. Ohtsuka T, Shimizu K, Yamamori B, Kuroda S, Takai Y (1996) Journal of Biological Chemistry 271:1258
102. York RD, Yao H, Dillon T, Ellig CL, Eckert SP, Mccleskey EW, Stork PJS (1998) Nature 392:622
103. Zwartkruis FJT, Wolthuis RMF, Nabben NMJM, Franke B, Bos JL (1998) EMBO Journal 17:5905
104. Bos JL, Franke B, Mrabet L, Reedquist K, Zwartkruis F (1997) FEBS Letters 410:59
105. Zwartkruis FJT, Bos JL (1999) Experimental Cell Research 253:157
106. de Rooij J, Zwartkruis FJT, Verheijen MHG, Cool RH, Nijman SMB, Wittinghofer A, Bos JL (1998) Nature 396:474
107. Franke B, Akkerman JWN, Bos JL (1997) EMBO Journal 16:252
108. Mcleod SJ, Ingham RJ, Bos JL, Kurosaki T, Gold MR (1998) Journal of Biological Chemistry 273:29,218
109. Kawasaki H, Sprihtgett GM, Toki S, Canales JJ, Harlan P, Blumenstiel JP, Chen EJ, Bany IA, Mochizuki N, Ashbacher A, Matsuda M, Housman DE, Graybiel AM (1998) Proceedings of the National Academy of Sciences of the United States of America 95:13,278
110. Rubinfeld B, Munemitsu S, Clark R, Conroy L, Watt K, Crosier WJ, McCormick F, Polakis P (1991) Cell 65:1033
111. Wang KL, Khan MT, Roufogalis BD (1997) Journal of Biological Chemistry 272:16,002
112. Clark GJ, Kinch MS, Gilmer TM, Burridge K, Der CJ (1996) Oncogene 12:169
113. Cox AD, Brtva TR, Lowe DG, Der CJ (1994) Oncogene 9:3281
114. Graham SM, Cox AD, Drivas G, Rush MG, D'Eustachio P, Der CJ (1994) Molecular & Cellular Biology 14:4108
115. Huff SY, Quilliam LA, Cox AD, Der CJ (1997) Oncogene 14:133
116. Fernandez-Sarabia MJ, Bischoff JR (1993) Nature 366:274
117. Lee CHJ, Della NG, Chew CE, Zack DJ (1996) Journal Of Neuroscience 16:6784
118. Clark GJ, Kinch MS, Rogers Graham K, Sebti SM, Hamilton AD, Der CJ (1997) Journal of Biological Chemistry 272:10,608
119. Yee WM, Worley PF (1997) Molecular Cellular Biology 17:921
120. Symons M (1996) Trends in Biochemical Sciences 21:178
121. Vanaelst L, Dsouzaschorey C (1997) Genes & Development 11:2295
122. Hoffman GR, Nassar N, Cerione RA (2000) Cell 100:345
123. Block C, Wittinghofer A (1995) Structure 3:1281
124. Vojtek AB, Cooper JA (1995) Cell 82:527
125. Mackay DJG, Hall A (1998) Journal of Biological Chemistry 273:20,685
126. Kjoller L, Hall A (1999) Experimental Cell Research 253:166
127. Rodriguezviciana P, Warne PH, Khwaja A, Marte BM, Pappin D, Das P, Waterfield MD, Ridley A, Downward J (1997) Cell 89:457
128. Schimmoller F, Simon I, Pfeffer SR (1998) Journal of Biological Chemistry 273:22,161
129. Novick P, Zerial M (1997) Current Opinion in Cell Biology 9:496
130. Martinez O, Goud B (1998) Biochimica et Biophysica Acta – Molecular Cell Research 1404:101
131. Zerial M (1998) Current Opinion in Cell Biology 10:475
132. Jahn R, Sudhof TC (1999) Annual Review of Biochemistry 68:863
133. Schimmoller F, Itin C, Pfeffer S (1997) Current Biology 7:R235
134. Goerlich D, Mattaj IW (1996) Science 271:1513

135. Bischoff FR, Ponstingl H (1991) Nature 354:80
136. Bischoff FR, Maier G, Tilz G, Ponstingl H (1990) Proc Natl Acad Sci USA 87:8617
137. Becker J, Melchior F, Gerke V, Bischoff FR, Ponstingl H, Wittinghofer A (1995) Journal of Biological Chemistry 270:11,860
138. Izaurralde E, Kutay U, Vonkobbe C, Mattaj IW, Goerlich D (1997) EMBO Journal 16:6535
139. Rexach M, Blobel G (1995) Cell 83:683
140. Goerlich D, Henklein P, Laskey RA, Hartmann E (1996) EMBO Journal 15:1810
141. Pollard VW, Michael WM, Nakielny S, Siomi MC, Wang F, Dreyfuss G (1996) Cell 86:985
142. Kutay U, Bischoff FR, Kostka S, Kraft R, Goerlich D (1997) Cell 90:1061
143. Bischoff FR, Krebber H, Smirnova E, Dong W, Ponstingl H (1995) EMBO J 14:705
144. Yokoyama N, Hayashi N, Seki T, Panté N, Ohba T, Nishii K, Kuma K, Hayashida T, Miyata T, Aebi U, Fukui M, Nishimoto T (1995) Nature 376:184
145. Sazer S (1996) Trends in Cell Biology 6:81
146. Chrambach A, Reisfeld RA, Wyckoff M, Zaccari J (1967) Analytical Biochemistry 20:150
147. Bradford MM (1976) Analytical Biochemistry 72:248
148. Crowther JR (1995) Methods in Molecular Biology 42:1
149. Chaiken I, Rose S, Karlsson R (1992) Analytical Biochemistry 201:197
150. Manavalan P, Johnson WC Jr (1987) Analytical Biochemistry 167:76
151. Greenfield NJ (1996) Analytical Biochemistry 235:1
152. Scatchard G (1949) Ann NY Acad Sci 51:660
153. Herrmann C, Nassar N (1996) Progress in Biophysics & Molecular Biology 66:1
154. Hill AV (1910) J Physiol 40:4
155. Wiseman T, Williston S, Brandts JF, Lin LN (1989) Analytical Biochemistry 179:131
156. Cush R, Cronin JM, Stewart WJ, Maule CH, Molloy J, Goddard NJ (1993) Biosensors & Bioelectronics 8:347
157. Karlsson R, Falt A (1997) Journal of Immunological Methods 200:121
158. Yarmush ML, Patankar DB, Yarmush DM (1996) Molecular Immunology 33:1203
159. Schuck P, Minton AP (1997) Trends in Biochemical Sciences 22:150
160. Schuck P, Minton AP (1996) Trends in Biochemical Sciences 21:458
161. Salzmann M, Wider G, Pervushin K, Senn H, Wuthrich K (1999) Journal of the American Chemical Society 121:844
162. Salzmann M, Pervushin K, Wider G, Senn H, Wuthrich K (1998) Proceedings of the National Academy of Sciences of the United States of America 95:13,585
163. John J, Frech M, Wittinghofer A (1988) J 263:11,792
164. Hiratsuka T (1983) Biochimica et Biophysica Acta 742:496
165. Neal SE, Eccleston JF, Hall A, Webb MR (1988) Journal of Biological Chemistry 263:19,718
166. McGrath JP, Capon DJ, Goeddel DV, Levinson AD (1984) Nature 310:644
167. Manne V, Bekesi E, Kung HF (1985) Proceedings of the National Academy of Sciences of the United States of America 82:376
168. Gibbs JB, Sigal IS, Poe M, Scolnick EM (1984) Proc Natl Acad Sci USA 81:5704
169. Schweins T, Geyer M, Scheffzek K, Warshel A, Kalbitzer HR, Wittinghofer A (1995) Nature Structural Biology 2:36
170. Neal SE, Eccleston JF, Webb MR (1990) Proc Natl Acad Sci USA 87:3562
171. Rensland H, Lautwein A, Wittinghofer A, Goody RS (1991) Biochemistry 30:11,181
172. Moore KJ, Webb MR, Eccleston JF (1993) Biochemistry 32:7451
173. Brownbridge GG, Lowe PN, Moore KJ, Skinner RH, Webb MR (1993) Journal of Biological Chemistry 268:10,914
174. Mittal R, Ahmadian MR, Goody RS, Wittinghofer A (1996) Science 273:115
175. Scheffzek K, Ahmadian MR, Wiesmuller L, Kabsch W, Stege P, Schmitz F, Wittinghofer A (1998) EMBO Journal 17:4313
176. Lenzen C, Cool RH, Wittinghofer A (1995) Methods in Enzymology 255:95
177. Lenzen C, Cool RH, Prinz H, Kuhlmann J, Wittinghofer A (1998) Biochemistry 37:7420
178. Feig LA (1999) Nature Cell Biology 1:E25
179. Cool RH, Schmidt G, Lenzen CU, Prinz H, Vogt D, Wittinghofer A (1999) Molecular & Cellular Biology 19:6297

180. Cool RH, Schmidt G, Lenzen CU, Prinz H, Vogt D, Wittinghofer A (1999) Molecular & Cellular Biology 19:6297
181. van den Berghe N, Cool RH, Wittinghofer A (1999) Journal of Biological Chemistry 274:11,078
182. Herrmann C, Martin GA, Wittinghofer A (1995) Journal of Biological Chemistry 270:2901
183. Block C, Janknecht R, Herrmann C, Nassar N, Wittinghofer A (1996) Nature Structural Biology 3:244
184. Herrmann C, Horn G, Spaargaren M, Wittinghofer A (1996) Journal of Biological Chemistry 271:6794
185. Linnemann T, Geyer M, Jaitner BK, Block C, Kalbitzer HR, Wittinghofer A, Herrmann C (1999) Journal of Biological Chemistry. 274:13,556
186. Nassar N, Horn G, Herrmann C, Block C, Janknecht R, Wittinghofer A (1996) Nature Structural Biology 3:723
187. Vetter IR, Linnemann T, Wohlgemuth S, Geyer M, Kalbitzer HR, Herrmann C, Wittinghofer A (1999) FEBS Letters. 451:175
188. Gorman C, Skinner RH, Skelly JV, Neidle S, Lowe PN (1996) Journal of Biological Chemistry 271:6713
189. Sydor JR, Engelhard M, Wittinghofer A, Goody RS, Herrmann C (1998) Biochemistry 37:14,292
190. Walker EH, Perisic O, Ried C, Stephens L, Williams RL (1999) Nature 402:313
191. Pai EF, Krengel U, Petsko GA, Goody RS, Kabsch W, Wittinghofer A (1990) EMBO J 9:2351
192. Jurnak F (1985) Science 230:32
193. Scheffzek K, Klebe C, Fritzwolf K, Kabsch W, Wittinghofer A (1995) Nature 374:378
194. Ostermeier C, Brunger AT (1999) Cell 96:363
195. Maesaki R, Ihara K, Shimizu T, Kuroda S, Kaibuchi K, Hakoshima T (1999) Molecular Cell 4(5):793–803
196. Rittinger K, Walker PA, Eccleston JF, Nurmahomed K, Owen D, Laue E, Gamblin SJ, Smerdon SJ (1997) Nature 388:693
197. Noel JP, Hamm HE, Sigler PB (1993) Nature 366:654
198. Prakash B, Praefcke GJK, Renault L, WittinghoferA, Herrmann C (2000) Nature 403:567
199. Nassar M, Horn G, Herrmann C, Scherer A, McCormick F, Wittinghofer A (1995) Nature 375:554
200. Vetter IR, Nowak C, Nishimoto T, Kuhlmann J, Wittinghofer A (1999) Nature. 398:39
201. Sprinzl M (1994) Trends in Biochemical Sciences 19:245
202. Goody RS, Pai EF, Schlichting I, Rensland H, Scheidig A, Franken S, Wittinghofer A (1992) Philosophical Transactions of the Royal Society of London – Series B: Biological Sciences 336:3
203. Schlichting I, Almo SC, Rapp G, Wilson K, Petratos K, Lentfer A, Wittinghofer A, Kabsch W, Pai EF, Petsko GA, Goody RS (1990) Nature 345:309
204. Scheidig AJ, Burmester C, Goody RS (1999) Structure. 7:1311
205. Cepus V, Scheidig AJ, Goody RS, Gerwert K (1998) Biochemistry 37:10,263
206. Boriacksjodin PA, Margarit SM, Barsagi D, Kuriyan J (1998) Nature 394:337
207. Sondek J, Lambright DG, Noel JP, Hamm HE, Sigler PB (1994) Nature 372:276
208. Rittinger K, Walker PA, Eccleston JF, Smerdon SJ, Gamblin SJ (1997) Nature 389:758
209. Scheffzek K, Ahmadian MR, Kabsch W, Wiesmuller L, Lautwein A, Schmitz F, Wittinghofer A (1997) Science 277:333
210. Scheffzek K, Ahmadian MR, Wittinghofer A (1998) Trends in Biochemical Sciences 23:257
211. Emerson SD, Madison VS, Palermo RE, Waugh DS, Scheffler JE, Tsao KL, Kiefer SE, Liu SP, Fry DC (1995) Biochemistry 34:6911
212. Terada T, Ito Y, Shirouzu M, Tateno M, Hashimoto K, Kigawa T, Ebisuzaki T, Takio K, Shibata T, Yokoyama S, Smith BO, Laue ED, Cooper JA (1999) Journal of Molecular Biology 286:219

213. Geyer M, Herrmann C, Wohlgemuth S, Wittinghofer A, Kalbitzer HR (1997) Nature Structural Biology 4:694
214. Huang L, Weng XW, Hofer F, Martin GS, Kim SH (1997) Nature Structural Biology 4:609
215. Huang L, Hofer F, Martin GS, Kim SH (1998) Nature Structural Biology 5:422
216. Kraulis PJ, Domaille PJ, Campbell-Burk SL, Van Aken T, Laue ED (1994) Biochemistry 33:3515
217. Geyer M, Schweins T, Herrmann C, Prisner T, Wittinghofer A, Kalbitzer HR (1996) Biochemistry 35:10,308
218. Ito Y, Yamasaki K, Iwahara J, Terada T, Kamiya A, Shirouzu M, Muto Y, Kawai G, Yokoyama S, Laue ED, Walchli M, Shibata T, Nishimura S, Miyazawa T (1997) Biochemistry 36:9109
219. Esser D, Bauer B, Wolthuis RMF, Wittinghofer A, Cool RH, Bayer P (1998) Biochemistry 37:13,453
220. Resh MD (1996) Cellular Signalling 8:403
221. Gelb MH (1997) Science 275:1750
222. Waddick KG, Uckun FM (1998) Biochemical Pharmacology 56:1411
223. Page MJ, Hall A, Rhodes S, Skinner RH, Murphy V, Sydenham M, Lowe PN (1989) Journal of Biological Chemistry 264:19,147
224. Porfiri E, Evans T, Chardin P, Hancock JF (1994) Journal of Biological Chemistry 269:22,672
225. Orita S, Kaibuchi K, Kuroda S, Shimizu K, Nakanishi H, Takai Y (1993) Journal of Biological Chemistry 268:25,542
226. Rubio I, Wittig U, Meyer C, Heinze R, Kadereit D, Waldmann H, Downward L, Wetzker R (1999) European Journal of Biochemistry 266:70
227. Silvius JR, Leventis R (1993) Biochemistry 32:13,318
228. Silvius JR, Zuckermann MJ (1993) Biochemistry 32:3153
229. Ghomashchi F, Zhang XH, Liu L, Gelb MH (1995) Biochemistry 34:11,910
230. Leventis R, Silvius JR (1998) Biochemistry 37:7640
231. Chaiken I, Rose S, Karlsson R (1992) Analytical Biochemistry 201:197
232. Plant AL, Brigham-Burke M, Petrella EC, O'Shannessy DJ (1995) Analytical Biochemistry 226:342
233. Schelhaas M, Nagele E, Kuder N, Bader B, Kuhlmann J, Wittinghofer A, Waldmann H (1999) Chem 5:1239
234. Bader B, Kuhn K, Owen DJ, Waldmann H, Wittinghofer A, Kuhlmann J (2000) Nature 403:223
235. Wittinghofer A, Waldmann H (2000) Angew Chem Int Ed (in press)
236. Pingoud A, Urbanke C (1997) Arbeitsmethoden der Biochemie
237. Cantor CR, Schimmel PR (1980) Biophysical chemistry, part 2/3. WH Freeman, NY (ISBN: 0-7167-1190-7, 0-7167-1192-3)
238. Marshall AG (1978) Biophysical chemistry. Wiley, NY (ISBN: 0-471-02718-9)
239. Glasel JA, Deutscher MP (1995) Introduction to biophysical methods for protein and nucleic acid research. Academic Press, San Diego (ISBN: 0-12-286230-9)
240. Gutfreund H (1995) Kinetics for the life sciences – receptors, transmitters and catalysts. Cambridge University Press (ISBN: 0-521-48586-X)
241. Blundell TL, Johnson LN (1976) Protein crystallography. In: Horecker B, Kaplan NO, Marmur J, Scheraga HA (eds) Molecular biology. International Series of Monographs and Textbooks. Academic Press

Ras-Farnesyltransferase-Inhibitors as Promising Anti-Tumor Drugs

Herbert Waldmann · Michael Thutewohl

Max-Planck-Institut für molekulare Physiologie, Otto-Hahn-Strasse 11, 44227 Dortmund, Germany
E-mail: herbert.waldmann@mpi-dortmund.mpg.de

Universität Dortmund, Fb. 3, Organische Chemie, Otto-Hahn-Strasse 6, 44227 Dortmund, Germany

The protein Ras, an important intracellular signal transducer, is crucially involved in the development of tumor growth. The farnesylation of Ras, catalyzed by the enzyme Ras-farnesyltransferase, is essential to its proper functioning in the normal and in the transformed state. Therefore, the inhibition of Ras lipidation has become a promising target for the development of new classes of anti-tumor agents. This review focuses on the different classes of Ras-farnesyltransferase inhibitors and compares their biological properties and modes of action in vitro as well as in vivo.

Keywords: Bioorganic chemistry, Cellular signal transduction, Enzyme-Inhibition, Ras-farnesyltransferase, Tumor-therapy

1
Introduction

The protein Ras plays a key role in the cellular transduction of mitogenic signals given by external growth factors [1]. It is a central molecular switch that regulates cell growth, cell division, and differentiation. False regulation of the Ras pathway can lead to transformation of the cell. In fact, point mutations in the corresponding Ras genes are found in approximately 30% of all human tumors, particularly in over 90% of human pancreatic carcinomas and 50% of human colon cancers [2]. The crucial role of the Ras proteins in maintaining the regular life cycle of cells and the extensive involvement of mutated Ras in the development of numerous cancers has sparked the idea that influencing the function of Ras might open up new opportunities for the development of alternative anti-tumor drugs [3,4]. Disruption of the aberrant growth signals of oncogenic Ras proteins provides an example of what is now recognized as 'signal transduction therapy' [2a]. Ras proteins must be localized to the plasma membrane in order to perform their biological function in both the normal and the transformed state. Covalently attached lipids are the major driving force for membrane localization. Non-lipidated Ras is cytosolic and biologically inactive. Several post-translational modifications are required at the carboxyl-terminal end of Ras before the initially biosynthesized precursor protein matures into a biologically active protein localized to the plasma membrane. Interfering with these processing steps might disrupt the aberrant growth signals of oncogenic Ras proteins.

2
Biochemical Lipid-Modification of Ras

The Ras proteins are synthesized as biologically inactive, cytosolic precursor proteins. They are then modified by several post-translational processing steps at the carboxyl terminal end and thereby converted into biologically active proteins localized at the plasma membrane. The cysteine of the C-terminal CAAX sequence (C is cysteine, A is generally an aliphatic amino acid, and X is methionine, serine, alanine, or glutamine) is first enzymatically S-farnesylated; the AAX part is then cleaved off by a specific protease, and the free C-terminal cysteine is finally converted into a methyl ester (Scheme 1).

In H- and N-Ras, additional cysteine residues in the immediate vicinity of the CAAX moiety are lipidated by formation of palmitic acid thioesters. Anchoring of the Ras proteins to the plasma membrane is mediated by the lipid groups. Genetic experiments have shown that farnesylation is essential to Ras function in the normal and the transformed state [5]. Palmitoylation appears to be not absolutely required for Ras cell-transforming activity but is important for conferring full biological activity on Ras [6]. The K4B-Ras protein compensates for the absence of a hydrophobic palmitoyl moiety with a positively-charged lysine-rich region near the CAAX box to enhance membrane interactions with negatively charged phospholipids.

The finding that lipid-modification of Ras, in particular farnesylation, is crucial to its biological activity led to the idea that inhibiting the enzymes re-

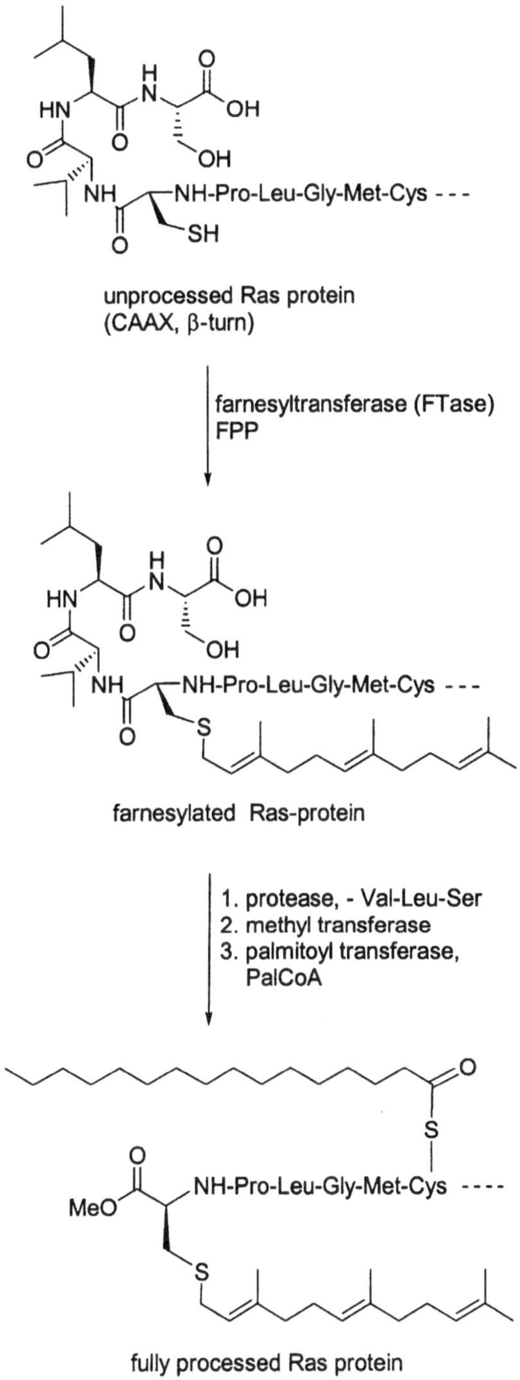

unprocessed Ras protein
(CAAX, β-turn)

farnesyltransferase (FTase)
FPP

farnesylated Ras-protein

1. protease, - Val-Leu-Ser
2. methyl transferase
3. palmitoyl transferase,
   PalCoA

fully processed Ras protein

Scheme 1. Posttranslational modification of Ras protein with lipid groups (FPP: Farnesyl-pyrophosphate, PalCoA: Palmitoyl CoA)

sponsible for the modification would in turn inhibit Ras function and provide new approaches for developing potent anti-tumor drugs.

3
Structure of Ras-Farnesyltransferase and Possible Opportunities for Its Inhibition

Farnesyltransferase is a heterodimeric protein composed of an α- and a β-sub-unit, and zinc and magnesium ions are required for its activity [7, 8]. A closely related heterodimeric enzyme is geranylgeranyltransferase I (GGTase I). It also recognizes proteins with a CAAX-box when X is leucine. Both enzymes share the same α-subunit and the requirement for Zn^{2+} and Mg^{2+} but the β-subunits are different. The similarity of these enzymes highlights the importance of selectivity of farnesyltransferase (FT)-inhibitors.

Since the discovery that the Ras proteins are farnesylated numerous inhibitors of FTase have been developed. This field has been reviewed in detail and only a brief overview is given here [3, 4, 9, 10–12]. Initial studies on inhibitor development relied on information about substrate specificity as well as structural studies of the substrate in solution and bound to the enzyme (obtained from NMR spectroscopic investigations) [13, 14]. The crystal structure of the free enzyme was reported at 2.25 Å resolution [15]. The active site consists of two clefts which are at the junction of a bound zinc ion. The zinc coordinates the thiol of cysteine in a ternary complex. One of the clefts was found to be a hydrophilic surface groove near the subunit interface. It could correspond to the site of binding of the Ras protein. The second cleft is a region of aromatic residues. It is of a size that would accommodate the farnesyl moiety but not the longer geranylgeranyl pyrophosphate. In addition, the structure of the rat enzyme in complex with farnesyl-pyrophosphate was solved at 3.41 Å resolution [16]. Recently, the analysis of four ternary complexes of rat-FTase co-crystallized with FPP analogs and KB4-Ras peptides was reported [17]. According to these studies based on crystal structures at 2 Å resolution, zinc is essential for productive peptide binding, whereby the conformation of the zinc bound ligand seems not to be a β-turn (see Scheme 1), in contrast to the results deduced from NMR-investigations [13, 14]. The three-dimensional information obtained from the X-ray structures should now spur the development of more efficient inhibitors.

4
Inhibitors of Ras-Farnesyltransferase

4.1
Inhibitors Based on the CAAX Motif

Based on the CAAX motif, peptide analogues were designed in which peptide amide bonds are replaced by amine and ether groups [18]. In particular β-turn mimetic 1 (Fig. 1) inhibits the FTase in vitro with an IC_{50} value of 1.8 nmol/l and shows highly specific activity in comparison to inhibition of GGTase I.

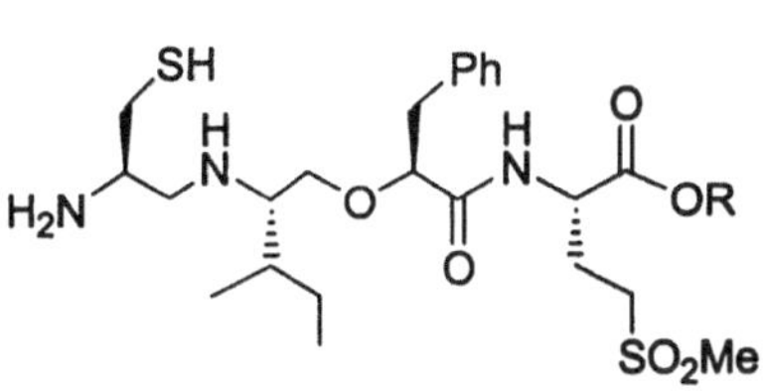

R = H L-739,749 **1**
R = CH_3 L-739,750 **2**

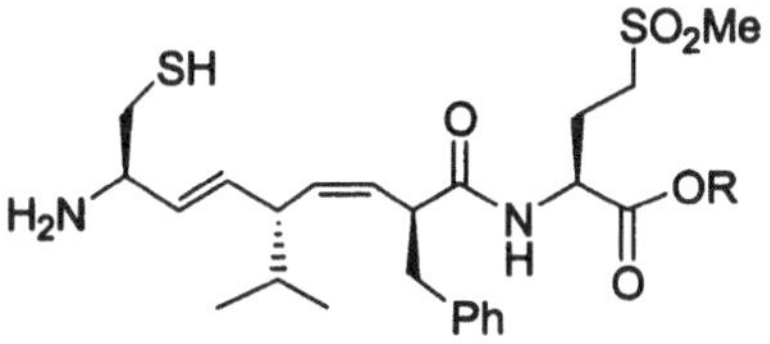

R = H B 956 **3**
R = CH_3 B 1086 **4**

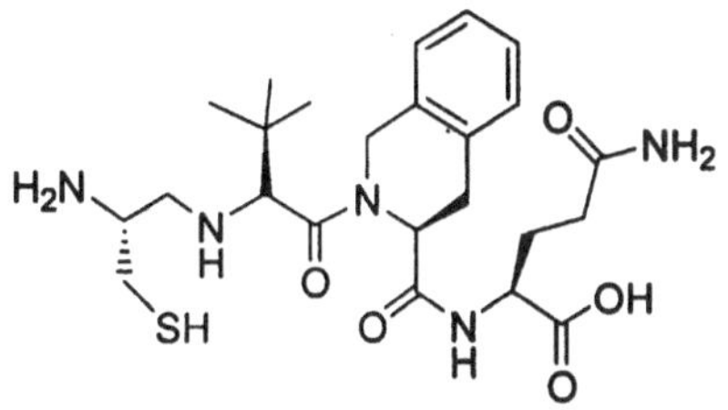

5

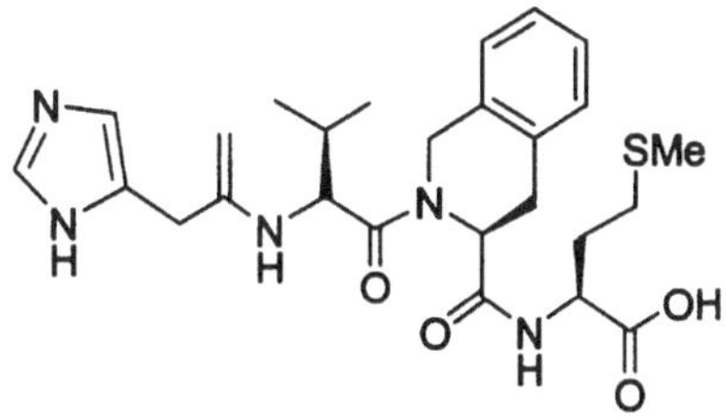

6

Fig. 1. FTase inhibitors in which amide bonds were replaced by isosteric amines and ethers, and which incorporate non-natural amino acids

The less polar methyl ester **2** as prodrug showed better results in vivo and inhibits both farnesylation of the Ras protein and growth of Ras-transformed cells, whilst proliferation of Raf- or Mos-transformed cells was not influenced. Growth of human pancreatic adenocarcinoma cells with mutated K-Ras, c-Myc and p53 genes was inhibited by application of **2**. If the compound is administered over a period of 5 days to mice with implanted Ras-dependent tumors, tumor growth can be reduced by up to 66% compared to untreated mice, whereas application of the antitumor antibiotic doxorubicin only resulted in 33% reduction under the same conditions. It is particularly noteworthy that treatment with the β-turn mimetic – in contrast to treatment with doxorubicin – was without any visible side effects, such as weight loss.

However, **2** also affected the regulation of actin stress fiber formation [19]. Rho proteins are involved in the regulation of various cytoskeletal structures, and RhoB is believed to be one of the prime targets of FTase inhibitors. Rho B is apparently both geranylgeranylated and farnesylated [20, 21]. If cells were treated with **2**, vesicular localization of Rho B was inhibited. Thus **2** may also inhibit the farnesylation of Rho B, thereby interfering with actin stress fiber formation [22].

Further modifications of the CAAX tetrapeptide structure led to inhibitor **3** which blocked H-Ras farnesylation with an IC_{50} of 11 nmol/l [23]. Tumor cell lines expressing mutant H- and N-Ras were most sensitive against this compound that inhibits tumor growth of EJ-1 human bladder carcinomas by about

60% at a dose of 100 mg/kg. The in vivo results and particularly tumor regression induced by the CAAX mimetics described thus clearly demonstrate the efficacy of FTase inhibitors as antitumor agents.

High activity and selectivity was also achieved by replacing the phenylalanine residue in Cys-Val-Phe-Met with 1,2,3,4-tetrahydroisoquinoline-3-carboxylate (Tic) and modifying the peptide backbone to give 5 (Fig. 1) [24]. This compound displayed an IC_{50} of 2.8 nmol/l against rat FTase and was 500-fold less potent against GGTase I. In compound 6 the terminal cysteine residue was replaced by 4-imidazole leading to an increase in activity to $IC_{50} = 0.79$ nmol/l [25].

Replacement of the two aliphatic amino acids in the CAAX motif was achieved with benzodiazepines such as 7 and 8 (Fig. 2).

The central unit of these peptidomimetics imitates a β-turn and brings the NH_2-terminus of the cysteine analogue and the COOH terminus of the methionine in spatial proximity; these can then complex the Zn^{2+} ion which is essential for activity of the FTase [26]. The free acid 7 inhibits the enzyme with an IC_{50} value of 1 nmol/l, whilst in intact cells the methyl ester 8, despite its weaker in vitro activity, is significantly more potent because it can penetrate the plasma membrane better due to its lower polarity. This property can be used to convert the morphology of H-Ras-transformed cells back to the normal form and to inhibit growth of these cells, whereas the substance shows no effect on Src-transformed and untransformed rat fibroblasts. The inhibitor therefore acts selectively on transformed cells and does not influence growth of normal cells. This result is noteworthy because farnesylation of the wild type H-Ras protein

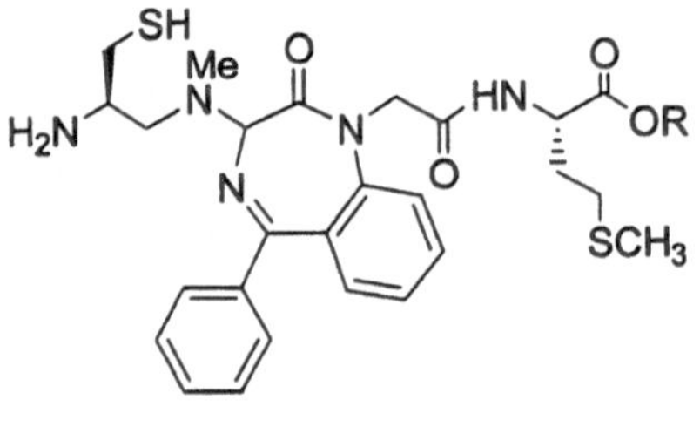

R = H   7
R = CH₃ 8

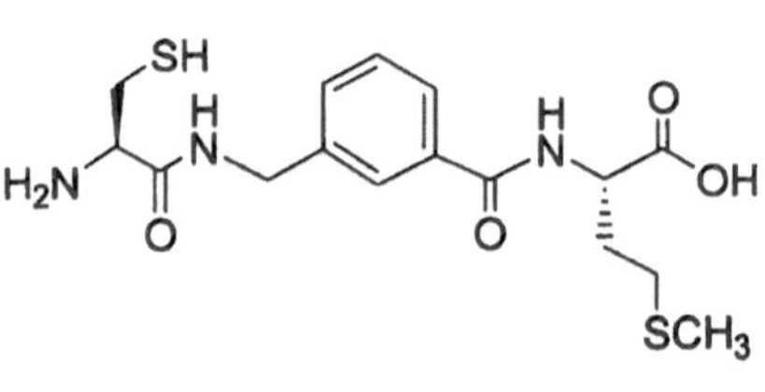

Cys-3-AMBA-Met 9

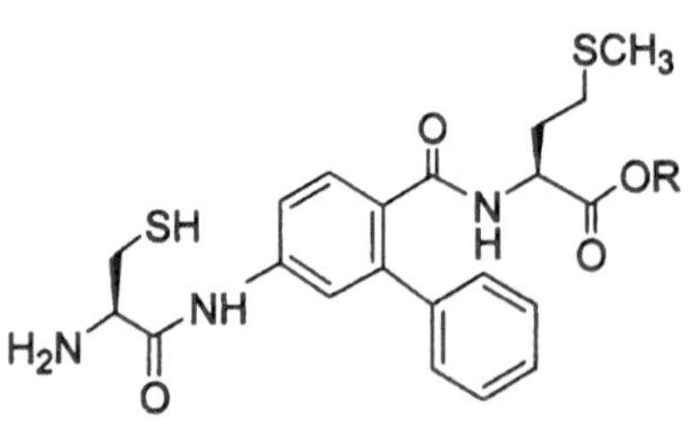

R = H, CH₃   10

Fig. 2. FTase inhibitors in which the AA dipeptide was replaced

is inhibited by the benzodiazepine and H-Ras is involved in growth of normal fibroblasts. Detailed examination of this unexpected result shows that inhibitor **8** reduces the amount of farnesylated H-Ras protein in both non-transformed and H-Ras transformed cells. However, *only* in the transformed cells does this lead to the expected reduction of enzyme activity in the Ras cascade: both the amount of phosphorylated Raf and the activities of MEK-1, MEK-2, and MAP kinase were reduced in the tumor cells [27]. In addition, stimulation of the Ras signal pathway was not influenced by EGF in normal cells or in Src-transformed cells. Normal cells apparently have the ability to activate the signal pathway independently of Ras inhibitory drugs. In fact, further experiments have shown that farnesylation of the K-Ras protein by the inhibitors is considerably less influenced than that of H-Ras. K-Ras is also a substrate of the GGTase I and can therefore be geranylgeranylated if activity of the FTase is reduced. Normal cells and Src-transformed cells, in contrast to H-Ras-transformed cells, are able to compensate for loss of a Ras protein or an enzyme important for normal cell growth, by activation of alternative pathways.

The two aliphatic amino acids have also been replaced by the hydrophobic spacers 3-(aminomethyl)benzoic acid (3-AMBA; see **9**, Fig. 2) [28] and 3- and 4-aminobenzoic acid (3- and 4-ABA) as well as 2-phenylaminobenzoic acid (see **10**, Fig. 2) [29, 30]. These compounds displayed IC_{50} values in the nanomolar range, and **10** (R = H) blocked the growth in nude mice of a human lung carcinoma expressing oncogenic Ras [31].

Further design led to replacement not only of the central AA-dipeptide but also of the AAX-tripeptide. Thus, biphenyl derivative **11** [32] and piperazine analogue **12** [33] were developed (Fig. 3). They display IC_{50} values in the nanomolar range, whereas geranylgeranyltransferase was blocked with micromolar IC_{50} values. Both compounds disrupt Ras processing in cells and piperazine derivative **12** suppresses tumor growth by 75% at a dose of 40 mg/kg.

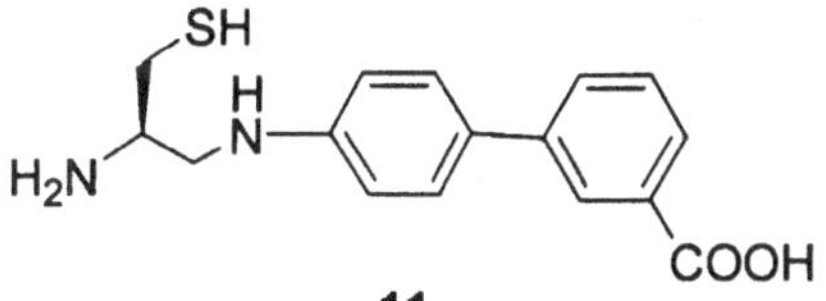

11 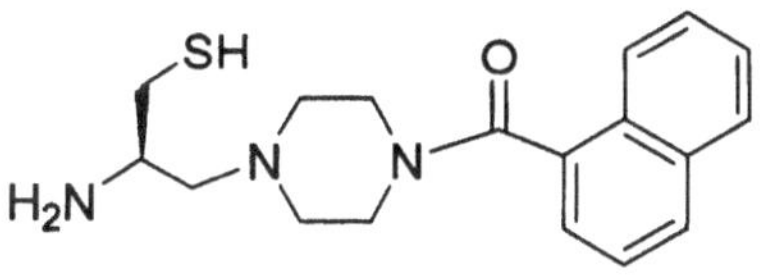   **12**

Fig. 3. FTase inhibitors in which the AAX tripeptide was replaced

4.2
Bisubstrate Inhibitors

The design of FTase inhibitors based on the structure of farnesyl pyrophosphate has been pursued with less intensity due to the possible nonselective effects of competing with other enzymes such as squalene synthetase that also accept farnesylpyrophosphate as substrate [3, 4, 9, 10–12].

However, bisubstrate inhibitors incorporating a farnesyl- and a CAAX mimetic are very promising since one can expect that they display enhanced activity and selectivity. The phosphinic acids 13 (R = H) and 14 (R = Me, Fig. 4) are examples of bisubstrate analogues.

Indeed, 13 is an effective in vitro inhibitor and the prodrug 14 has activity in H-Ras- and – to a lesser extent – K-Ras-transformed cells [34]. Prodrug 14 also inhibits growth of malignant cells of the neurofibromatosis type I which presumably contain overactive wild type Ras due to the absence of active neurofibromin GAP [35]. These highly promising results are tempered, however, by the lower bioavailability and in vivo activity in comparison to the peptidomimetics. Recently, benzyloxycinnamoylamide 15 and two analogs thereof were reported as bisubstrate inhibitors of yeast FTase [36] where the benzoylcinnamoyl group mimics the farnesyl moiety. These compounds displayed IC_{50} values in the low micromolar range.

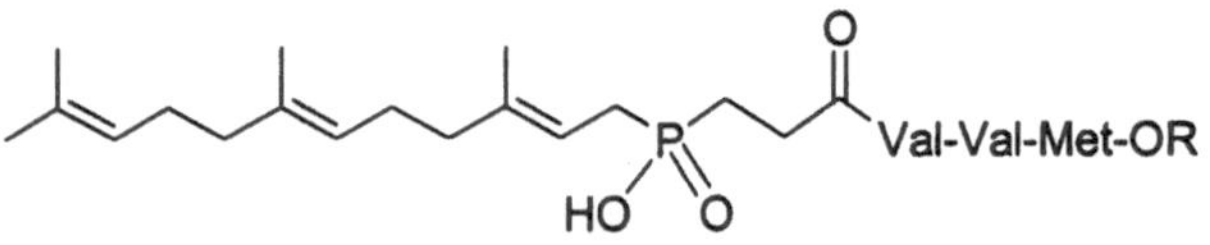

R = H   **13**
R = CH₃   **14**

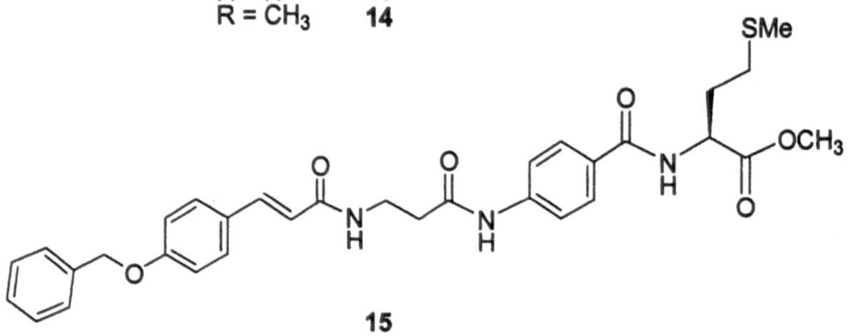

15

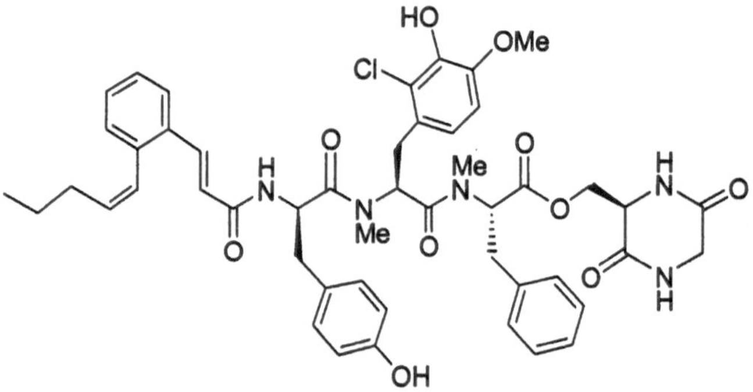

pepticinnamin E **16**

Fig. 4. Bisubstrate inhibitors of FTase

The natural product pepticinnamin E (16) is also considered a bisubstrate inhibitor of FTase. The successful synthesis of pepticinnamin E sets the stage to prepare analogs of this peptidic compound and to delineate the important structural parameters responsible for its FTase-inhibiting activity. Initial studies revealed that the central tripeptide part and, in particular, the absolute configuration of the central chlorinated amino acid are decisive for the inhibitory activity [37]. The bisubstrate inhibitors developed so far (as well as some of the CAAX mimetics, see above) have the advantage of not containing a free cysteine SH-group and thus being more stable.

4.3
Inhibitors from Screening Natural Sources and Compound Libraries

In addition to pepticinnamin E several natural products have been identified as FTase inhibitors, including, e.g., manumycin (17) and analogues [38] fusidienol (18) [39] and the preussomerins, e.g., 19 [40] (Fig. 5; for further examples see [10]). In general, these compounds have been less potent than the CAAX-based mimetics typically displaying IC_{50} values in the low micromolar or the high nanomolar range.

High-throughput screening of compound libraries was used to identify further inhibitors of farnesyltransferase. Thus a highly potent pentapeptide 20 was identified (Fig. 6) that inhibited FTase with $IC_{50} = 17$ nmol/l and that antagonized Ras in *Xenopus oocytes* [41]. Systematic derivatization and truncation of the peptide backbone finally led to a dipeptide derivative 21 that inhibited Ras processing in cells and increased the life span of tumor-bearing nude mice by 35% at 200 mg/kg [10, 42].

In particular, a series of completely non-peptidic nonsulfhydryl FTase inhibitors 22 and 23 was uncovered (Fig. 6).

These chlorobenzocycloheptapyridines display pronounced selectivity for FTase over GGTase I. They show improved in vivo anti-tumor activity and pharmacokinetic profiles in mice when administered orally and inhibit H-Ras processing in Cos monkey kidney cells [43, 44]. Extensive variation of the substituents on the piperidine moiety and the tricyclic ring system led to the devel-

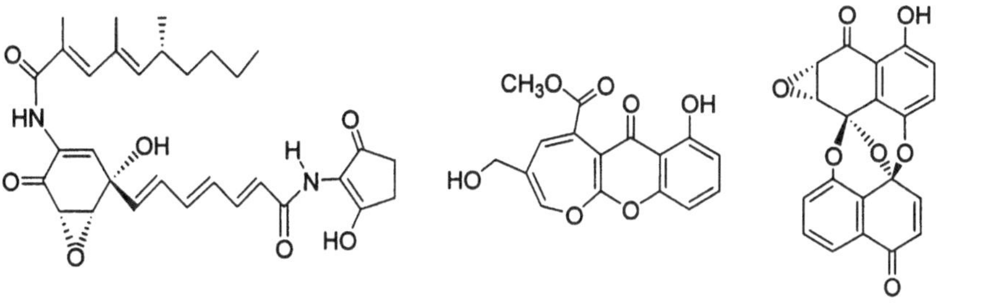

manumycin A **17**   fusidienol **18**   preussomerin G **19**

Fig. 5. FTase inhibitors from natural sources

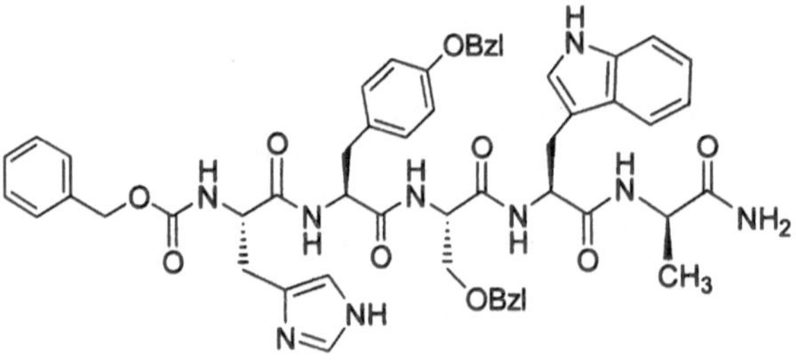

Fig. 6. FTase inhibitors identified from compound libraries

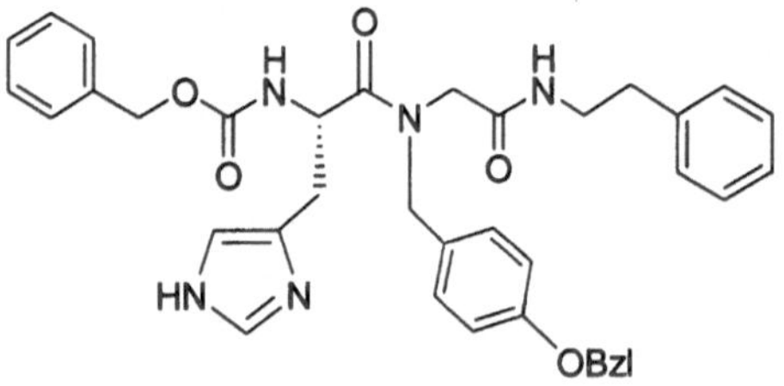

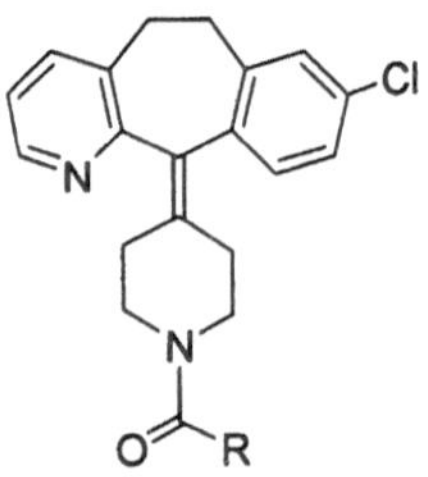

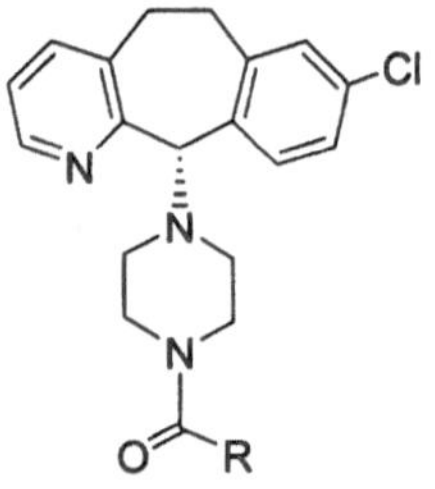

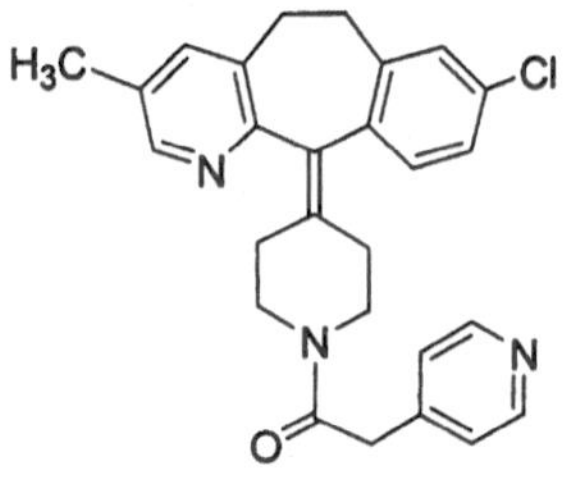

opment of compound **24** which inhibits FTase with $IC_{50} = 40$ nmol/l and is inactive toward GGTase I.

5
Lack of Toxicity to Normal Cells

Although FTase inhibitors influence the farnesylation of Ras they are likely to
interfere with the posttranslational modifications of other CAAX-containing
proteins as well. Apart from the approximately 20 farnesylated proteins that are
known today, farnesylation is also required for normal Ras function which in
turn is critical for normal cell viability. For these reasons farnesyltransferase

inhibitors were thought to be potentially toxic. However, these drugs proved to be surprisingly non-toxic and did not display inhibitory activity against rodent fibroblast cells [45]. Farnesyltransferase inhibitors not only inhibited the growth of transformed cells in culture far more than the growth of normal cells, but also showed surprisingly few side effects in the treatment of tumor-bearing animals, with no obvious toxicity in normal tissues following treatments of up to six weeks.

There are several possible explanations to account for this apparent lack of toxicity. Some geranylgeranylated Ras-related proteins might compensate for the loss of Ras function (see, e.g., [46]). Alternatively inhibition of farnesyl transferase may reduce Ras activity below the level required for transformation, yet allow sufficient Ras activity for maintaining normal cell viability [47]. Alternatively, a different signaling pathway may be activated when Ras is not anchored to the plasma membrane.

Toxicity and effectivity studies have often been performed in rodent fibroblast cells containing oncogenic H-Ras. However, prenylation of K-Ras B and N-Ras are not as effectively blocked by the farnesyltransferase inhibitors as H-Ras [48] (see below). Thus normal cells may be less sensitive to these drugs because they express K-Ras 4B and N-Ras. In this context it should be noted that H-Ras mutations are relatively uncommon in human tumors [49]. Rather, the K-Ras gene is the most frequently mutated in solid human cancers, whereas N-Ras is prevalent in leukemias. Thus the preclinical evaluation of the farnesylation inhibitors has yet to be critically re-evaluated for trials in humans.

6
Mechanism of Action

Intensive investigations of a variety of FTase inhibitors have clearly revealed that these drugs demonstrate antiproliferative activity in cell culture against a variety of transformed cell lines and tumors in animals. In particular, cells with a mutant H-Ras protein are sensitive against these drugs. Although the growth of K-Ras 4B transformed cells is also blocked, much higher concentrations are needed for this effect since this protein is a much better substrate for FTase than H-Ras [50]. Furthermore, the K4B-Ras protein can more easily be geranylgeranylated by GGTase, thus compensating for the lack of farnesylation [51, 52]. Since geranylgeranylated forms of Ras proteins can also potently transform cells, farnesyltransferase inhibitors are not expected to inhibit the growth of tumors with mutant K4B-Ras. However, despite this alternative prenylation growth inhibition is observed. Obviously blocking the farnesylation of one or more proteins other than Ras is responsible for or at least contributes to the observed effect. This is supported by the finding that growth of fibroblast cells transformed by an N-terminally myristoylated (and thereby membrane-anchored) Ras that cannot become farnesylated or geranylgeranylated is inhibited by FTase inhibitors as well [53].

These findings have raised questions of which additional proteins are the molecular targets of FTase inhibitors. One class of candidates involves members of the Rho family of proteins, in particular Rho B (see Sect. 4.1). Rho B is a particularly attractive target as Rho proteins have been shown to be required for Ras transformation and Rho B is an immediate early gene response protein activated by growth factor signaling. Rho B is targeted by FTase inhibitors (see above and [54]). Another novel Ras-related farnesylated protein, Rheb, may also be a target. Rheb modulates Ras function and is an antagonist of Ras transformation. Its processing and membrane localization can be inhibited by a FTase inhibitor [55].

Many of the biological studies with FTase inhibitors suggested that these agents are cytostatic and suppress cell growth rather than being cytotoxic in patients. However, farnesyltransferase inhibition was also observed to induce tumor regression of mammary and salivary carcinomas in a mouse tumor model [45]. The mechanism of shrinkage was unclear, but a recent study showed that an FTase inhibitor can induce apoptosis (cellular suicide) in cell culture, independent of the p53 tumor suppressor protein (a mediator of apoptosis) [56]. These observations may be clinically important because resistance to standard anti-cancer agents is often conferred by alterations in p53. Thus FTase inhibitors may still be efficacious against tumors containing mutant p53.

Many of the questions concerning the biological activity of FTase inhibitors have been addressed, although numerous questions remain to be answered. In particular the consequences of the K4B-Ras geranylgeranylation bypass and the uncertainty about the cellular targets of these drugs deserve intensive investigation. Nonetheless it is clear and accepted that FTase inhibitors have proven effective and shown significant anti-tumor activity against Ras-dependant tumors with little or no whole animal toxicity in many animal tumor modes. They have now reached the stage of human clinical evaluation [42, 57]. The clinical trials will provide answers to questions about efficacy, toxicity, development of resistance, and mechanism of action of these compounds such that their full therapeutic potential can be appreciated.

7
Outlook

Since Ras is the most frequent oncogene found in human tumors it has long been considered an attractive target for anti-cancer therapy. One promising and advanced approach is the inhibition of the enzyme farnesyl transferase, which attaches the lipophilic isoprenyl group to the C-terminal cysteine of the mature protein and is thus vital for Ras function. A number of efficient compounds have been developed, showing inhibitory activities in vitro and in vivo up to the low nanomolecular range, accompanied by high selectivities. The inhibitory properties, chemical stabilities and bioavailabilities of these substances could be optimized by chemical modification of the appropriate molecular structure. Furthermore, FTase-inhibitors show considerable tumor suppression properties that are, in some cases, even superior to conventional anti-tumor agents. These FTase inhibitors are non-toxic to normal cells and have proved their efficacy in various phases of clinical trials. Because of the fact that neither the exact mechanisms of action nor the affected molecular targets are well understood yet, it remains an important task to investigate the pathways of signal transduction in order to open new, more effective ways in anti-tumor therapy.

Acknowledgement. M. T. is grateful to the Fonds der Chemischen Industrie for a Kekulé-Stipendium.

8
References

1. (a) Barbacid M (1987) Ann Rev Biochem 56:779; (b) Lodish H, Baltimore D, Berk A, Zipursky SL, Matsudeira P, Darnell J (1995) Molecular Cell Biology, 3rd edn. Freeman, New York, Chap 20; (c) Hinterding K, Alonso-Díaz D, Waldmann H (1998) Angew Chem Int Ed 37:688
2. (a) Levitzki A (1994) Eur J Biochem 226:1; (b) Pai AF, Krengel U, Goody RS, Kabsch W, Wittinghofer A (1990) EMBO J 9:2351
3. Buolamwini KJ (1999) Curr Op Chem Biol 3:500
4. Waddick KG, Uckun FM (1998) Biochem Pharmacol 56:1411
5. Hancock JF, Paterson H, Marshall CJ (1990) Cell 63:133
6. Dudler T, Gelb MH (1996) J Biol Chem 271:11,541
7. Moomaw JF, Casey PJ, (1992) J Biol Chem 267:17,438
8. Reiss Y, Brown MS, Goldstein JL (1992) J Biol Chem 267:6403
9. Cox AD, Der CJ (1997) Biochim Biophys Acta 1333:F51
10. Leonard DM (1997) J Med Chem 40:2971
11. Buss JE, Marsters JC Jr (1995) Chem Biol 2:787
12. Gibbs JB, Graham SL, Hartmann GD, Voblau KS, Vohl NE, Omer CA, Oliff A (1997) Curr Op Chem Biol 1:197
13. Stradley SJ, Rizo J, Gierasch LM (1993) Biochemistry 32:12,586
14. Koblan KS, Bogusky MJ (1995) Protein Sci 4:681
15. Park HW, Boduhire SR, Momaw JF, Casey PJ, Beese LS (1997) Science 275:1800
16. Long SB, Casey PJ, Beese LS (1998) Biochemistry 37:9612
17. Long SB, Casey PJ, Beese LS (2000) Structure 8:209
18. Vohl NE, Wilson FR, Mosser SD, Guiliani E, Solms SJ, Conner MW, Anthony NJ, Holtz WJ, Gomez RP, Lee TJ, Smith RL, Graham SL, Hartman GD, Gibbs JB, Oliff A (1994) Proc Natl Acad Sci 91:9141
19. Prendergast GC, Davide JP, de Solms SJ, Giuliani E, Graham SL, Gibbs JB, Oliff A, Vohl NE (1994) Mol Cell Biol 14:4193
20. Adamson P, Marshall JC, Hall A, Tilbook PA (1992) J Biol Chem 267:20,033
21. Du W, Lebowitz PF, Prendergast GC (1999) Mol Cell Biol 19:1831
22. Lebovitz PF, Davide JP, Prendergast GC (1995) Mol Cell Biol 15:6613
23. Nagasu T, Yoshimatsu K, Rowell C, Lewis MD, Garcia AM (1995) Cancer Res 55:5310
24. Leftheris K, Kline T, Vite GD, Cho YH, Bhide RS, Patel DV, Manorama M, Schmidt RJ, Weller HN (1996) J Med Chem 39:224
25. Hunt JT, Lee VG, Leftheris K, Seizinger B, Carboni J, Mabus J, Ricca C, Yan N, Manne V (1996) J Med Chem 39:353
26. James GL, Goldstein JL, Brown MS, Rawson TE, Somers TC, McDowell RS, Crowley CW, Lucas BK, Levinson AD, Marsters JC Jr (1993) Science 260:1937
27. James GL, Brown MS, Cobb MH, Goldstein JL (1994) J Biol Chem 269:27,705
28. Nigam M, Seong C-M, Qion Y, Hamilton AD, Sebti SM (1993) J Biol Chem 268:20,695
29. Qian Y, Blaskovitch MA, Sahem M, Seong CM, Wathen SP, Hamilton AD, Sebti SM (1994) J Biol Chem 269:12,410
30. Lerner EC, Qian Y, Blaskovitch MA, Fossum RD, Vogt A, Sun J, Cox AD, Der CJ, Hamilton AD, Sebti SM (1995) J Biol Chem 270:26,802
31. Sun J, Qian Y, Blaskovitch MA, Hamilton AD (1995) Cancer Res 55:4243
32. Qian Y, Vogt A, Sebti SM, Hamilton AD (1996) J Med Chem 39:217

33. Williams TM, Ciccarone TM, MacTough SC, Bock RL, Conner MW, Davide JP, Hamilton K, Koblan KS, Kohl NE (1996) J Med Chem 39:1345
34. (a) Patel DV, Gordon EM, Schmidt RJ, Weller HN, Young MG, Zahler R, Barbacid M, Carboni JM, Gullo-Brown JL, Hunikan L, Ricca C, Robinson S, Seizinger BR, Tuormari AV, Manne V (1995) J Med Chem 38:435; (b) Manne V, Yan N, Carboni JM, Tuomari AV, Ricca C, Gullo-Brown J, Andahazy ML, Schmidt RJ, Patel D, Zahler R, Weinmann R, Der CJ, Cox AD, Hunt JT, Gordon EM, Barbacid M, Seizinger BR (1995) Oncogene 10:1763
35. Yan N, Ricca C, Fletcher J, Glover T, Seizinger BR, Manne V (1995) Cancer Res 55:3569
36. Schlitzer M, Sattler J (1999) Angew Chem Int Ed 38:2032
37. (a) Hinterding K, Hagenbuch P, Rétey J, Waldmann H (1998) Angew Chem Int Ed 37:1236; (b) Hinterding K, Hagenbuch P, Rétey J, Waldmann H (1999) Chem Eur J 5:227
38. Hara M, Akasaya K, Akinaya S, Ohabe M, Nakano H, Gomez R, Wood D, Uh M, Tamanoi F (1993) Proc Natl Acad Sci 90:2281
39. Singh SB, Jones ET, Goetz MA, Bills GF, Nallin-Omstead M, Jenkins RG, Kingham RB, Silverman KC, Gibbs JB (1994) Tetrahedron Lett 35:4963
40. Singh SB, Zink DL, Liesch JM, Ball RG, Goetz MA, Bolessa EA, Giacobbe RA, Silverman KC, Bills GF (1994) J Org Chem 59:6269
41. Leonard DM, Schuber KR, Poulter CJ, Eaton SR, Sawyer TK, Hodges JC, Su T-Z, Scholten JD, Gowan R, Sebold-Leopold JS, Doherty AM (1997) J Med Chem 40:192
42. Rawls RL (1998) Chem Eng News 67
43. Mallams AK, Njoroge FG, Doll RJ, Snow ME, Kaminski JJ, Rossmann RR, Vibulbhan B, Bishop WR, Kirschmeier P, Carruthers NI, Wong JK (1997) Bioorg Med Chem 5:93
44. Njoroge FG, Doll RJ, Vibulbhan B, Alvarez CS, Bishop WR, Petrin J, Kirschmeier P, Carruthers NI, Wong JK (1997) Bioorg Med Chem 5:101
45. Kohl NE, Omer CA, Conner MW, Anthony NJ, Davide JP, deSolms SJ, Guiliani EA, Gomez RP, Graham SL (1995) Nature Medicine 1:792
46. Graham SM, Vojtek AB, Huff SY, Cox AC, Clark J, Woper JA, Der CJ (1996) Mol Cell Biol 16:6132
47. Finney RE, Bishop JM (1993) Science 260:1524
48. James GL, Goldstein JL, Brown MS (1996) Proc Antl Acad Sci USA 93:4454
49. Lavell F (1997) CR Seances Soc Biol Fil 191:211
50. Zhang FL, Kirschmeier CD, James L, Bond RW, Wang L, Patton R, Windsor WT, Syto R, Zhang RM, Bishop WR (1997) J Bio Chem 272:10,232
51. Whyte DB, Kirschmeier P, Hockenberry NN, Nunez-Olivia I, Catino JJ, Bishop WR, Pai J-K (1997) J Biol Chem 272:14,459
52. Powell CA, Kowalczyk JJ, Lewis MD, Garcia AM (1997) J Biol Chem 272:14,093
53. Lebowitz PF, Davide JP, Prendergast GC (1995) Mod Cell Biol 10:2289
54. Lebowitz PF, Casey PJ, Prendergast GC, Thiessen JA (1997) J Biol Chem 272:15,591
55. Clark GJ, Knich MS, Rogers-Graham K, Sebti SM, Hamilton AD, Der CJ (1997) J Biol Chem 272:10,608
56. Lebowitz PF, Sakamuro D, Prendergast GC (1997) Cancer Res 57:708
57. Lovell RB, Kohl NE (1998) Cancer Metastasis Rev 17:203

Phosphatidylcholine-Preferring Phospholipase C from *B. cereus*. Function, Structure, and Mechanism

Paul J. Hergenrother · Stephen F. Martin

Department of Chemistry and Biochemistry, The University of Texas, Austin, TX 78712 USA
E-mail: *sfmartin@mail.utexas.edu*

The PLC class of enzymes has been studied extensively over the past 15–20 years because of their involvement in signaling pathways in which extracellular messages are delivered to the cell to induce a response. Of the PLC isoenzymes, the PI-PLCs have perhaps been examined in the greatest detail because of their key role in initiating cellular response by hydrolyzing the phosphodiester bond of phosphatidylinositols and their phosphorylated derivatives to release the second messengers IP_3 and DAG. However, the extended release of DAG that is critical to maintaining the stimulatory response arises from hydrolysis of the more abundant phosphatidylcholine by PC-PLC or by PLD followed by phosphatidic acid phosphatase. Because no eukaryotic PC-PLC has been cloned or isolated in pure form, the phosphatidylcholine-preferring PLC from *B. cereus* (PLC_{Bc}) has emerged as a focal point for investigation and as a putative model for mammalian PC-PLCs. The similarity of the active site of PLC_{Bc} with other phosphoryl transfer enzymes has also served as a stimulus for mechanistic studies. The present account details recent studies of this important member of the PLC superfamily of enzymes.

Keywords: Phospholipase, Phospholipid, Mechanism, Structure

Topics in Current Chemistry, Vol. 211
© Springer-Verlag Berlin Heidelberg 2000

1
Introduction

The view of phospholipids and phospholipases has evolved considerably over the course of the last 50 years. After their initial discovery, phospholipids were accorded the important, if somewhat pedestrian, role of maintaining the structural integrity of cell membranes. This perception of phospholipids as passive gatekeepers of the cell persisted for decades. However, in the early 1980s, it was discovered that the hydrolysis of phospholipids in cellular membranes provided products that stimulated a host of cellular functions, including those of calcium release and enzyme activation [1]. Further experimentation indicated that some phospholipases, which are the enzymes that catalyze the hydrolysis of phospholipids, are activated in response to various extracellular signals [2]. Therefore to send a message from outside the cell to inside the cell, Mother Nature chose phospholipids and their associated phospholipases as one means of transmitting the signal.

Although there are several classes of phospholipases, the phospholipase C (PLC) family is the one that has recently received intense scrutiny in the general context of signal transduction. The present account therefore details recent structural and mechanistic studies of one member of this important super family of enzymes, the phosphatidylcholine-preferring phospholipase C from B. cereus (PLC$_{Bc}$).

2
Classification and Importance of Phospholipases

With their newly accorded significance in signal transduction and membrane studies, the increase in knowledge about phospholipases has been remarkable.

Phospholipases are found in virtually every living organism, from bacteria to plants to mammals, and a wealth of information is now available about the various classes of phospholipases, which are categorized based upon which ester or phosphodiester bond is hydrolyzed (Fig. 1). For example, phospholipase D is specific for hydrolyzing the P-O bond to the polar headgroup of the phospholipid to furnish an alcohol and phosphatidic acid (PA) (Fig. 1, Path 1). Phospholipase C hydrolyzes the other phosphodiester bond to give diacylglycerol (DAG) and a phosphorylated headgroup (Path 2). Phospholipase A_1 (PLA$_1$) and phospholipase A_2 (PLA$_2$) catalyze the hydrolyses of the ester side chains at the *sn*-1 and *sn*-2 positions, respectively, of the glycerol backbone (Paths 3 and 4).

For a complex multicellular organism to function, cells must rapidly respond to messages from other regions of the organism. These messages, often in the form of hormones, arrive at the cell membrane through the aqueous environ-

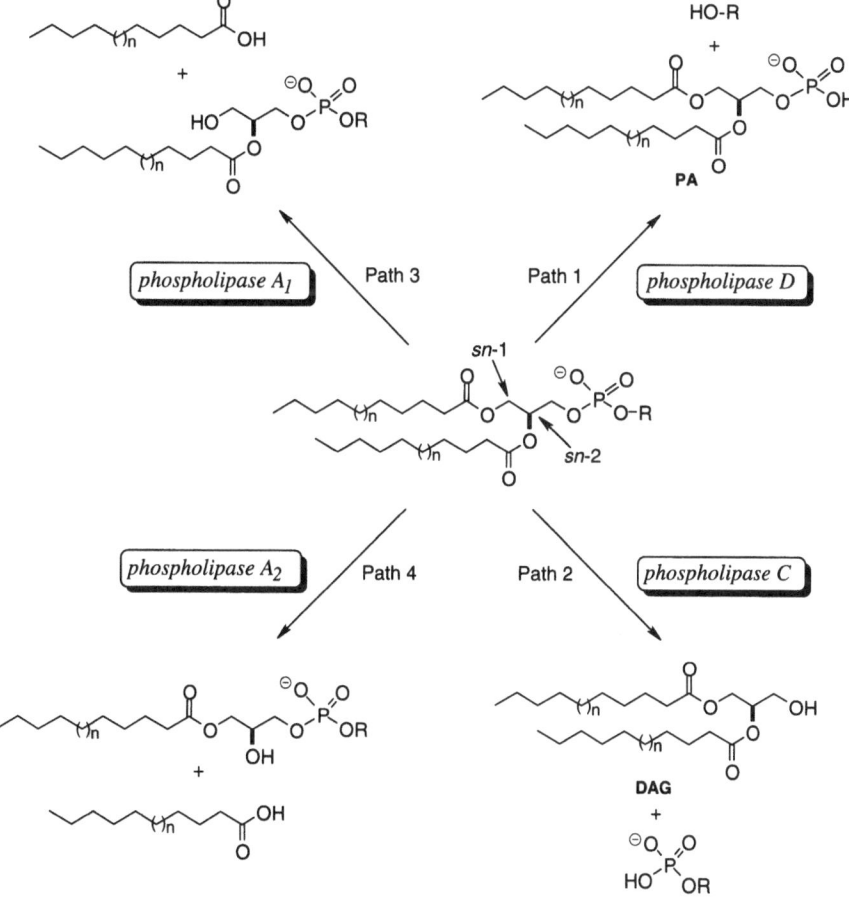

Fig. 1. Cleavage sites and products for the hydrolysis of phospholipids by PLD, PLC, PLA$_1$, and PLA$_2$

ment that comprises the bulk of living things. The message must then be relayed into the cell to effect the appropriate response. Because this signal is transduced across the cell membrane, the process is known as signal transduction, and the entire sequence of interdependent events is referred to as the signal transduction cascade. Thus, the cascade is typically initiated with binding of the hormone to a receptor protein imbedded in the membrane (Fig. 2). These receptors typically span the membrane and have localized areas of hydrophilicity and hydrophobicity. Upon binding the signaling entity, which can be a small molecule (e. g., catecholamines) or a protein (e. g., insulin), a conformational change is induced in the receptor. This reorganization allows an intracellular loop portion of the receptor to contact a guanine-nucleotide-binding regulatory protein, or G protein [3], which is comprised of α, β, and γ subunits [4].

When the receptor interacts with its associated G protein, the conformation of the guanine-nucleotide-binding site is altered. The subunits then dissociate, and a phosphatidylinositol-specific phospholipase C (PI-PLC) is activated [5]. The subsequent hydrolysis of phosphatidylinositol bisphosphate then produces inositol triphosphate (IP_3) and diacylglycerol (DAG), which are known to be secondary messengers. For example, the water soluble IP_3 is released into the cell where its ultimate targets are the calcium storage organelles from which Ca^{2+} is released [3]. The presence of DAG in cells is known to activate the cellular enzyme protein kinase C (PKC) [6, 7], which phosphorylates a number of cellular

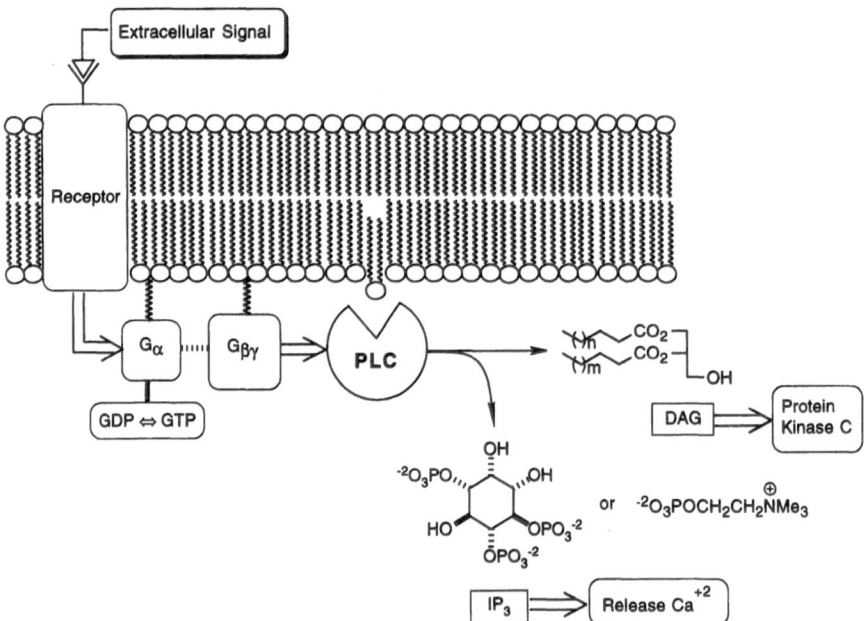

Fig. 2. Activation of an intracellular PLC by an extracellular signal (primary messenger) transduced across the membrane via a receptor and G-protein. The products DAG and IP_3 that are produced by the PLC catalyzed hydrolysis of phospholipids are secondary messengers

proteins, thereby leading to their activation [2, 8]. Thus, the G-protein coupled activation of PLC leads to the production of secondary messengers that relay the extracellular message to the interior of the cell, thereby effecting a response.

The rate of production of DAG in the cell does not occur linearly with time, but rather it is biphasic. The first peak is rapid and transient and coincides with the formation of IP_3 and the release of Ca^{2+}; this DAG is therefore derived from the PI-PLC catalyzed hydrolysis of phosphatidylinositols [1]. There is then an extended period of enhanced DAG production that is now known to be derived from the more abundant phospholipid phosphatidylcholine (PC), which has a different composition of fatty acid side chains [9]. Although DAG may be generated directly from PC through the action of PC-PLC, it can also be formed indirectly from PC. In this pathway, PC is first hydrolyzed by PLD to give choline and phosphatidic acid, which is then converted to DAG by the action of a phosphatidic acid phosphatase [10, 11].

Owing to their pivotal role in mammalian signal transduction, there has been an intense interest in the enzymes of the PLC superfamily. Progress toward understanding the mechanism, structure, and function of PI-PLCs from both bacterial and mammalian sources has been particularly impressive [12-15]. Several PI-PLCs have been isolated and cloned, and a number of high resolution, three-dimensional X-ray structures are available [16-19]. In contrast to the advances that have been made with mammalian PI-PLC isoenzymes, their PC-PLC counterparts are poorly characterized. Studies with mammalian PC-PLCs have typically been conducted with partially purified enzymes, and there has not been a report of the isolation of a pure, eukaryotic PC-PLC. To circumvent the currently intractable problems associated with mammalian PC-PLCs, PLCs from bacterial sources have been sought as potentially useful models.

The history of bacterial phospholipase C activity dates back to 1948 when bacterial strains from the *Bacillus* genus were screened for the hydrolysis of phosphatidylcholine on agar plates [20]. *B. cereus* gave a strong positive response, whereas *B. anthracis* and *B. mycoides* exhibited weaker activity. Subsequent culturing of *B. cereus* yielded an active phospholipase that is now known as the phosphatidylcholine-preferring phospholipase C from *Bacillus cereus* (PLC_{Bc}) [21]. This enzyme was ideal for further study because it was readily available in pure form. Although the in vivo function of PLC_{Bc} has not been established, it appears, like alkaline phosphatase and sphingomyelinase, to be involved in phosphate retrieval when *B. cereus* is grown under conditions of limited phosphate [22]. The discovery that PLC_{Bc} had an antigenic similarity with a mammalian PC-PLC prompted a number of experiments in which PLC_{Bc} was exploited to simulate PC-PLC activity [23]. For example, PLC_{Bc} can stimulate DAG production in vivo [24], and PLC_{Bc} has been employed to investigate the mechanisms of oncogenesis [25, 26] and membrane damage [27].

Thus, PLC_{Bc} has emerged as a potentially useful model for the presently inaccessible mammalian PC-PLCs and as a tool for studying different facets of PLC-mediated cell signaling. The present account will summarize some of the recent advances toward understanding the function, structure, and mechanism of PLC_{Bc}.

3
Aspects of Phospholipid Hydrolysis by PLC$_{Bc}$

One of the first steps toward developing an understanding of the functional features of an enzyme catalyzed process is to study the efficiency with which it processes different substrates. In the case of PLC$_{Bc}$ there are a number of variables in substrate structure whose effect upon catalytic activity and efficiency must be examined. These include: (1) the importance of the stereochemistry in the glycerol backbone; (2) the nature of the functional links between the side chains and the glycerol backbone; (3) the effect of varying the length of the acyl side chains; (4) the effect of varying the headgroup; and (5) the effect of altering the nature of the phosphodiester moiety. During the course of such studies, it is often possible to identify substrate analogues that are inhibitors, and these may be useful tools for studying the mechanism of the enzymatic reaction and for determining the biological effects upon inhibiting the enzyme in vivo.

3.1
Assays for PLC$_{Bc}$

Prior to being able to study the function and mechanism of an enzyme, it is essential that suitable assays be available to monitor enzyme activity toward different substrates and to determine the kinetic parameters k_{cat} and K_m for the reactions. A brief overview of the known assays for the evaluation of PLC$_{Bc}$ activity is thus appropriate. The ideal assay for a phospholipase C would utilize a phospholipid substrate, not an analogue with a modified headgroup or side chains. Such an assay should be sensitive to minimize the quantities of enzyme and substrates that would be required, and it should be convenient to implement so that analyses may be readily performed.

A titrametric assay of PLC$_{Bc}$, alternatively called the pH-stat method, was the workhorse in early studies [28]. This method simply involves titrating the acidic product of the PLC reaction as it is formed with a solution of standard base. An advantage of this continuous assay is that it can be used to detect the turnover of both synthetic and natural substrates, and its sensitivity has been estimated to be in the 20–100 nmol range. However, the pH-stat assay has low throughput capability, and it cannot be easily performed in a parallel fashion with multiple substrate concentrations. It is also necessary to exclude atmospheric carbon dioxide from the aqueous media containing the enzyme and substrate.

Because neither product in the PLC$_{Bc}$ catalyzed hydrolysis of natural phospholipids is chromogenic, non-natural substrates have been designed and developed explicitly for the spectrophotometric analysis of enzyme activity (Fig. 3). Hydrolysis of p-nitrophenyl phosphorylcholine (NPPC) by PLC$_{Bc}$ produces the p-nitrophenolate anion, which gives a yellow color at pH = 8.0. Although using NPPC as a substrate allows for the facile evaluation of PLC$_{Bc}$ activity, NPPC is a very poor substrate for PLC$_{Bc}$ [29, 30]; hence the sensitivity of this assay is low. Another non-natural substrate is 1,2-dioctanoyl-sn-(3-thioglycero)-phosphocholine [31], hydrolysis of which by PLC$_{Bc}$ liberates a free thiol that may be easily quantitated with 5,5′-dithio-bis(2-nitrobenzoic acid) (DTNB,

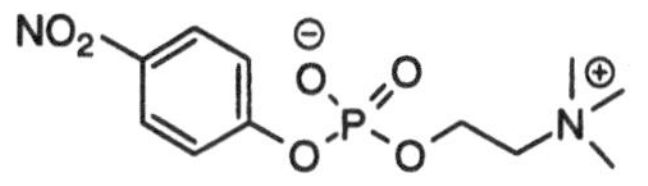

p-Nitrophenyl phosphorylcholine

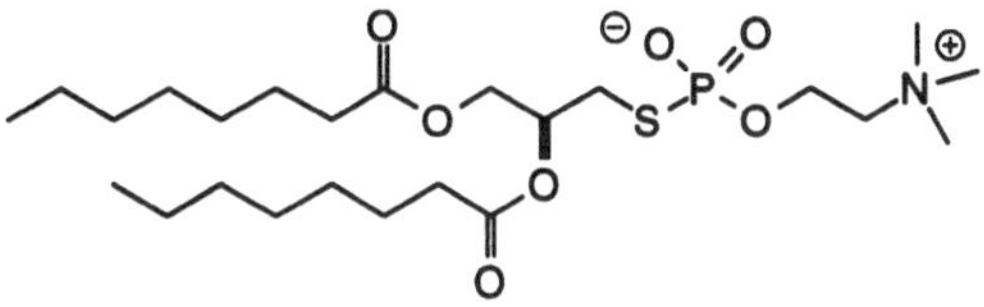

1,2-Dioctanoyl-*sn*-(3-thioglycero)phosphocholine

Fig. 3. Non-natural substrates for PLC$_{Bc}$

Ellman's Reagent) [32]. Unfortunately, PLC$_{Bc}$ processes this thiophospholipid 100 times slower than a natural phosphatidylcholine, so this assay is also of limited utility.

A more successful strategy for developing sensitive and facile assays to monitor PLC$_{Bc}$ activity involves converting the phosphorylated headgroup into a color-imetric agent via a series of enzyme coupled reactions. For example, phos-phatidylcholine hydrolysis can be easily monitored in a rapid and sensitive manner by enzymatically converting the phosphorylcholine product into a red dye through the sequential action of alkaline phosphatase, choline oxidase, and peroxidase [33]. This assay, in which 10 nmol of phosphorylcholine can be readily detected, may be executed in a 96-well format and has been utilized in deuterium isotope and solvent viscosity studies [34] and to evaluate inhibitors of PLC$_{Bc}$ [33] and site-directed mutants of PLC$_{Bc}$ [35, 36].

Because the preceding chromogenic assay rely on choline quantitation, the hydrolysis of substrates with headgroups other than choline cannot be followed. To circumvent this problem, another useful protocol was devised whereby the phosphorylated headgroup produced by the PLC$_{Bc}$ hydrolysis is treated with APase, and the inorganic phosphate (Pi) that is thus generated is quantitated by the formation of a blue complex with ammonium molybdate/ascorbic acid; 5 nmol of phosphate may be easily detected. This assay, which may also be per-formed in a 96-well format, has been utilized to determine the kinetic param-eters for the hydrolysis of a number of substrates by PLC$_{Bc}$ [37, 38].

Several other methods, including a ^{31}P NMR [39, 40] and radiometric assay [28, 41] have also been used to monitor the PLC$_{Bc}$ catalyzed hydrolysis of phospholipids. The assays that are currently available for PLC$_{Bc}$ are compared in Table 1.

Table 1. Comparison of common assays for phospholipase C

Assay method	Natural substrate?	Rapid and simple?	Sensitive and accurate?	Use with variety of headgroups?
Titrametric	Yes	No	Yes	Yes
NPPC	No	Yes	No	No
Thiophosphate	No	Yes	No	Yes
Choline quantitation	Yes	Yes	Yes	No
Pi quantitation	Yes	Yes	Yes	Yes
Radiometric	Yes	No	Yes	Yes
^{31}P NMR	Yes	No	No	Yes

3.2
Substrate Structure-Activity Relationships

A number of experiments have been conducted to establish various aspects of structure-activity relationships (SAR) of substrates. Although the greatest emphasis has been upon the nature and orientation of the side chains on the glycerol backbone, variations in the phosphate headgroup have also been explored.

3.2.1
The Side Chains

The glycerol side chains of the phospholipid substrate play an important role in recognition and binding to the enzyme, and variations in the orientation, nature and the length of these side chains can have dramatic effects on PLC_{Bc} activity. Natural phospholipids exist in the R stereochemical configuration at the sn-2 center, and the corresponding S-isomer is hydrolyzed 40 times slower than the natural phospholipid (Table 2) [42]. The introduction of methyl groups adjacent to the ester carbonyl group provides phospholipid derivatives that are poorer substrates [43]. Replacing the sn-1 and especially the sn-2 ester linkages with an ether linkage gives analogues with low levels of enzymatic activity (Table 2) [43, 44]. Although it is at first somewhat surprising that removal of a carbonyl group would so drastically affect activity, the crystal structure of PLC_{Bc} complexed with a phosphonate inhibitor provides a rationale for the observed activity: The sn-2 carbonyl group of the inhibitor is engaged in a hydrogen bond to the backbone N-H of Asn134 [45]. Substrates lacking the sn-2 carbonyl function cannot form this hydrogen bond, thereby resulting in lower kinetic activity. Thus, it follows that the lyso analogue, which lacks the sn-2 side chain, is processed by PLC_{Bc} approximately 1000 times slower than its diacyl parent [46]. On the other hand, when the sn-1 side chain is removed, the resulting phospholipid analogue is turned over at only a slightly reduced rate relative to the diacyl parent (Table 2) [42, 44]. Sphingomyelin, in which the sn-2 ester is an amide and the sn-1 carbonyl removed, is not a good substrate for PLC_{Bc} [47].

The length of the acyl side chains of a phosphatidylcholine derivative have significant effects upon the rate with which PLC_{Bc} catalyzes the hydrolysis of

Table 2. Relative rates of hydrolysis of phosphatidylcholine derivatives by PLC_{Bc}

Substrate	Relative activity	Reference
(structure)	100	[42]
(structure)	2.5	[42]
(structure)	1	[43, 44]
(structure)	~10	[42, 44]

phospholipids [48]. As the length of the side chains increases, the critical micelle concentration (CMC) of the substrate decreases, and there is a discontinuity in the rate progression curve at the CMC. Thus, phosphatidylcholine derivatives in micellular form are hydrolyzed about two to five times faster than when they are in monomeric form [49–51]. The reason for this interfacial activation is not fully understood, but several theories have been advanced. One possibility is that the release of DAG is more facile into a lipophilic micelle than into aqueous solution, so by accelerating the rate of departure of this leaving group, the overall rate of turnover increases. This interpretation requires that DAG release is rate-limiting in the PLC_{Bc} reaction of micellular substrates [49]. However, it has been shown that DAG release is *not* rate determining in reactions of soluble substrates at concentrations below their CMCs [34].

Another interesting observation is that as the length of the side chain decreases the K_ms of monomerically-dissolved phosphatidylcholines increase significantly, while V_{max} changes very little (Table 3) [43, 52]. NMR studies of phosphatidylcholines possessing three, four, and six carbons in each of the acyl side chains have been conducted to determine whether these variations in K_m might arise from differences in the solution conformation of the substrate [52]. However, the length of the side chain had little effect upon the relative orientations of the substituents on the glycerol backbone, so it appears that factors other than ground state conformational preferences of the phospholipid are responsible for the observed variations in K_m. Such factors might involve decreas-

Table 3. The effect of acyl chain length on PLC_{Bc} activity toward soluble phosphatidylcholine derivatives. The differences in the experimental data are likely due to the use of different assay conditions

Substrate	V_{max} (μmol/min · mg)	K_m (mM)	K_{cat}/K_m (1/s · mM)	Reference
	2290	61	18	[52]
	1430 (750)	1.2 (40)	32	[43, 52]
	2100 (1000)	2.4 (0.36)	417	[43, 52]
	(1340)	(0.20)		[43]

ed hydrophobic interactions between the enzyme and the substrate and/or differential solvation/desolvation effects associated with substrate binding.

3.2.2
Modifications of Polar Headgroup

Although the major focus of substrate SAR studies has been on phosphatidylcholine derivatives, phospholipids with other headgroups have also been analyzed. In a study of naturally occurring phosphatidylcholines, phosphatidylethanolamines (PE), and phosphatidylserines (PS) having acyl side chains ranging in length from 12 to 18 carbon atoms each, PE and PS exhibited V_{max}s that were approximately 25% that of PC [53]. However, it should be emphasized that these compounds had different side chain lengths, so it was not possible to compare directly the effect of having different headgroups.

Recently, a series of soluble phospholipids with n-hexanoyl side chains and a variety of headgroups were assayed as substrates for PLC_{Bc} [37]. The results (Table 4) indicate that both 1,2-dihexanoyl-sn-glycero-3-phosphocholine (C6PC) and 1,2-dihexanoyl-sn-glycero-3-phosphoethanolamine (C6PE) are processed with similar catalytic efficiencies (k_{cat}/K_m) by the enzyme, while 1,2-dihe-

Table 4. Kinetic values for the hydrolysis of phospholipids with various headgroups by PLC_{Bc}

Substrate	V_{max} (μmol/min · mg)	k_{cat} (1/s)	K_m (mM)	k_{cat}/K_m (1/(s · mM))	Reference
C6PC	2100 ± 150	1000	2.4 ± 0.4	417	[38]
C6PE	1130 ± 100	540	1.8 ± 0.2	300	[38]
C6PS	435 ± 85	210	4.5 ± 1.0	47	[38]
C6PDB	11	5.2	14	0.37	[37]

xanoyl-*sn*-glycero-3-phosphoserine (C6PS) has a roughly tenfold lower k_{cat}/K_m. The unnatural substrate 1,2-dihexanoyl-*sn*-glycero-3-phospho-3′,3′-dimethyl-1-butane (C6PDB), which has a headgroup isosteric with that of C6PC but lacking the positive charge, is hydrolyzed by PLC_{Bc} about 1000 times less efficiently than C6PC [38]. Thus, a positive charge on the phospholipid headgroup is important for recognition and catalysis.

3.2.3
Modifications of Phosphate Group

The constellation of atoms about the phosphorus of the phosphodiester linkage has been varied by replacing the bridging and non-bridging oxygens with sulfur, carbon, and nitrogen to give compounds that may be either substrates or inhibitors of PLC_{Bc}. For example, substitution of the bridging oxygen in the scissile oxygen-phosphorus bond with a sulfur atom gives a substrate that is turned over very slowly, whereas replacing this oxygen with NH, NBn, CH_2, or CF_2 provides competitive inhibitors (Table 5) [42, 54]. The structure of PLC_{Bc} complexed with a phosphonate inhibitor was elucidated by X-ray crystallographic analysis [45]. When both non-bridging oxygens are replaced with sulfur, a potent inhibitor of PLC_{Bc} activity is obtained [54].

Table 5. The effect upon PLC_{Bc} activity of replacing bridging and non-bridging oxygens around phosphorus in soluble phosphatidylcholine derivatives

Phospholipid	K_i (mM)	K_m (mM)	Reference
		0.023	[42]
		0.02	[54]
	1.15		[54]
	0.15		[54]
	0.01		[54]

3.3
Zinc Binding Inhibitors

There are inhibitors of PLC_{Bc} that are substrate analogues and non-substrate analogues. As evidenced in the preceding section, inhibitors that are substrate analogues contain non-hydrolyzable replacements of the phosphate group. One group of non-substrate analogues that are PLC_{Bc} inhibitors exploit their zinc binding ability as the basis for PLC_{Bc} inhibition; such inhibitors interact with one or more of the three active site zinc ions. One important inhibitor of this type is tricyclodecan-9-yl-xanthogenate (D609) (Fig. 4), a commercially available inhibitor that inhibits PLC_{Bc} ($K_i = 6.4 \ \mu mol \ l^{-1}$) [55]. D-609 is known to prevent the activation of protein kinase C and have anti-tumor and anti-viral properties [56], and it has recently been used in numerous studies that implicate mammalian PC-PLC participation in various signaling pathways such as those involving interleukin production and apoptosis [57–59]. However, experiments

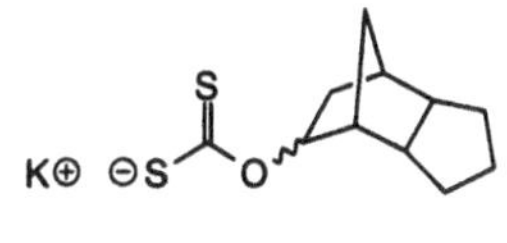

D609

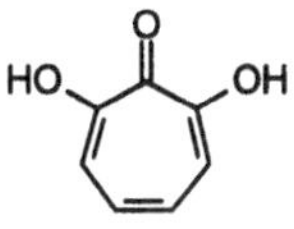

2,7-Dihydroxytropolone

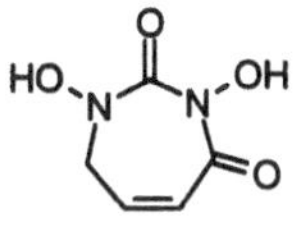

1,3-*N*,*N'*- Dihydroxy-5,6-dihydro-
1,3-diazepin-2,4-dione

Fig. 4. Non-substrate analogues as inhibitors of PLC$_{Bc}$

using D-609 must be interpreted with caution because it also affects PLD activity [24].

The zinc binding ability of small molecules has been exploited as a strategy for discovering other PLC$_{Bc}$ inhibitors, although a potential limitation of this approach is that inhibitors so identified may not be selective. For example, it was recently found that 2,7-dihydroxytropolone (Fig. 4), which is an inhibitor of the bimetallic hydrolases alkaline phosphatase and inositol monophosphatase [60], is also a strong inhibitor of PLC$_{Bc}$, having a K_i of ~ 16 µmol l^{-1} at pH 7.3 [61]. Cyclic *N*,*N'*-dihydroxy ureas also weakly inhibit PLC$_{Bc}$ as illustrated by 1,3-*N*,*N'*-dihydroxy-5,6-dihydro-1,3-diazepin-2,4-dione (Fig. 4), which exhibited a K_i of 388 µmol l^{-1} at pH 7.3 and 53 µmol l^{-1} at pH 9.5 [61]. The pK_as of the *N*-hydroxyl groups in this and other cyclic *N*,*N'*-dihydroxy ureas played a key role in inhibitor potency, presumably because an ionized *N*-hydroxyl function would be expected to bind to a zinc ion better than a neutral hydroxyl group. Whether tropolones or cyclic *N*,*N'*-dihydroxy ureas will prove to be useful and selective inhibitors of PLC$_{Bc}$ remains to be demonstrated.

4
Structural Features of PLC$_{Bc}$ and its Complexes

Protein X-ray crystallography is one of the best means of gaining insights that are necessary to begin to understand the function, substrate specificity and mechanism of a protein. Hence, the X-ray analysis of PLC$_{Bc}$ set the stage for numerous structural and mechanistic studies. The high resolution (1.5 Å) structure of PLC$_{Bc}$ showed it to be a monomeric protein containing three zinc ions in its active site [62]. The enzyme is roughly ellipsoidal in shape, 40 × 30 × 20 Å, and about 66% of the amino acids are located in ten α-helical regions with the remainder of the residues being in flexible loops connecting the helices. The surface of the protein is relatively smooth except for an 8 Å deep by 5 Å wide cleft at the active site, which is readily apparent in the complex of PLC$_{Bc}$ with a phosphonate inhibitor (Fig. 5) [45]. These and other structures of PLC$_{Bc}$ complexes

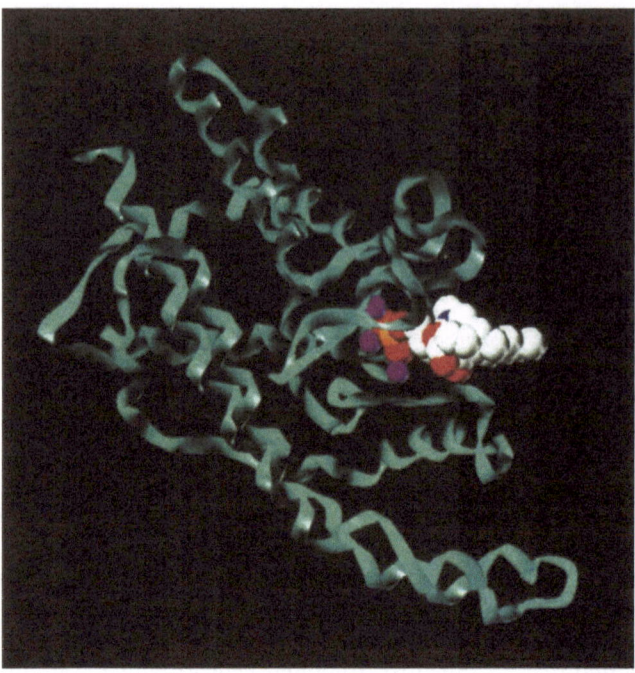

Fig. 5. View of PLC$_{Bc}$ bound to a phosphonate substrate analogue showing protein backbone (*cyan*), zinc ions (*magenta*), and inhibitor with carbons (*white*), oxygens (*red*), phosphorus (*orange*), and nitrogen (*blue*) [45]

have revealed a number of important features of the enzyme and the manner in which it binds substrate analogues. Of particular interest are the novel metal binding site, the choline binding pocket and the interactions of the enzyme with the lipid side chains.

4.1
Nature of Zinc Center

Because of the unusual cluster of three zinc ions at its active site, the tri-metal center of PLC$_{Bc}$ has been the subject of considerable interest. The amino acid side chains and water molecules that comprise the first coordination sphere of these zinc ions in the native structure are represented schematically in Fig. 6. All of the zincs are pentacoordinate, and each adopts an approximate trigonal-bipyramidal geometry. The distance between Zn1 to Zn3 is 3.6 Å, qualifying it as a bi-metallic center [63]. The side chain carboxyl group of Asp122 and a water molecule bridge these two ions. The remainder of the Zn3 coordination sphere is filled by the amino group and backbone carbonyl of the N-terminal Trp1 and the ε-nitrogen of His14. The Zn1 ion in turn is ligated to His69, His118, and Asp55. The constellation of two histidines (His128 and His142) and a glutamic acid (Glu146) about Zn2 resembles that of a classic zinc site. The importance of

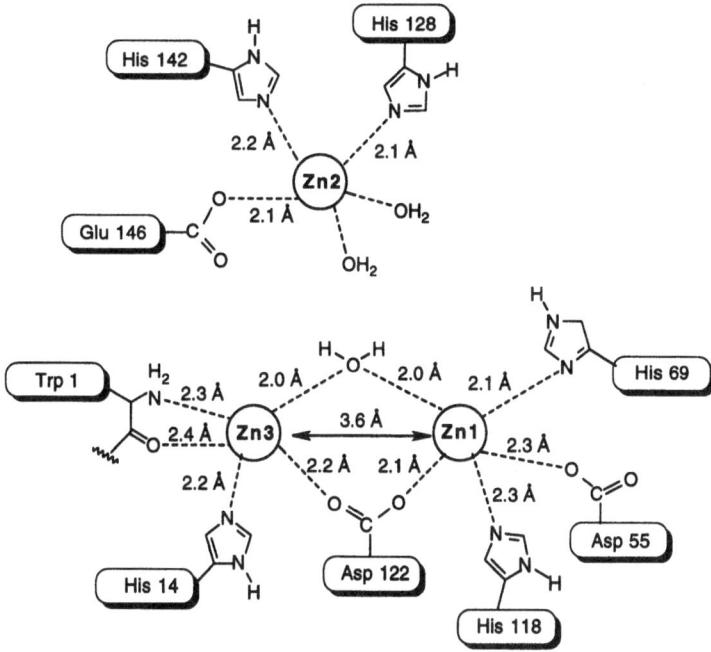

Fig. 6. Zinc ligands in the native PLC$_{Bc}$ structure [62]

His118, Glu146, His128, His142, Trp1, and His14 as zinc binding ligands has been supported by a series of site-directed mutagenesis studies in which each of these residues was replaced with Ala; each mutant bound only two zinc ions [35, 64].

Inasmuch as structures of enzymes complexed with substrate-like or product-like ligands can reveal useful information about the mechanism of the enzyme catalyzed reaction, crystals of PLC$_{Bc}$ were soaked with phosphorylcholine, phosphorylethanolamine, and phosphorylserine, but no complexes were formed [65]. Tris, which is a common biological buffer, inhibits PLC$_{Bc}$ activity by 50% at 5 mmol l^{-1} and 100% at 50 mmol l^{-1}, and the structure of a complex of PLC$_{Bc}$ with Tris has been obtained. In this complex, the amine nitrogen of Tris is coordinated to Zn2, thereby replacing a water molecule, and the three hydroxyl groups of Tris are involved in a hydrogen bonding network with the protein [66]. This binding motif places the Tris molecule at the opening of the PLC$_{Bc}$ active site blocking access of substrate to Zn1 and Zn3.

From the mechanistic viewpoint, a more informative structure is the complex of PLC$_{Bc}$ with inorganic phosphate (Pi), which also inhibits the enzyme (50% inhibition at 50 mmol l^{-1} Pi) [65]. In this complex, which was obtained at 2.1 Å resolution, the three zinc ions were coordinated to the two non-bridging phosphate oxygens (Fig. 7a). Comparison of this structure with the native one (Fig. 6) reveals that one of the non-bridging oxygens on the phosphate replaced the bridging water molecule between Zn1 and Zn3, while the other oxygen displaced one of the waters coordinated to Zn2.

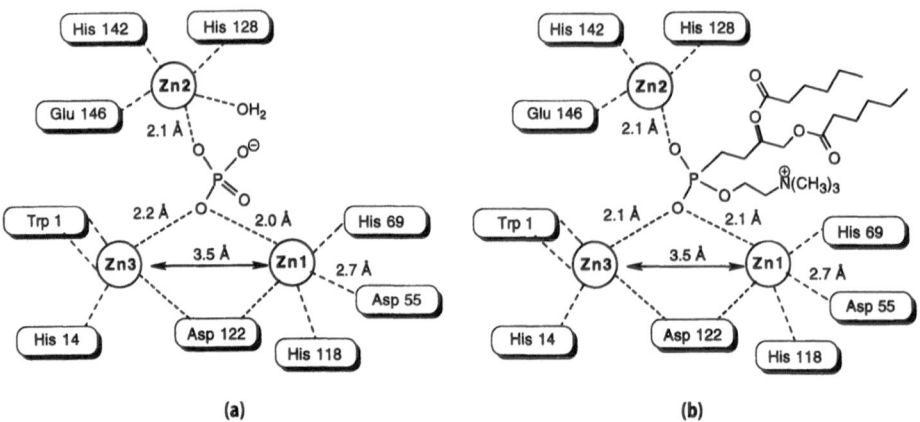

Fig. 7a, b. Active site showing Zn1–Zn3 and Zn3–Asp55 distances in complexes of PLC$_{Bc}$ with: **a** inorganic phosphate [65]; **b** a phosphonate inhibitor [45]

A major advance that led to a much better understanding of how substrate might bind to PLC$_{Bc}$ was a seminal report detailing the structure of the complex of PLC$_{Bc}$ with the phosphonate inhibitor 3(S),4-dihexanoylbutyl-1-phosphonyl-choline ($K_i = 1.15$ mM) (Fig. 5) [45]. This structure revealed that the phosphonate moiety bound to the active site zinc ions in a manner similar to inorganic phosphate (Fig. 7b). The non-bridging oxygens of phosphorus were coordinated to all three zinc ions, replacing the Zn1–Zn3 bridging water and one water on Zn2. In this complex, however, the second water that was bound to Zn2 in the native enzyme is absent making this ion tetracoordinate. The coordination state of Zn2 may provide a clue to how PLC$_{Bc}$ catalyzes the hydrolysis of phospholipids, and this issue will be revisited in the context of mechanistic studies (see Sects. 5.2.1.4 and 5.2.2.1).

There are only minor differences between the structures of native PLC$_{Bc}$ and its complexes with inorganic phosphate and the phosphonate inhibitor. However, one of these may be significant. In the X-ray structure of the uncomplexed enzyme, one of the carboxyl oxygens of Asp55 is 2.3 Å from Zn1 (Fig. 6), whereas it is about 2.70 Å in both the phosphate and phosphonate complexes (Fig. 7). This shift is accompanied by a decrease in the distance between Zn1 and Zn3 from about 3.6 Å in the native structure to about 3.5 Å in the complexed structures. Although the reasons for these changes are unknown, the observation that Asp55 is more loosely coordinated to Zn1 in the complexes than in the native enzyme may have a bearing on the potential role of Asp55 as the general base in the mechanism of PLC$_{Bc}$ catalyzed hydrolysis of phospholipids (see Sect. 5.2.1.3).

While having three metal ions in an enzyme active site is uncommon, it is not unique to PLC$_{Bc}$. The well-known alkaline phosphatase from *E. coli* (APase) contains two zinc ions and a magnesium ion [67], whereas the α-toxin from *Clostridium perfringens* [68] and the P1 nuclease from *Penicillium citrinum* [69] each contain three zinc ions. Indeed, the zinc ions and coordinating ligands of P1 nuclease bear an uncanny resemblance to those of PLC$_{Bc}$ as the only differ-

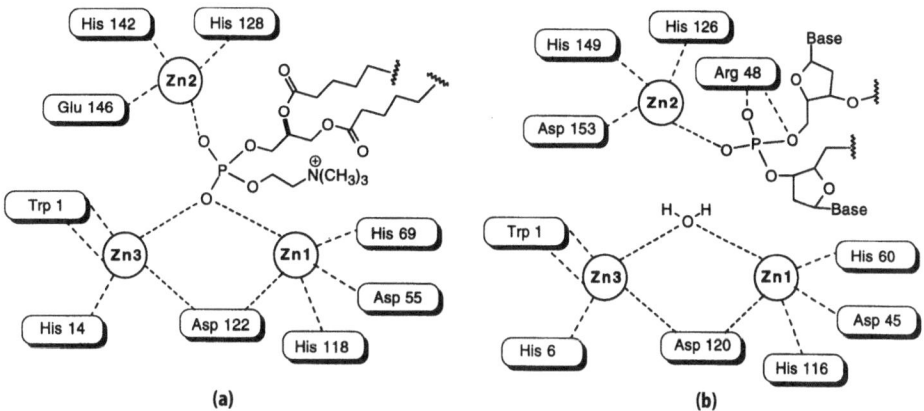

Fig. 8a, b. Proposed modes of substrate binding to: **a** PLC_{Bc} [45]; **b** P1 nuclease [70], based on X-ray structures with non-hydrolyzable substrate analogues

ence is that the Glu ligand for Zn2 in PLC_{Bc} is an Asp in P1 nuclease. Because PLC_{Bc} and P1 nuclease have virtually identical active sites, it is tempting to invoke similar modes of binding and catalysis. However, there are significant structural differences in these enzymes, and they bind substrate analogues in very different fashions. For example, in a complex with the uncleavable diastereomer of a thio-phosphorylated dinucleotide, only Zn2 of P1 nuclease is coordinated with a phosphate oxygen. The binuclear zinc site of Zn1 and Zn3 had a bridging water molecule and appeared inaccessible [69]. In a subsequent X-ray study of P1 nuclease with other substrate analogues, one of the non-bridging phosphate oxygens was coordinated to Zn2 while the other was electrostatically engaged with the side chain of Arg48; again the bridging Zn1–Zn3 water remained in place [70].

Based upon these findings, it appears that P1 nuclease binds substrates in a markedly different manner (Fig. 8b) than PLC_{Bc} (Fig. 8a), and the enzymatic mechanism for the hydrolysis of phosphodiester bonds by the two is likely different. PLC_{Bc} contains no residue analogous to Arg48 in P1 nuclease, so PLC_{Bc} binds and neutralizes the negative charge of the substrate with three zinc ions. On the other hand, P1 nuclease uses only one zinc and the guanidino group of an arginine; this binding mode frees the remaining two zincs of P1 nuclease to perform another function, possibly to activate the bridging hydroxyl group for nucleophilic attack on the phosphodiester bond.

4.2
The Choline Binding Site

The substrate binding cleft of PLC_{Bc} that accommodates the choline headgroup is lined on one side with Glu4, Asp55, Tyr56, and Glu146 and on the other with Ser64, Thr65, Phe66, Ile80, Thr133, Leu135, and Ser143. Of particular interest is the arrangement of Glu4, Tyr56 and Phe66 about the positively charged trimethylammonium moiety. Namely, the side chain carboxyl group of Glu4 is 3.9 Å away from the partially-positive polarized hydrogens on the methyl groups, whereas the centroids of the aromatic rings of Tyr56 and Phe66 are approximately 4.7 Å and 4.2 Å, respectively, from these methyl groups (Fig. 9) [45]. The questions that immediately arise are what are the roles of these different amino acid side chains in stabilizing the positive charge on the phospholipid headgroup and what contributions do these residues make to substrate binding and specificity?

The choline binding pocket of PLC_{Bc} provides an interesting case study of how the side chains of amino acids interact with ammonium ions in biological systems. It seems reasonable to assume that the carboxyl group of Glu4 stabilizes the positive charge by an electrostatic interaction. Both Tyr56 and Phe66 are positioned so that one π-face of each of the respective aromatic rings, which have partial negative character because of the quadrupole moment of the rings [71], is coordinated with the trimethylammonium ion to stabilize the positive charge through electrostatic effects. Such π-cation interactions represent a genre of protein-ligand recognition motifs that was not widely recognized until about a decade ago [72], but there have since been a number of studies documenting their importance [73–77].

A question at this stage is how do individual changes in the polar/hydrophilic nature of the side chains at residues 4, 56, and 66 influence the kinetic parameters k_{cat} and K_m and catalytic efficiency, k_{cat}/K_m, for the hydrolyses of PC, PE, and PS? To address this issue, a series of single mutants of Glu4 (E4Q, E4D, E4L), Tyr56 (Y56W, Y56R, Y56A), and Phe66 (F66Y, F66W, F66A) were prepared by PCR mutagenesis [78]. The mutant proteins, which were determined to be struc-

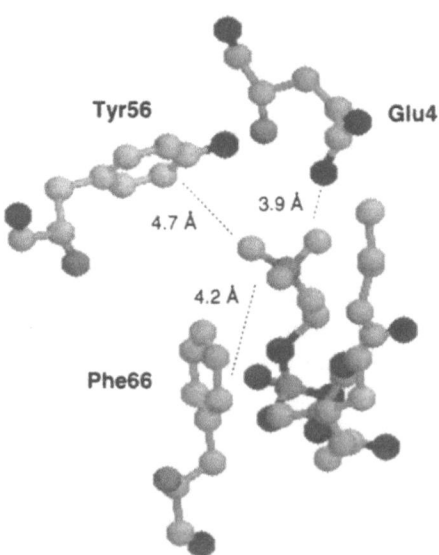

Fig. 9. Stabilizing interactions of trimethylammonium ion of a substrate analogue with Glu4, Tyr56, and Phe66 of PLC_{Bc} [45]

Table 6. Qualitative assessment of the contributions of Glu4, Tyr56, and Phe66 to the hydrolysis of soluble phospholipid substrates PC, PE, and PS by PLC$_{Bc}$

PLC$_{Bc}$ residue	Important for PC processing?	Important for PE processing?	Important for PS processing?
Glu4	++	++	–
Tyr56	++	+	+
Phe66	+++	+++	+++

+++, mutants retain <5% wild-type activity.
++, most mutants retain <15% wild-type activity.
+, most mutants retain <30% wild-type activity.
–, mutants retain >50% wild-type activity.

turally similar to the wild-type by temperature dependent CD and atomic absorption, were assayed for their ability to catalyze the hydrolysis of the soluble phospholipid substrates C6PC, C6PE, and C6PS.

The relative importance of Glu4, Tyr56, and Phe66 in the PLC$_{Bc}$ catalyzed hydrolyses of different phospholipids is qualitatively summarized in Table 6, but a few more specific comments are warranted. The biggest changes in the kinetic parameters were in k_{cat}, which varied over nearly three orders of magnitude, whereas changes in K_m varied by less than a factor of ten. Mutations of Glu4 had little effect on k_{cat} and K_m for the hydrolysis of C6PS, but they did influence the kinetic parameters for C6PC and C6PE. Thus, a carboxyl group at the E4 position is beneficial for the hydrolyses of C6PC and C6PE, which each have a positive charge on the headgroup of the substrate, but not for the hydrolysis of C6PS, which has no net charge on the headgroup. Interestingly, E4Q hydrolyzes C6PS *faster* than wild-type. An aromatic residue at position 56 appears to confer some substrate specificity on the enzyme as Y56W is *more selective* toward C6PC and C6PE relative to C6PS than the wild-type enzyme; this selectivity is largely a consequence of a significant drop in k_{cat} for C6PS. Mutations of Phe66 to a non-aromatic residue led to dramatic drops in catalytic efficiencies that were manifested in large decreases in k_{cat} for each of the substrates. Although these experiments have revealed some interesting relationships between "choline" binding residues and *catalysis*, they are basically silent on *binding*, and additional experiments, such as microcalorimetry of protein-ligand complexes, will be required to gain insights on this important issue.

4.3
The Binding of Lipid Side Chains

Although the stabilizing interactions between the amino acid side chains of PLC$_{Bc}$ and the choline headgroup are readily apparent in the PLC$_{Bc}$-phosphonate inhibitor complex, it is more difficult to identify contacts between the protein and the acyl chains of the inhibitor [45]. In part this is because thermal motion in the acyl side chains, especially the *sn*-1 chain, renders them somewhat disordered. Consequently, the measured distances between the side chain carbons and enzyme residues cannot be taken as evidence for or against favorable hydrophobic interactions. With this caveat in mind, there appear to be possible contacts at distances in the range 3.9 – 4.5 Å between the side chains of Leu135, Phe66, and Phe70 and the C(2) – C(6) segment of the *sn*-2 side chain. These interactions may be important as determinants of enzyme activity because, in their absence, K_m might be expected to increase. Indeed, as the data in Table 3 indicate, decreasing the length of the acyl side chains in phosphatidylcholine derivatives does result in significant increases in K_m.

One interaction between the enzyme and the phosphonate inhibitor that is unambiguously significant involves the strong hydrogen bond (3.0 Å) between the backbone amide of Asn134 and the carbonyl group of the *sn*-2 ester function. This interaction provides a rationale for the observation that a simple ether linkage results in a greatly diminished rate of PLC_{Bc} processing (see Sect. 3.2.1). Presumably because phospholipid analogues lacking the *sn*-2 carbonyl group cannot make this hydrogen bond, they are poorly recognized by the enzyme and hence are poor substrates.

5
PLC_{Bc} Mechanism

Before discussing the mechanistic aspects of the PLC_{Bc} catalyzed hydrolysis of phospholipids, a brief survey of the manner in which metal ions in enzyme active sites participate in catalysis is warranted.

5.1
Mechanisms of Metalloenzymes

Two major roles that the Lewis acidity of metal ions can play include: (1) coordination with the substrate to increase its electrophilicity and render it more susceptible to nucleophilic attack and (2) coordination with an active site water, thereby lowering its pK_a. The subclass of metallohydrolyases to which PLC_{Bc} belongs is characterized by a bi-metallic or bi-nuclear center wherein two metal ions in the active site are relatively close to one another (3 – 4 Å) [63]. These two ions are bridged by a single amino acid ligand (typically a carboxylate) and a water molecule. Due to the strong Lewis acidity of the two metal ions, this bridging water likely exists as a hydroxide ion at physiological pH.

There are three mechanistic possibilities for catalysis by two-metal ion sites (Fig. 10). The first of these is the classic *two-metal ion catalysis* in which one metal plays the dominant role in activating the substrate toward nucleophilic attack, while the other metal ion furnishes the bound hydroxide as the nucleophile (Fig. 10a). Upon substrate binding, the previously bridged hydroxide shifts to coordinate predominately with one metal ion. Enzymes believed to function through such a mechanism include a purple acid phosphatase [79], DNA polymerase I [80], inositol monophosphatase [81], fructose-1,6-bisphosphatase [82], *Bam* HI [83], and ribozymes [63].

In a second scenario, the substrate is not necessarily coordinated to the metal ions but rather is positioned by amino acid side chains at the active site. The

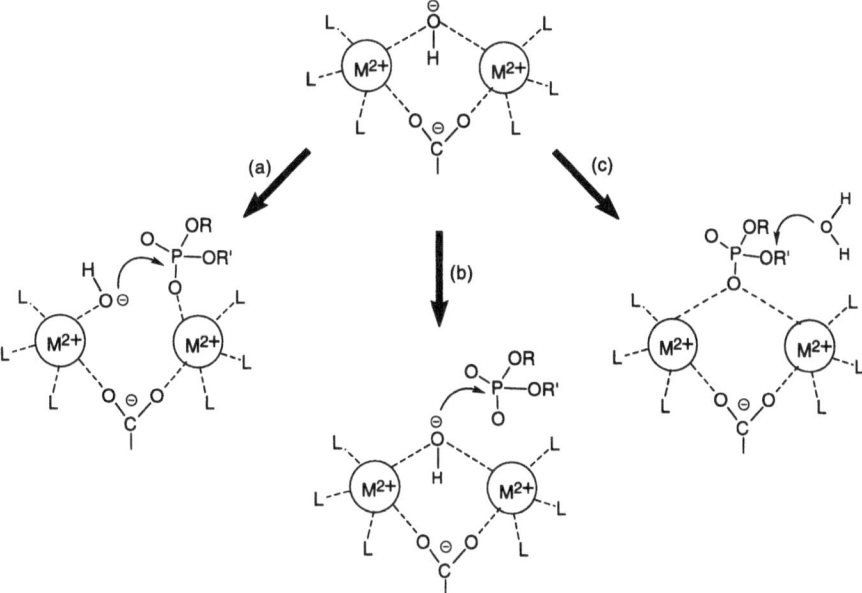

Fig. 10. Participation of a bi-nuclear metal site in enzymatic catalysis: (a) the substrate coordinates to one metal ion while the other activates the nucleophile; (b) the substrate is stabilized by amino acid side chains, and the metal ions provide the bridging hydroxide as the nucleophile; (c) the bridging hydroxide is displaced by the substrate, and the nucleophile is another water, which may be activated by an amino acid side chain in the active site

bi-metallic site then lowers the pK_a of the bridging water to generate the nucleophilic hydroxide ion, even at low pHs (Fig. 10b). Enzymes believed to use this tactic of nucleophile activation include P1 nuclease [69, 70], leucine aminopeptidase [84], and inorganic pyrophosphatase [85].

In the final case, the bridging hydroxide is displaced upon substrate binding (Fig. 10c). The metal ions would then act as classic Lewis acids to polarize the substrate, thereby activating it for attack. The nucleophilic water is not co-ordinated with a metal ion in the bi-metallic center, but rather it would be activated by an amino acid side chain that serves as a general base. Even though this latter mode of substrate and nucleophile activation is possible, there presently appears to be no well-documented case of a metalloenzyme with a bi-metallic center that catalyzes hydrolysis in this fashion. It should be noted however that an enzyme like PLC_{Bc} having three metal ions in the active site might operate by a related mechanistic pathway in which the water is activated by the third metal ion and/or an amino acid side chain.

A comparison of the X-ray crystal structures of the enzyme in its native state with that of a complex with an inhibitor can provide clues as to which of these three mechanistic possibilities is operative. In most cases the location of the bridging water in the inhibitor complex is diagnostic of the role of the metal ions in catalysis. For example, if this water is present in the inhibitor complex and if the inhibitor is coordinated to the metal, mechanism (a) is probable. However, if the water remains in place and the inhibitor is *not* coordinated to a metal ion, mechanism (b) is more likely operative. Conversely, if the water is displaced upon inhibitor binding and the inhibitor is coordinated to one or both metal ions, then the reaction probably proceeds via mechanism (c). Of course, the usual caveat that such complexes are catalytically inactive should be borne in mind during this analysis.

5.2
The Catalytic Mechanism of PLC$_{Bc}$

Because of its unusual active site and its potential as a model for mammalian PC-PLC enzymes, there has been considerable interest in the mechanism of the PLC$_{Bc}$ catalyzed hydrolysis of phospholipids. The few mechanistic points that are commonly accepted are that a water must be activated for attack on the phosphodiester linkage, the leaving group must be protonated, and the products must leave the enzyme active site (Fig. 11). These issues will now be examined.

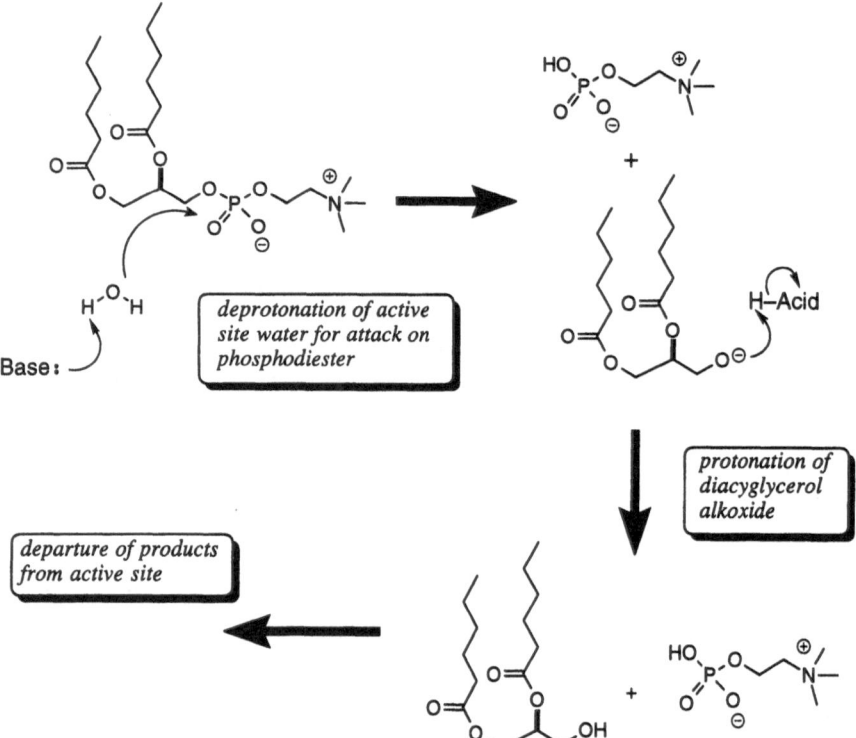

Fig. 11. Necessary elements of the PLC$_{Bc}$ mechanism. An active site water is activated for attack on the phosphodiester, a proton is transferred to the departing DAG, and the products leave the active site

5.2.1
Activation of the Nucleophile

One of the questions that is commonly addressed in mechanistic proposals is how is the active site water activated for nucleophilic attack on the phosphodiester bond? Numerous combinations of amino acid side chains and zinc ions have been proposed for this role, but there has been little consensus. Critical to all the general base hypotheses is a quite reasonable assumption about catalysis by PLC_{Bc}: The nucleophilic attack on the phosphodiester moiety proceeds via an in-line mechanism resulting in stereochemical inversion of configuration at phosphorus [86]. While this assumption is consistent with the position of the active site water molecules in the PLC_{Bc}-phosphonate inhibitor complex [45], it has not yet been established experimentally. This structure provides a detailed picture of how the amino acid side chains of Glu146, Glu4, Asp55, and the zinc ions interact with the phosphonate inhibitor (Fig. 12), so mechanistic hypotheses now have a structural basis.

5.2.1.1
Glu146 as a General Base

One mechanism that was proposed for the PLC_{Bc} reaction was based upon the observation that an active site water (W1) was 2.5 Å from the side chain carboxyl group of Glu146, which would then serve as the putative general base [45, 87]. In this scenario Glu146 deprotonates W1, which is 4.4 Å from the phosphorus of the inhibitor (Fig. 12). This water is opposite the DAG leaving group and in position to perform an in-line, $S_N2(P)$-type attack on the phosphodiester via an associative mechanism [88]. However, activation of this water by Glu146 is problematic because it requires that Glu146 function both as a ligand for Zn2

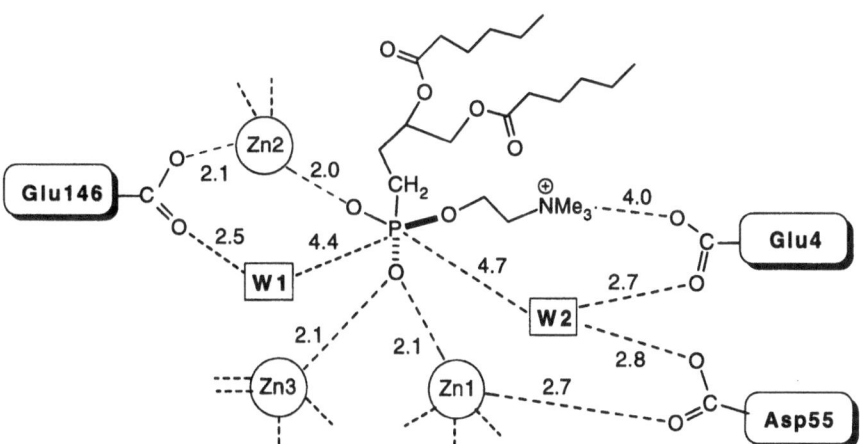

Fig. 12. Distances (Å) of PLC_{Bc} active site waters from selected amino acid side chains and a phosphonate substrate analogue [45]

and as the general base. Although a similar dual role for a carboxyl group has been proposed for DNA polymerase-β [80] and inositol monophosphatase [81], the ability of a carboxyl group to serve both functions has been openly question-ed because coordination with a zinc ion would lower its basicity [86].

To test experimentally whether Glu146 might be the general base, the single mutants E146L, E146D, and E146Q of PLC$_{Bc}$ were expressed in *E. coli*, isolated, purified, and characterized [35]. The kinetic parameters k_{cat} and K_m were deter-mined by the choline quantitation method [33]. The mutants were found to retain 0.2–1.6% of the wild-type activity with the major difference in activity being in k_{cat}; the K_m values were all similar to that of the wild type. Although there was a significant decrease in catalytic efficiency (60–500-fold) in these mutants, the loss in activity did not approach the 10,000–100,000-fold decrease often observed upon removal of the general base. For example, removal of the putative general base in the phosphoryltransfer enzymes triosephosphate iso-merase [89] and fructose-1,6-bisphosphatase [82] gave enzymes with a 10^6 and 10^4 reduction in activities, respectively.

Additional evidence that Glu146 is not the general base was obtained by determining the pH dependence on activity of the E146Q mutant. If E146 were responsible for the ascending acidic limb in the bell-shaped pH dependence/activity curve for the wild-type enzyme, then exchanging this residue for a non-ionizable one should abolish this limb, and a flat line at acidic pH would be observed. Because both the wild-type and E146Q mutants exhibit similar pH/activity profiles at pH 5–8 (Fig. 13), a general base with a similar pK_a is being

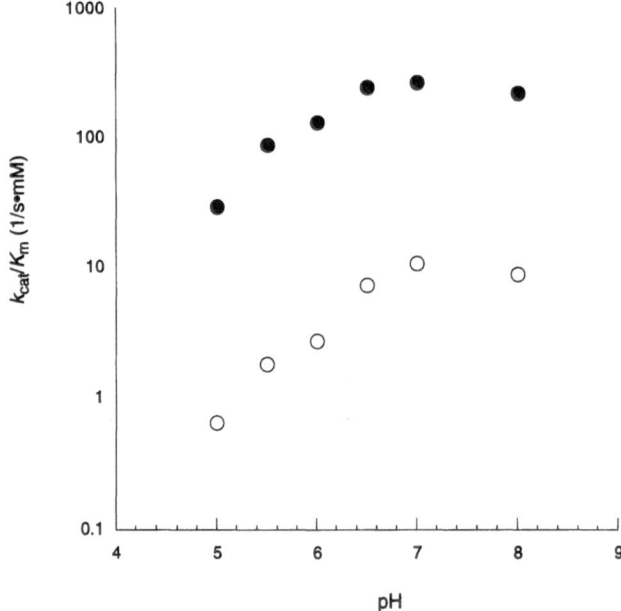

Fig. 13. k_{cat}/K_m as a function of pH for wild-type recombinant PLC$_{Bc}$ (*shaded circles*) and the E146Q mutant (*open circles*) [35]

ionized in both cases [35]. The initial portion of both curves has a slope of one, consistent with a single residue undergoing ionization.

Since it appears that the E146 side chain plays no role in activating the nucleophilic water, it remained to establish the role of E146. As noted previously, the crystallographic evidence also implicated E146 as a Zn2 ligand. Removal of metal ligands in metalloenzymes often affects their structure and stability, so the structures and zinc occupancies of the E146 mutants were examined. Indeed, the circular dichroism (CD) spectra of the E146 mutants differed markedly from the wild-type, and the active sites in each of these mutants were occupied by only two zinc ions as measured by atomic absorption.

The weight of the available evidence therefore supports the role of Glu146 as a Zn2 ligand. As the primary functions of Zn2 are presumably to organize the active site and help neutralize the negative charge on the phospholipid substrate, its removal is consistent with the observed loss of enzymatic activity in the E146 mutants. Similar catalytic changes have been documented in aminopeptidase A [90], alkaline phosphatase [91], and carbonic anhydrase II [92] where mutations of Zn ligands resulted in proteins with 0.1–5% of the activity of the wild type.

5.2.1.2
Glu4 as a General Base

Inspection of Fig. 12 shows that Glu4 of PLC_{Bc} is also suitably positioned to serve as a general base. One of the carboxylate oxygens of this residue is 2.7 Å from an active site water (W2) that is located 4.7 Å from the phosphorus atom of the inhibitor. Like W1, this water is opposite the DAG leaving group and in position to attack the phosphorus atom via an associative transition state and an in-line, $S_N2(P)$ type mechanism [88]. However, the other carboxyl oxygen of Glu4 is 4.0 Å from the methyl groups of the choline headgroup, suggesting that this residue could also stabilize the positive charge of the trimethylammonium moiety.

The case for Glu4 as the general base was supported by one molecular mechanics study using the program GRID, which is designed to calculate favorable binding sites on macromolecules, to dock the phosphatidylcholine substrate into the active site of PLC_{Bc} [93]. This calculation, which was performed before the availability of the structure of the complex of PLC_{Bc} with the phosphonate inhibitor, predicted that the phosphate oxygens would coordinate with all three zinc ions, leaving the three acidic residues Glu4, Asp55, and Glu146 within the requisite 6.5 Å of the phosphorus required for them to function as the general base. Because earlier crystallographic data suggested Asp55 and Glu146 served as zinc ligands, Glu4 seemed the rational choice as the general base.

There have been two independent mutagenesis studies that have been directed toward probing the role of E4 in the PLC_{Bc} reaction [36, 94]. In the first of these, the kinetic parameters k_{cat} and K_m of the E4L, E4D, and E4Q mutants, which each gave CD spectra similar to wild-type, were determined by the choline quantitation method [33], and these mutants were found to retain 6–60% of the catalytic efficiency (i.e., k_{cat}/K_m) of wild-type [36, 64]. Furthermore, the pH-dependence with activity curve of the E4L mutant was virtually identical with that for E146Q (Fig. 13) and similar to that of wild-type. In the other investigation, the E4A mutant was observed to have greater phosphomonoesterase activity than wild-type PLC_{Bc} [94].

Inasmuch as these experiments eliminated Glu4 as a contender for the general base, the question remains of the function of Glu4. All Glu4 mutants retain ~50% of the wild type k_{cat}, so the loss in catalytic efficiency is due primarily to increases in K_m. This observation coupled with the relationship of the side chain carboxyl group in the PLC_{Bc}-inhibitor complex suggests that Glu4 is involved in substrate binding and recognition. It is known that mutations of Glu4 affect the processing of C6PC and C6PE but not C6PS (see Table 3), so it appears that Glu4 may play a role in dictating the substrate specificity of PLC_{Bc}.

5.2.1.3
Asp55 as a General Base

One of the carboxylate oxygens of the Asp55 side chain is 2.8 Å from an active site water molecule (W2), while the other oxygen is 2.7 Å from Zn1 (Fig. 12). It is perhaps significant that this Zn1–O distance is about 0.6 Å longer than other Zn-ligand distances in this complex, and it is about 0.4 Å longer than the same Zn1–O distance in the uncomplexed enzyme (see Figs. 6, 7b) [62]. The side chains of Asp122, His118, and His69 are also ligands for Zn1, so the importance of Asp55 as a ligand for Zn1 during catalysis is uncertain. Although Asp55 had been suggested as a candidate for the general base [93], it was never actually proposed to perform this role until a recent molecular mechanics and dynamics study identified the position of the putative nucleophilic water that would be activated by Asp55 [95].

In order to elucidate the function of Asp55 in the PLC_{Bc} catalyzed reaction, the site-specific mutants D55L, D55N, and D55E were isolated and purified [36]. Even the conservative replacement of Glu for Asp gave a protein with a 4000-fold lower catalytic efficiency, whereas the D55N and D55L mutants exhibited reductions in activity of 10^4- to 10^6-fold! The primary changes were in k_{cat} with K_m remaining essentially unperturbed. The structure of the D55L mutant was essentially identical to wild-type PLC_{Bc} on the basis of comparisons of the CD, temperature dependent CD, and atomic absorption spectra. Because of its low catalytic activity, it has not been possible to determine the pH dependence of activity for D55N. Thus, the preponderance of available evidence supports the role of Asp55 as the general base in the PLC_{Bc} catalyzed hydrolysis of phospholipids.

If Asp55 is the general base, pH dependence curves suggest that it has a pK_a of 5.3–5.5 [96], which is substantially higher than the pK_a of 3.9 that is measured for aspartic acid in solution. However, specific protein environments are able to alter significantly the observed pK_as of amino acids. For example, Glu35 in lysozyme has the unusually high pK_a of ~6.2, a value that has been attributed to hydrophobic and electrostatic effects [97, 98]. Similar effects on other enzymatic carboxylate residues have been commonly recorded [99]. Examination of the immediate environment of Asp55 in various crystal structures reveals several factors that could contribute to its elevated pK_a [45, 62, 65]. For example, Asp55 is proximal to the hydrophobic side chains of Phe66 and Ala3. Moreover, the permanent dipole in α-helical structures has been found to raise the pK_as of side chains at the C-terminus of the helix, while the pK_as of these side chains are lowered when they are at the N-terminus [100, 101]. Inasmuch as Asp55 is located at the extreme C-terminal end of α-helix B, one would expect an increase in its pK_a. The combination of these effects appears to override the proximity effect of Zn1 that would tend to lower the pK_a of Asp55.

5.2.1.4
Metal Ion Activation of Water

With three zinc ions in the active site, it might seem reasonable that PLC_{Bc} would utilize a zinc-bound hydroxide as the attacking nucleophile. Indeed, zinc bound hydroxides are routinely invoked as nucleophiles for zinc peptidases, amidases, and zinc phosphoryl transfer enzymes [86]. For PLC_{Bc}, any one of the three zinc ions could individually activate a water molecule, or a hydroxide bound to Zn1 and Zn3, which comprise the bi-nuclear site, could be the attacking nucleophile. However, activation of water by an active site zinc ion seems irreconcilable with the crystallographic data available for PLC_{Bc} and its complex with a phosphonate inhibitor (see Fig. 7b). In this complex, the non-bridging oxygens on phosphorus are tightly associated (2.1 Å) with the three zinc ions, and the bridging water associated with Zn1 and Zn3 and both waters that were coordinated with Zn2 in the native structure (see Fig. 6) have been displaced. Thus, in the PLC_{Bc}-phosphonate complex, no zinc bound water molecule that can be activated is visible. This situation should perhaps be contrasted with the structure of the complex of PLC_{Bc} with inorganic phosphate wherein two of the zinc-bound waters have been displaced, but one water remains coordinated to Zn2 (see Fig. 7a).

An enzyme inhibitor complex represents a static picture of an enzyme engaged in a non-catalytic event, so care should be exercised in drawing conclusions based upon such structures. Recognizing this caveat, a second theoretical study of the PLC_{Bc} mechanism was conducted using the GRID program [102]. Modeling the substrate into the active site led to the prediction that the phosphate moiety would bind solely to Zn2 as observed in P1 nuclease (cf Fig. 8b) [69]. This mode of complexation then allows Zn1 and Zn3 to bind an active site water, thereby facilitating its deprotonation and activation. This version of the PLC_{Bc} reaction mechanism has been propagated in several recent reviews [63, 86, 103], but there is no direct experimental evidence that bears on this proposal. Moreover, use of a zinc-bound hydroxide as an attacking nucleophile does not appear to be consistent with the observed normal isotope effect of 1.9 (see Sect. 5.3.2) [34].

5.2.2
Protonation of Diacylglycerol Leaving Group

The nature of the reaction catalyzed by PLC_{Bc} in which phosphatidylcholine is split into diacylglycerol and phosphorylcholine (Fig. 11) requires two proton transfer steps: The first is the deprotonation of an active site water to generate the attacking hydroxide nucleophile, and the second is the protonation of the alkoxide leaving group. Although analyses of the X-ray structures of PLC_{Bc} and its complexes provided some initial insights about the general base, they offered little information regarding the identity of the general acid that must furnish the proton for the departing diacylglycerol. Despite this dearth of crystallographic evidence, certain scenarios involving a zinc ion in concert with bulk solvent, a zinc bound water, or an amino acid side chain appear feasible.

5.2.2.1
Zinc Ion or Zinc-Bound Water as General Acid

Owing to the absence of any amino acid residues within 4 Å of the DAG-oxygen in the crystal structure of the PLC_{Bc}-phosphonate inhibitor complex [45], the likelihood that one of the zinc ions could participate as the general acid must be considered, and there are several possibilities. In the first of these, either Zn2, which is located 3.9 Å away from the DAG methylene in the inhibitor complex, or Zn3, which is 2.9 Å from this methylene group, could coordinate with the leaving oxygen of the diacylglycerol moiety. Inasmuch as Zn1 is tetracoordinate in this complex, it perhaps seems a more likely candidate. The transient Zn-O interaction would persist until the alkoxide was protonated, presumably by bulk solvent, as no amino acids or active site waters seem properly positioned. Such a role for a zinc ion is precedented in the alkaline phosphatase reaction where a similar function has been assigned to a zinc ion [67]. However, this reaction proceeds through a covalent enzyme-substrate intermediate, and it is thus quite different from the PLC_{Bc} reaction.

The second way in which a zinc ion might be involved in facilitating departure of the diacylglycerol group is that a zinc-bound water molecule could serve as the general acid [104]; the pK_a of a Zn-bound water is typically in the range 7–9 [105]. In this scenario, a proton on a zinc-bound water would be transferred to the departing DAG alkoxide. Although the X-ray structure of the complex of PLC_{Bc} with inorganic phosphate reveals a water molecule coordinated with Zn2, there is no such water in the complex of the phosphonate inhibitor with PLC_{Bc} (Fig. 7). The absence of any zinc-bound water molecule renders this scenario somewhat problematic, but it is premature to draw any definite conclusions as structures of additional complexes of PLC_{Bc} are needed to resolve this dilemma.

5.2.2.2
Amino Acid Side Chain as General Acid

In the absence of compelling evidence supporting a zinc ion or a zinc-bound water as the general acid, an amino acid side chain should still be considered as a candidate for the general acid. The difficulty is selecting a suitable active site residue. Consequent to a molecular mechanics simulation of the PLC_{Bc} reaction, an intriguing proposal has recently been advanced that purports Asp55 as both the general base and as an assistant in the alkoxide protonation (Fig. 14) [95]. In this mechanism, Asp55 first deprotonates an active site water, and the resulting hydroxide attacks the phosphodiester function (Step a). The DAG leaving group is then protonated by another water molecule, and the resulting hydroxide

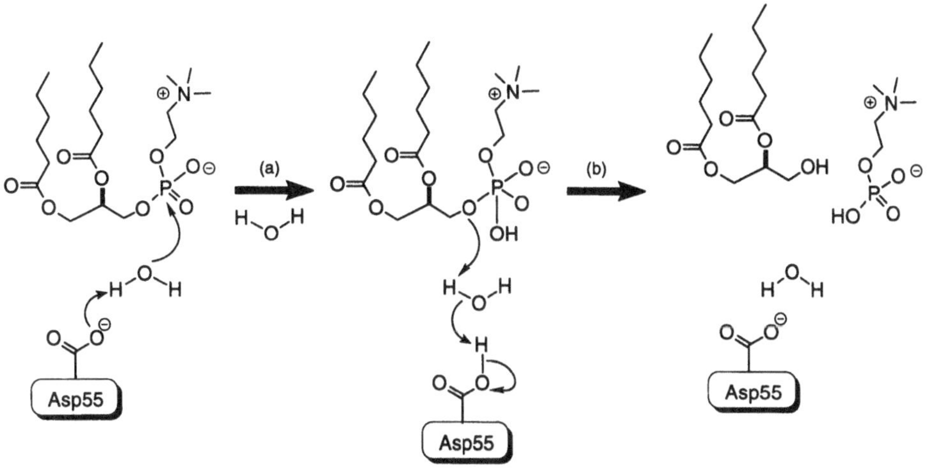

Fig. 14. Proposed mechanism for PLC_{Bc} whereby Asp55 serves a dual role as general acid/general base

removes the proton from Asp55, leaving the active site ready for the next catalytic event (Step b).

Although this is an interesting proposal, there are some potential problems that argue against its likelihood. Analysis of the activity vs pH data indicates that the pK_a of the general acid would be approximately 10 [48, 96], and such an elevated pK_a would seem unlikely for an aspartic acid residue. Moreover, the ability of a single amino acid residue to serve the roles of general base and general acid is inconsistent with the bell-shaped plot of activity of PLC_{Bc} vs pH. Assuming that pH changes are not inducing unforeseen effects upon enzyme activity, such a bell-shaped curve indicates that the general base and the general acid are being titrated on the corresponding acidic and basic limbs. If Asp55 performed in the proposed dual capacity, the pH dependence curve would have an ascending acidic limb without a descending basic limb, because the general acid would be continuously formed irrespective of pH. In light of this analysis and the experimental evidence that supports the role of Asp55 as the general base (see Sect. 5.2.1.3), it is kinetically impossible for Asp55 to serve as the general acid.

With the exception of Asp122 and His69, we have replaced all of the active site-residues by site-specific mutagenesis [64], and we have not yet observed the expected thousand-fold or greater decrease in activity unaccompanied by a significant structural change that would typically be associated with removal of the general acid [106]. Because Asp122 bridges Zn1 and Zn3, it does not seem a likely candidate for the general acid. In the PLC_{Bc}-phosphonate complex, the NH of His69 is positioned about 5.2 Å from the methylene group that would correspond to the diacylglycerol oxygen [45], so it might serve as the general acid. The ability of histidine side chains to function as general acids has been well documented, and the pK_a of the N-H on a histidine that is coordinated to a metal ion is about 11–12 [105], a value close to that estimated from pH vs activity plots for the general acid in the PLC_{Bc} reaction. Thus, the possibility that an amino acid side chain is the general acid has not yet been rigorously excluded.

5.3
Catalytic Cycle of PLC$_{Bc}$

The reactions of all hydrolytic enzymes proceed through the basic steps of sub-strate binding, hydrolysis, product release, and regeneration of the active form of the enzyme. This sequence of events is known as the catalytic cycle, and its elucidation is a critical element in understanding the function and mechanism of an enzyme. With respect to the PLC_{Bc} catalyzed hydrolysis of phospholipids, there were two questions of interest: What is the rate-determining step (RDS), and what is the order of product release? The rate-determining step may involve substrate binding or product release, which are external steps, or the hydrolysis reaction itself, which is an internal step. Solvent viscosity experiments attest to whether the RDS is an internal or external step, whereas deuterium isotope effects indicate if a proton transfer is involved in the RDS. If a proton transfer occurs in the RDS, a proton inventory plot can give insights as to how many protons are in flight. Product inhibition studies indicate which product is the last to leave the enzyme.

5.3.1
Viscosity Effects

If substrate binding or product release is rate-determining in an enzymatic reaction, then the reaction will be slower in media of increased viscosity. Conversely, if the RDS is hydrolysis, then changes in viscosity would not affect the reaction rate. Inasmuch as the catalytic efficiency of PLC_{Bc} (335 mM^{-1}s^{-1}) [35] does not approach the diffusion controlled limit of $10^8 – 10^{10}$ M^{-1}s^{-1}, substrate binding was not expected to be rate-limiting; however, this hypothesis had to be verified. Thus, if a diffusive step were rate limiting, product release seemed the only reasonable possibility. Indeed, release of DAG from the active site had been suggested previously as the likely rate-determining step [49].

To determine the effect of microviscosity on the PLC_{Bc} catalyzed hydrolysis of the soluble substrate 1,2-dihexanoyl-*sn*-glycero-3-phosphocholine (C6PC), the kinetic parameters were obtained in solutions containing sucrose as the viscogen. These experiments indicated that there was little change in reactivity as the viscosity was increased (Fig. 15), and hence a diffusion controlled process such as substrate binding or product release is not rate limiting in the PLC_{Bc} catalyzed hydrolysis of C6PC [34].

5.3.2
Deuterium Isotope Effects

The O-D bond has a lower zero point energy than an O-H bond, so it requires a higher activation energy to reach the transition state (TS) for bond breaking.

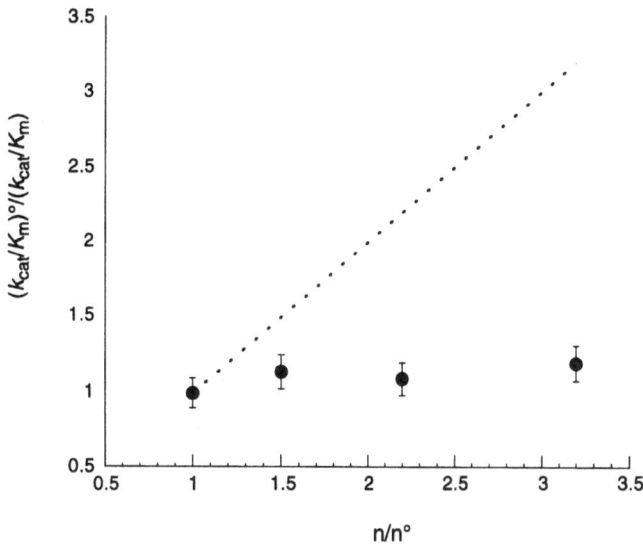

Fig. 15. Effect of various concentrations of the microviscogenic agent sucrose on the PLC_{Bc} catalyzed hydrolysis of C6PC. The *dashed line* has a slope of one and represents data expected if an external or diffusive step were rate-determining [34]

Thus, if an O-H(D) bond is broken during the rate-determining step of a reaction, there is a difference in the rates of the two reactions, the magnitude of which varies with the nature of the TS. If the proton being transferred is bound equally between two atoms in the TS, then the effect is maximized, whereas if the TS is product- or reactant-like, primary isotope effects closer to one will be observed.

Because solvent viscosity experiments indicated that the rate-determining step in the PLC_{Bc} reaction was likely to be a chemical one, deuterium isotope effects were measured to probe whether proton transfer might be occurring in this step. Toward this end, the kinetic parameters for the PLC_{Bc} catalyzed hydrolysis of the soluble substrate C6PC were determined in D_2O, and a normal primary deuterium isotope effect of 1.9 on k_{cat}/K_m was observed for the reaction [34]. A primary isotope effect of magnitude of 1.9 is commonly seen in enzymatic reactions in which proton transfer is rate-limiting, although effects of up to 4.0 have been recorded [107–110].

The observed normal isotope effect of 1.9 provides further evidence supporting the role of Asp55 as the general base. Namely, a normal isotope effect of 1.9 is most consistent with general base catalysis by an amino acid side chain, as inverse isotope effects are commonly observed when a zinc-bound water molecule, or hydroxide, is the attacking nucleophile. For example, the zinc-containing enzymes AMP deaminase [111], thermolysin [112], stromelysin [113], and a desuccinylase [114] are each believed to utilize a zinc-bound water as the nucleophile, and all of these reactions are characterized by an inverse deuterium isotope effect. This inverse isotope effect is thought to result from a dominant equilibrium solvent isotope effect for proton transfer at the metal-bound water. If a zinc-bound water were utilized by PLC$_{Bc}$, a normal isotope effect would be expected only if the proton transfer step were not at equilibrium, a situation markedly different from the four known cases. Indeed, we are unaware of any enzyme that utilizes a zinc-bound water as the nucleophile and shows a normal deuterium isotope effect. Thus, the observation of a normal isotope effect for the PLC$_{Bc}$ reaction is *prima facie* evidence that a zinc-bound water or hydroxide is not the attacking nucleophile (see Sect. 5.2.1.4).

5.3.3
Proton Inventory

Having established that a proton transfer is involved in the rate-determining step of the PLC$_{Bc}$ catalyzed reaction, the question of how many protons were "in flight" in this step remained. This issue is most easily addressed by a proton inventory experiment in which the enzyme catalyzed reaction is conducted in buffers containing different mole fractions of D$_2$O [115]. The shape of the resulting plot of activity vs mole fraction of D$_2$O provides information regarding the number of protons that are being transferred during the rate-determining step of the reaction. When a straight line is obtained, the activity decreases proportionally as the mole fraction of D$_2$O is increased, thus indicating that one proton is being transferred in the rate-determining step. Thus, the proton inventory data, which are depicted in Fig. 16, are consistent with the transfer of a single

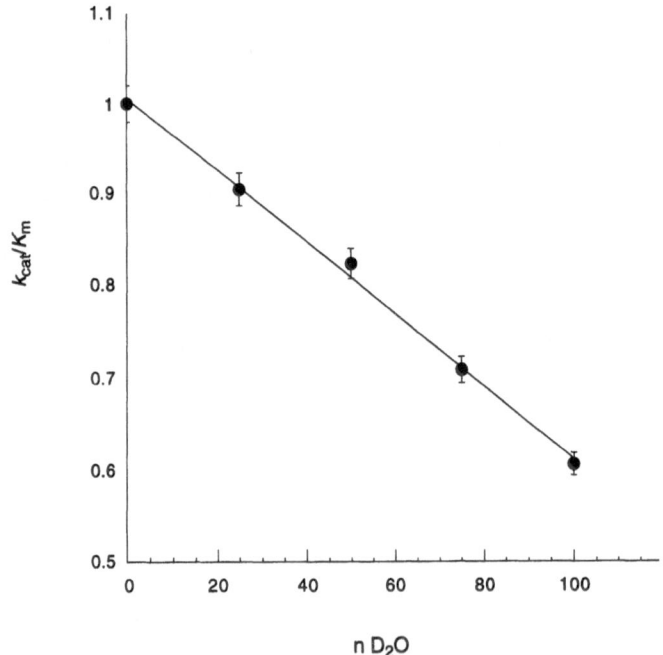

Fig. 16. Proton inventory data for PLC$_{Bc}$ [34]

proton in the rate-determining step for the PLC_{Bc} catalyzed hydrolysis of phospholipids [34].

5.3.4
Product Inhibition

Diacylglycerol has long been known to be a weak competitive inhibitor of PLC_{Bc}, whereas phosphorylcholine shows very little inhibition [40, 49, 116]. Recent kinetic assays of PLC_{Bc} activity in the presence of DAG indicate that it is a competitive inhibitor with a K_i of the order of 10 mM, whereas phosphorylcholine was found to be an extremely weak ($K_i = 30-50$ mM), mixed inhibitor of PLC_{Bc} [34]. Because diacylglycerol is a competitive inhibitor of the enzyme, the nature of the catalytic cycle dictates that it must be the last product to leave the enzyme active site.

5.4
Mechanistic Summary of PLC_{Bc}-Reaction

Although uncertainties remain, the available structural and experimental data now enable the proposal of a more comprehensive view of the mechanism and the catalytic cycle of the PLC_{Bc} catalyzed hydrolysis of phospholipids than has heretofore been possible. The process commences when the phospholipid substrate binds to the PLC_{Bc} active site (Fig. 17). The resulting enzyme-substrate complex is stabilized by interactions between: (1) the non-bridging oxygens of the phosphate group and the three zinc ions; (2) the positively charged choline moiety and the amino acid residues at Glu4, Tyr56, and Phe66; (3) the *sn*-2 ester carbonyl group and the NH of Asn134; and perhaps (4) hydrophobic interactions between the *sn*-2 acyl side chain and Leu135, Phe66, and Phe 70 [45, 65]. Due to the fixed geometry of the phosphate moiety, Zn1 and Zn3 move about 0.2 Å closer to one another upon substrate binding, a reorganization that is accompanied by an increase of about 0.4 Å in the distance between the side chain carboxyl group of Asp55 and Zn1 (see Figs. 6, 7 b). It is perhaps this reduced interaction with Zn1 that now enables Asp55 to act as the general base.

The catalytic reaction is then initiated by the deprotonation of an active site water by the general base Asp55, and the resulting hydroxide attacks the phosphorus of the phosphodiester group. In this step, the zinc ions stabilize the negative charge on the phosphodiester moiety, thereby rendering it more susceptible to nucleophilic attack. The position of the water in the active site relative to the phosphorus atom of the phosphodiester moiety is consistent with an in-line attack via an associative mechanism, [86, 88] but there is no experimental evidence to support this hypothesis. The details of how the diacylglycerol leaving group acquires a proton are unknown, but the possibilities include: (1) transient stabilization of the DAG alkoxide by Zn2 followed by protonation by bulk water; (2) a zinc-bound water; and (3) an amino acid side chain. Based upon the X-ray structural data of PLC_{Bc} complexes, the former two possibilities appear more likely, but additional structural and experimental data are required before this important issue can be resolved. Solvent viscosity experiments coupled with the

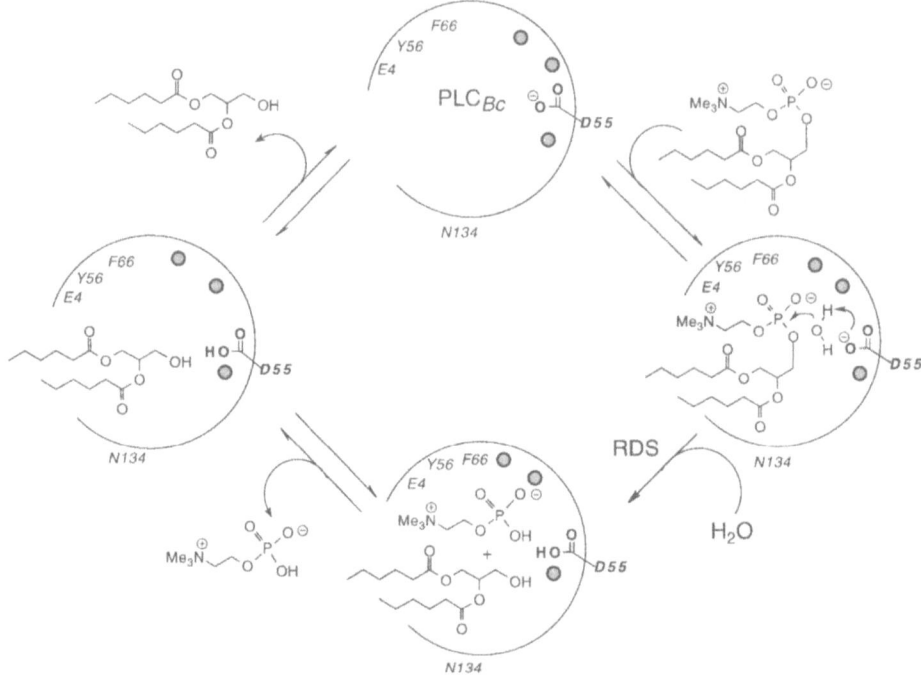

Fig. 17. Catalytic cycle for PLC$_{Bc}$. After substrate binding, hydrolysis occurs in the rate-determining step, followed by the sequential release of phosphorylcholine and diacylglycerol. Amino acids known to be involved in substrate binding are shown, and zinc ions appear as filled circles

observed normal deuterium isotope effect indicate that an internal, proton transfer step is the rate-determining step in the reaction. The two steps in the mechanism that involve proton transfer are the deprotonation of the active site water by the general base and the protonation of the DAG leaving group by the general acid, but it is not presently possible to determine which of these is the RDS. Product inhibition experiments indicate that DAG is a competitive inhibitor, necessitating that it is the last product released by the enzyme; phosphorylcholine is therefore the first to be released.

6
Perspectives

While much is now known about the function, structure, and mechanism of PLC$_{Bc}$, there remain numerous unanswered questions. For example, it is well documented that PLC$_{Bc}$ preferentially hydrolyzes micellular substrates and monomeric substrates with longer acyl side chains. However, the basis for the discontinuity in the rate of hydrolysis that is observed at the critical micelle concentration of the substrate is unknown as is the reason for the increase in the K_m for soluble substrates as the length of the acyl side chains decrease. What is the structural basis for the observed substrate specificity (i. e., PC > PE > PS) of PLC_{Bc}, and what roles do the three amino acid residues Glu4, Tyr56, and Phe66, which form the choline binding pocket, play in substrate recognition, binding, and catalysis? What are the energetics of the stabilizing interactions between amino acids in the choline binding pocket and the headgroups of different phospholipid substrates? Mechanistic questions also remain unresolved. Is deprotonation of the attacking water molecule or protonation of the departing diacylglycerol alkoxide the rate-determining, proton transfer step? What is the general acid? Inhibitors may often be utilized to gain insights into many of these and other questions, but better inhibitors must still be identified. Indeed, potent inhibitors might be utilized to isolate a mammalian PC-PLC and enable a more detailed study of these enzymes whose existence relies to date only on circumstantial evidence. These and related questions will serve as the basis for future investigations in the exciting area of the enzymes of the PLC family.

Acknowledgment. We thank the National Institutes of Health and the Robert A. Welch Foundation for their generous support of our research in the PLC area.

7
References

1. Liscovitch M (1992) TIBS 17:393
2. Asaoka Y, Nakamura S, Yoshida K, Nishizuka Y (1992) TIBS 17:414
3. Exton JH (1997) Eur J Biochem 243:10
4. Lambright D, Sondek J, Bohm A, Skiba N, Hamm H, Sigler P (1996) Nature 379:311
5. Lee S, Rhee S (1995) Curr Opin Cell Biol 7:183
6. Kishimoto A, Takai Y, Mori T, Kikkawa U, Nishizuka Y (1980) J Biol Chem 255:2273
7. Bell R (1986) Cell 45:631
8. Nishizuka Y (1989) Nature 334:661
9. Billah M, Anthes J (1990) Biochem J 269:281
10. Exton JH (1994) Biochim Biophys Acta 1212:26
11. Exton J (1998) Biochim Biophys Acta 1436:105
12. Bruzik KS, Tsai M-D (1994) Bioorg Med Chem 2:49
13. Gassler CS, Ryan M, Liu T, Oh G, Heinz DW (1997) Biochemistry 36:12,802
14. Hondal RJ, Zhao Z, Kravchuk AV, Liao H, Riddle SR, Yue X, Bruzik KS, Tsai M-D (1998) Biochemistry 37:4568
15. Heinz DW (1999) Angew Chem Int Ed 38:2348
16. Heinz DW, Ryan M, Smith MP, Weaver LH, Keana JF, Griffith OH (1996) Biochemistry 35:9496
17. Williams RL, Katan M (1996) Current Biol 4:1387
18. Grobler JA, Essen LO, Williams RL, Hurley JH (1996) Nat Struct Biol 3:788
19. Essen LO, Perisic O, Katan M, Wu Y, Roberts MF, Williams RL (1997) Biochemistry 36:1704
20. McGaughey C, Chu H (1948) J Gen Microbiol 2:334
21. Little C, Aurebekk B, Otnaess A-B (1975) FEBS Letters 52:175
22. Guddal PH, Johansen T, Schulstad K, Little C (1989) J Bacteriol 171:5702
23. Clark MA, Shorr RGL, Bomalaski JS (1986) Biochem Biophys Res Commun 140:114
24. van Dijk M, Muriana F, de Widt J, Hilkmann H, van Blitterswijk W (1997) J Biol Chem 272:11,011
25. Mulcahy L, Smithe M, Stacey D (1985) Nature 313:241
26. Garcia de Herreros A, Dominguez I, Diaz-Meco MT, Graziani G, Cornet ME, Guddal PH, Johansen T, Moscat J (1991) J Biol Chem 266:6825

27. Forsdahl K, Larsen T (1995) J Mol Cell Cardiol 27:893
28. Reynolds LJ, Washburn WN, Deems RA, Dennis EA (1991) Methods Enzymol 197:3
29. Kurioka S (1968) J Biochem 63:678
30. Kurioka S, Matsuda M (1976) Anal Biochem 75:281
31. Snyder WR (1987) Anal Biochem 164:199
32. Cox J, Snyder W, Horrocks L (1979) Chem Phys Lipids 25:369
33. Hergenrother PJ, Spaller MR, Haas MK, Martin SF (1995) Anal Biochem 229:313
34. Martin SF, Hergenrother PJ (1999) Biochemistry 38:4403
35. Martin SF, Spaller MR, Hergenrother PJ (1996) Biochemistry 35:12,970
36. Martin SF, Hergenrother PJ (1998) Biochemistry 37:5755
37. Hergenrother PJ, Martin SF (1997) Anal Biochem 251:45
38. Martin SF, Hergenrother PJ (1998) Bioorg Med Chem Lett 8:593
39. Bhamidipati SP, Hamilton JA (1993) J Biol Chem 268:2431
40. Tan CA, Roberts MF (1996) Biochim Biophys Acta 1298:58
41. Bartlett GR (1959) J Biol Chem 234:466
42. Snyder W (1987) Biochim Biophys Acta 920:155
43. El-Sayed MY, DeBose CD, Coury LA, Roberts MF (1985) Biochim Biophys Acta 837:325
44. Massing U, Eibl H (1994) Substrates for phospholipase C and sphingomyelinase from
    Bacillus cereus. In: Woolley P, Petersen SB (eds) Lipases. Their structure, biochemistry
    and application. Cambridge University Press, Cambridge, p 225
45. Hansen S, Hough E, Svensson LA, Wong Y-L, Martin SF (1993) J Mol Biol 234:179
46. Ries U, Fleer EAM, Unger C, Eibl H (1992) Biochim Biophys Acta 1125:166
47. Little C (1981) Acta Chem Scan Ser B 35:39
48. Little C (1977) Acta Chem Scand B31:267
49. El-Sayed MY, Roberts MF (1985) Biochim Biophys Acta 831:133
50. Gabriel NE, Agman NV, Roberts MF (1987) Biochemistry 26:7409
51. Lewis K, Bian J, Sweeney A, Roberts M (1990) Biochemistry 29:9962
52. Martin SF, Pitzer GE (2000) Biochim Biophys Acta 1464:104
53. Roberts MF, Otnaess A-B, Kensil CA, Dennis EA (1978) J Biol Chem 253:1252
54. Martin SF, Wong Y-L, Wagman AS (1994) J Org Chem 59:4821
55. Amtmann E (1996) Drugs Exp Clin Res 22:287
56. Muller-Decker K (1989) Biochem Biophys Res Commun 162:198
57. Kozawa O, Suzuki A, Kaida T, Tokuda H, Uematsu T (1997) J Biol Chem 272:25,099
58. Miao JY, Araki S, Hayashi H (1997) Endothelium 5:297
59. Li Y, Maher P, Schubert D (1998) Proc Natl Acad Sci USA 95:7748
60. Piettre SR, Ganzhorn A, Hoflack J, Islam K, Hornsperger J-M (1997) J Am Chem Soc 119:
    3201
61. Martin SF, Follows BC, Hergenrother PJ, Franklin CL (2000) J Org Chem 65:4509
62. Hough E, Hansen LK, Birkness B, Jynge K, Hansen S, Hordvik A, Little C, Dodson E,
    Derewenda Z (1989) Nature 338:357
63. Wilcox D (1996) Chem Rev 96:2435
64. Hergenrother PJ (1999) PhD Dissertation. The University of Texas
65. Hansen S, Hansen LK, Hough E (1992) J Mol Biol 225:543
66. Hansen S, Hansen LK, Hough E (1993) J Mol Biol 231:870
67. Kim EE, Wyckoff HW (1991) J Mol Biol 218:449
68. Steinthorsdottir V, Fridriksdottir V, Gunnarsson E, Andresson O (1998) FEMS Microbiol
    Lett 158:17
69. Volbeda A, Lahm A, Sakiyama F, Suck D (1991) EMBO J 10:1607
70. Romier C, Dominguez R, Lahm A, Dahl O, Suck D (1998) Proteins Struct Func Gen 32:414
71. Dougherty DA (1996) Science 271:163
72. Burley SK, Petsko GA (1988) Adv Protein Chem 39:125
73. Ma JC, Dougherty DA (1997) Chem Rev 97:1303
74. Basran J, Mewies M, Mathews FS, Scrutton N (1997) Biochemistry 36:1989
75. Ting A, Shin I, Lucero C, Schultz PG (1998) J Am Chem Soc 120:7135
76. Wouters J (1998) Protein Sci 7:2472

77. Gallivan JP, Dougherty DA (1999) Sci Natl Acad Sci USA 96:9459
78. Martin SF, Follows BC, Hergenrother PJ, Trotter BK (2000) Biochemistry 39:3410
79. Klabunde T, Strater N, Frohlich R, Witzel H, Krebs B (1996) J Mol Biol 259:737
80. Beese LS, Steitz TA (1991) EMBO J 10:25
81. Pollack S, Atack J, Knowles M, McAllister G, Ragan C, Baker R, Fletcher S, Iversen L, Broughton H (1994) Sci Natl Acad Sci USA 91:5766
82. Kelly N, Giroux EL, Guqiang L, Kantrowitz ER (1996) Biochem Biophys Res Commun 219:848
83. Viadiu H, Aggarwal AK (1998) Nature Struct Biol 5:910
84. Strater N, Lipscomb WN (1995) Biochemistry 34:9200
85. Heikinheimo P, Pohjanjoki P, Helminen A, Tasanen M, Cooperman B, Goldman A, Baykov A, Lahti R (1996) Eur J Biochem 239:138
86. Strater N, Lipscomb WN, Klabunde T, Krebs B (1996) Angew. Chem Int Ed Engl 35:2024
87. Hough E, Hansen S (1994) Structural aspects of phospholipase C from *Bacillus cereus* and its reaction mechanism. In: Woolley P, Petersen SB (eds) Lipases. Their structure, biochemistry and application. University Press, Cambridge, Cambridge, p 95
88. Mildvan AS (1981) Phil Trans R Soc Lond B 293:65
89. Knowles JR (1991) Nature 350:121
90. Vazeux G, Wang J, Corvol P, Llorens-Cortes C (1996) J Biol Chem 271:9069
91. Ma L, Kantrowitz ER (1996) Biochemistry 35:2394
92. Kiefer L, Fierke C (1994) Biochemistry 33:15,233
93. Byberg JR, Jørgensen FS, Hansen S, Hough E (1992) Proteins Struct Func Gen 12:331
94. Tan CA, Roberts MF (1998) Biochemistry 37:4275
95. da Graca Thrige D, Buur JR, Jorgensen FS (1997) Biopolymers 42:319
96. Ikeda K, Inoue S, Amasaki C, Teshima K, Ikezawa H (1991) J Biochem 110:88
97. Inoue M, Yamada H, Yasukochi T, Kuroki R, Miki T, Horiuchi T, Imoto T (1992) Biochemistry 31:5545
98. Bartik K, Redfield C, Dobson CM (1994) Biophys J 66:1180
99. McIntosh LP, Hand G, Johnson PE, Joshi MD, Korner M, Plesniak LA, Ziser L, Wakarchuk WW, Withers SG (1996) Biochemistry 35:9958
100. Sancho J, Serrano L, Fersht AR (1992) Biochemistry 31:2253
101. Joshi HW, Meier MS (1996) J Am Chem Soc 118:12,038
102. Sundell S, Hansen S, Hough E (1994) Protein Eng 7:571
103. Lipscomb W, Strater N (1996) Chem Rev 96:2375
104. Cha J, Pedersen M, Auld D (1996) Biochemistry 35:15,831
105. Christianson D (1991) Structural biology of zinc. In: Anfinsen C, Richards F, Edsall J, Eisenberg D (eds) Advances in protein chemistry. Metalloproteins: structural aspects. vol 42. Academic Press, p 281
106. Mildvan AS (1997) Proteins: Struct Func Gen 29:401
107. Adams JA (1996) Biochemistry 35:10,949
108. Leichus BN, Blanchard JS (1992) Biochemistry 31:3065
109. Xiang B, Markham GD (1997) Arch Biochem Biophys 348:378
110. Cook PF, Yoon M-Y, Hara S, McClure GD (1993) Biochemistry 32:1795
111. Merkler DJ, Schramm VL (1993) Biochemistry 32:5792
112. Izquierdo M, Stein R (1990) J Am Chem Soc 112:6054
113. Harrison R, Chang B, Niedzwiecki L, Stein R (1992) Biochemistry 31:10,757
114. Born T, Zheng R, Blanchard J (1998) Biochemistry 37:10,478
115. Venkatasubban KS, Schowen RL (1984) CRC Critical Reviews in Biochemistry 17:1
116. Burns RA Jr, Friedman JR, Roberts MF (1981) Biochemistry 20:5945

Photoaffinity Labeling in Biological Signal Transduction

György Dormán

ComGenex Inc, H-1027 Budapest, Bem rkp. 33–34
E-mail: dgyorgy@comgenex.hu

This paper is dedicated to Prof. Glenn D. Prestwich (University of Utah) on the occasion of his 50th birthday.

Biological signal transduction pathways control the essential functions in all cells and tissues. Understanding the mechanisms at the molecular level accelerates the development of new selective medicines. The determination of the disease related molecular targets is also essential in these efforts. Photoaffinity labeling provides a well-established technique for identification, localization of binding proteins, and studying intracellular protein-protein interactions. The application of photoaffinity labeling in biological signal transduction is discussed and illustrated with a rich collection of specific examples.

Keywords: Signal transduction, Photoaffinity labeling, G-Protein coupled receptors, Protein kinase, Tyrosine kinase, Aryl azide, Benzophenone, Diazirine

Topics in Current Chemistry, Vol. 211
© Springer-Verlag Berlin Heidelberg 2000

List of Abbreviations

[^{3}H]-BZDC	[^{3}H](benzoyldihydrocinnamyl)-aminopropyl
5-LO	5-lipoxygenase
AA	arachidonic acid
α-MSH	Alpha-melanocyte-stimulating hormone
ANP	Atrial natriuretic peptide
Apa	azido-phenyalanine
ASA	azido salicylic acid
AT	Angiotensin
AZ	aryl azide
BP	benzophenone
Bpa	benzoyl-phenylalanine
CAAX	C – cystein, AA – aliphatic amino acids, X – methionine
CaM	calmodulin
cAMP	cyclic adenosine monophosphate
CCK	Cholecystokinine

CD2	'complementary determining'
CK2	Protein kinase casein kinase
COX	cyclooxygenase
CRF	Corticotropin-releasing factor
CsA	Cyclosporin A
DAG	diacylglycerol
DZR	diazirines
EGFR	epidermal growth factor receptor
FKBP	FK506 binding protein
FLAP	5-lipoxygenase activating protein
FP	N-formyl-peptide
GDP	guanosine diphosphate
GPCR	G-protein coupled receptor
GTP	guanosine triphosphate
H4Bip	tetrahidro-L-biopterin
IL	interleukin
IP_3	D-myo-inositol-1,4,5-trisphosphate
LT	leukotriene
MAP-kinase	mitogen activated protein kinase
MHC	major histocompatibility complex
NAD	nicotinamide adenine dinucleotide phosphate
NFκB	nuclear transcription factor
NHS	N-hydroxy succinimide
NOS	nitric oxide synthase
NPY	Neuropeptide Y
NSAID	non-steroid anti-inflammatory drugs
PAL	photoaffinity labeling
PGI_2	prostacyclin
PH	pleckstrin homology
PI	phosphatidyl inositol
PIP_2	phosphatidyl-inositol 4,5-bisphosphate
PK	Protein kinase
PL	phospholipase
PTH	Parathyroid hormone
PTK	protein tyrosine kinase
Ras	ras gene encoded proteins
SH2	sarcoma homology
SP	Substance P
TMDPhe	trifluorometil-diazirine-phenylalanine
VP	Vasopressin receptor

1
Introduction

Cells in multicellular organisms communicate with each other in order to harmonize their individual actions. These signal transduction pathways control

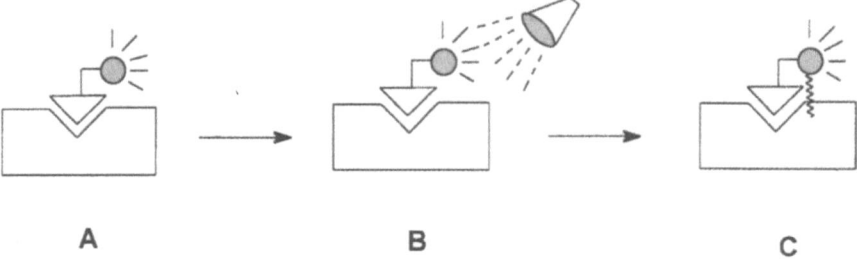

Fig. 1 A–C. General feature of photoaffinity labeling

several essential functions in all cells or tissues. In photoaffinity labeling (PAL) [1–9] a new covalent linkage is created between a light-sensitive, detectable ligand and the target biopolymer upon irradiation in a reversibly bound state (Fig. 1.). As a result of the photoinitiated covalent coupling reaction, the functional biopolymer (receptor protein or enzyme) undergoes irreversible activation or inactivation, and at the same time the occupied functional site is labeled with a detectable tag which allows easy determination of its location. Numerous ligand-protein or protein-protein interactions are involved in signal transduction and PAL represents a remarkably simple and efficient method for studying those interactions in many aspects.

2
Biological Signal Transduction: Basics and Opportunities for Photoaffinity Labeling Studies

2.1
The Diversity of Signaling Pathways

The major intracellular signaling pathways [10–13] are shown in the radically simplified Scheme 1. The simplest way of signaling utilizes lipophilic substances like steroid hormones which directly pass through the cell membrane and induce a response inside the cell without the assistance of second messenger molecules. The other fairly simple way of delivering a message into the cell utilizes ions as direct messengers. Cations or anions use different membrane-bound ion-channels that open up after binding of an agonist molecule on a specific site of the channel. On the other hand, the majority of the extracellular chemical messengers are fairly big and hydrophilic, which prevents direct crossing of the membrane, and consequently other ways were evolved. Communication through the cell membrane utilizes either membrane embedded receptor-enzymes or membrane-coupled receptors, which upon binding of the chemical messenger stimulates conformational change. That alteration initiates a series of intracellular events including enzyme activation, protein-protein interaction and translocation, or release of second messengers.

One important class of these systems is the membrane embedded enzyme-receptors such as tyrosine kinases. The binding of the extracelullar agonists

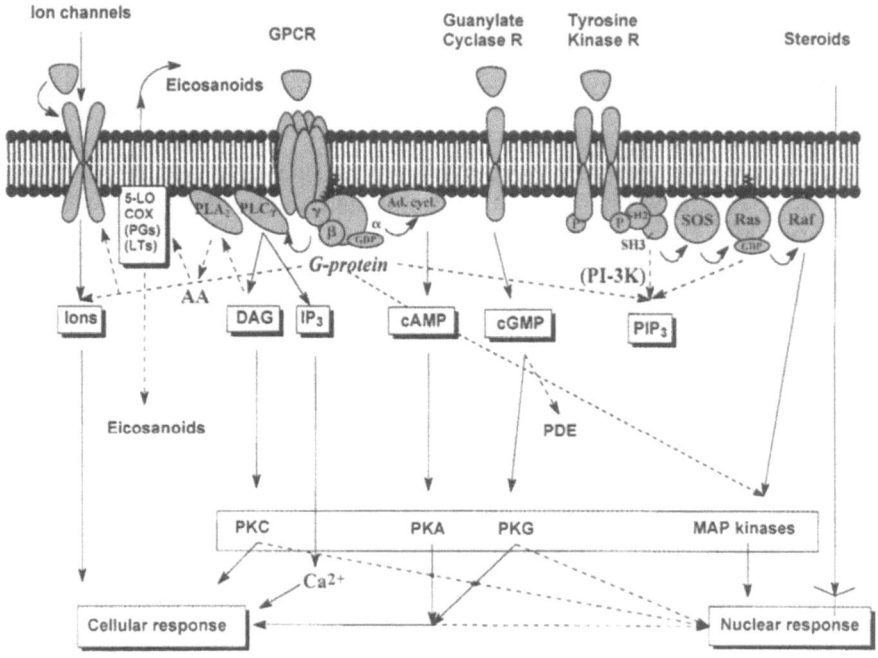

Scheme 1. A simplified overview of the intracellular signaling pathways

to the receptor-enzyme modulates the activity of the biocatalyst causing phosphorylation of the tyrosine residues on certain target proteins within the cell. This type of signal transduction is used by growth hormones and insulin etc.

Membrane-associated receptors are linked to transducing proteins (like G-proteins) in the inner portion of the membrane. G-protein coupled receptor (GPCR) families comprise a major class of the receptors that are pharmacologically relevant, such as muscarinic acetyl choline receptors, adrenoceptors, dopamine receptors, serotonine, opiate, peptide hormone, purinerg receptors, and also sensory chemoreceptors. A large variety of subtypes are described in the pharmacological literature.

Many of those receptors have already been cloned, which revealed a consistent pattern (40 %) with conserved homologies. The majority of GPCR receptors consist of a single polypeptide chain of 400–500-amino acid residues, and all possess seven transmembrane α-helices. A highly conserved region, the long cytoplasmic loop enables the receptor to couple to a heterotrimeric guanine-nucleotide-binding transducer protein (G-protein). This protein mediates the signaling from the cell surface receptor via conformational changes upon binding to a variety of intracellular effectors.

G-proteins consist of three characteristic subunits: α-subunit (36–52 kDa), β-subunit (35–36 kDa), and γ-subunit (8–10 kDa). The β- and γ-subunits form a strong dimer, while the α-subunit is able to bind guanine nucleotides. Upon the agonist activation GDP is replaced for GTP, which induces the disintegration of the G-protein trimer liberating the α-subunit. As a result, it interacts with effector systems, which can be either enzyme controlled or simply an ion-channel. The intrinsic GTPase activity of the α-subunit slowly converts GTP back to GDP, completing a signaling cycle.

In studying G-proteins, several tools were developed. For example, certain toxins are able to modify covalently the G-proteins, others inhibit the GTPase activity, which allows a permanent activation. PAL can provide the same perspective, whereas photolabile agonists or GTP analogues covalently modify the receptor or G-proteins.

The γ-subunit contains a characteristic CAAX motif, which allows covalent attachment of prenyl groups and anchors the protein to the inner surface of the membrane.

Independently the β- and γ-subunits have accompanying signaling function. As recently shown they participate in the activation of mitogen activated kinases (MAP-kinases).

Much knowledge has been gathered about signaling mechanisms of G-protein to date, but many questions remained unanswered and new areas of investigation have opened. PAL is a particularly valuable tool for studying intracellular protein-protein interactions and the diverse function of second messengers.

Two major second messenger systems have been evolved. When the α-subunit is temporarily linked to adenylate cyclase, the enzyme is activated and catalyzes the transformation of ATP to cAMP. This second messenger activates a specific protein kinase, which in turn phosphorylates target proteins in the cell leading to the overall physiological response. Such a pathway is utilized by the β-adrenoceptor, dopaminergic, and prostaglandin receptors, etc.

In an alternative system the ultimate cellular response is the increase of the intracellular Ca^{2+}-level. A minor phospholipid (PI, phosphatidyl inositol) is sequentially phosphorylated to PIP_2 (phosphatidyl-inositol 4,5-bisphosphate) and that serves as a substrate for the effector enzyme (phospholipase C, PLCγ) regulated by cell-surface receptors via G-protein α-subunit. PLCγ cleaves PIP_2 yielding the water soluble D-myo-inositol-1,4,5-trisphosphate (IP_3) and a lipid soluble diacylglycerol (DAG). IP_3 is a major Ca^{2+}-releasing second messenger, which activates a Ca-channel in the endoplasmic reticulum. The other second messenger, DAG activates a specific protein kinase, PKC, promoting phosphorylation of target proteins in the cell.

In summary, it is characteristic of the GPCR signaling system, that the ligand binding biopolymer and the effector enzyme are separate proteins and the enzyme is activated by the assistance of a receptor coupled guanylate binding-protein.

Conversely, in the tyrosine kinase pathway the effector enzyme is either an intracellular portion of the cell-surface receptor or directly coupled to it without the assistance of a transducer protein.

2.2
Signal Transmission from Seconds to Hours

Interestingly, the signaling speed is relatively fast in the G-protein coupled systems. The signal is transmitted in seconds from the surface receptor to the destination inducing the cellular changes; however, the simpler direct phosphorylation pathway requires minutes to get changes in the cellular properties. In the tyrosine kinase pathway the chemical messenger binds to the large extracellular domain, which is connected via a single α-helix to the intracellular catalytic region. Upon binding the receptor dimerizes, followed by autophophorylation of their tyrosine residues. The emerging phosphotyrosine residues act as acceptor sites for a unique peptide sequence (SH2 domain) of a distinct protein family. That binding initiates activation of various cell control functions through enzyme or transcription factor activation.

Another receptor-enzyme system was elucidated, which forms a guanylate cyclase. This enzyme converts GTP to cGMP upon receptor binding on the outer surface. The resulting cGMP activates either a protein kinase or a phosphodiesterase.

The previously discussed simplified signaling systems all point in one direction [10]. Beside several cell controlling functions they initiate gene transcription/expression. That general phenomenon allows detection of the binding to the cell surface receptors at genetic level, establishing pharmacogenetics. On the other hand, this new approach enables easy identification of any gene-product (signaling proteins and enzymes) involved in the signaling cascade and allows one to study the pathways as a whole.

Finally, steroids and thyroid hormones interact directly with the ultimate target, and stimulate transcription factors and gene expression. These lipophilic ligands readily cross the membrane and bind to nuclear proteins, which consist of a highly conserved DNA binding domain, and recognize a particular base-sequence. Upon agonist binding they activate the corresponding gene leading to increased protein synthesis. These events frequently require hours.

In summary, the duration of the different signaling pathways can range from seconds to hours. Since irreversible activation and inactivation can be controlled by photocovalent attachment these versatile tools are very useful in studying the signaling pathways as noted. The traditional photoexcitation enabled only investigating processes taking place within several minutes to hours. Now new laser pulse techniques [14] allow the establishing of photocrosslinking within *nanoseconds*, opening up brand new perspectives for studying very fast signaling cascades. Laser flash photolysis has already been widely used in initiating signals in several biological systems using the inverse technology to PAL. In that methodology a photoreactive protecting group masks the pharmacophore in agonist or messenger molecules, causing a loss of biological activity (since the biologically active species are trapped in an inactive state they are frequently referred to as 'caged' compounds). Upon irradiation, the bond between the protecting group and the ligand is cleaved and the biologically active species is recovered within a very short time, allowing one to detect the fast kinetic consequences of the signal initiation [8, 15, 16]. PAL and photorelease of bioactive compounds from "caged" compounds can be combined in many creative ways and their potential has not been fully exploited. One recent report demonstrates it. One of the photophores creates a cross-link between the ligand and the receptor, while activation of the other photoreacting group under different condition enables one to remove the ligand with the label from the protein-fragment after separation and digestion. Finally, the released peptide fragment can be identified with MS/MS sequence analysis [17].

2.3
Identification of New Drug Targets in the Signaling Pathways: Another Important Role for Photoaffinity Labeling

Several disease states are directly linked to the malfunction of the receptor or the signaling system. Identifying the most specific disease-related targets in the signaling machinery and gaining information about the binding domain are very important for selective drug design.

Since these proteins are difficult to crystallize and site-directed mutagenesis has certain limitations, PAL can find much use in exploring pathways, transducers, and effectors in signal transduction as well as some of their main structural features.

The potential of photoprobes is increasingly utilized as a remotely controllable tool in the drug discovery process. The difference between the dark state stability and light sensitivity allows the separation of two actions:

1. Introduction of a photoreactive group to biologically active molecules and setting all the parameters for the experiment.
2. Specific photochemical reaction can be initiated at will by remote irradiation activating a photoreactive structural unit (photophore), which results in the formation of a new stable covalent linkage.

PAL has been known for 30 years, but a new era of applications has appeared recently through the evolution of more efficient photophores and activation, as well as new high-resolution separation and detection techniques. The photocovalent modification of the binding site, which has an irreversible effect on activity and places a (radio)label, enables a multilevel analysis and identification on the biological target.

2.4
Different Levels of Protein Identification in PAL

The result of a standard photoaffinity experiment can provide an output in three levels [8] as shown in Scheme 2. The photolytically induced targeted introduction of a (radio)label onto the target biopolymer helps to identify any specific binding proteins (*first level of identification*) responsible for a particular action in the signaling cascade or to identify the binding domain within the target protein (*second level of identification*) or allows one to identify the amino acid sequence of the binding protein (*third level of identification*).

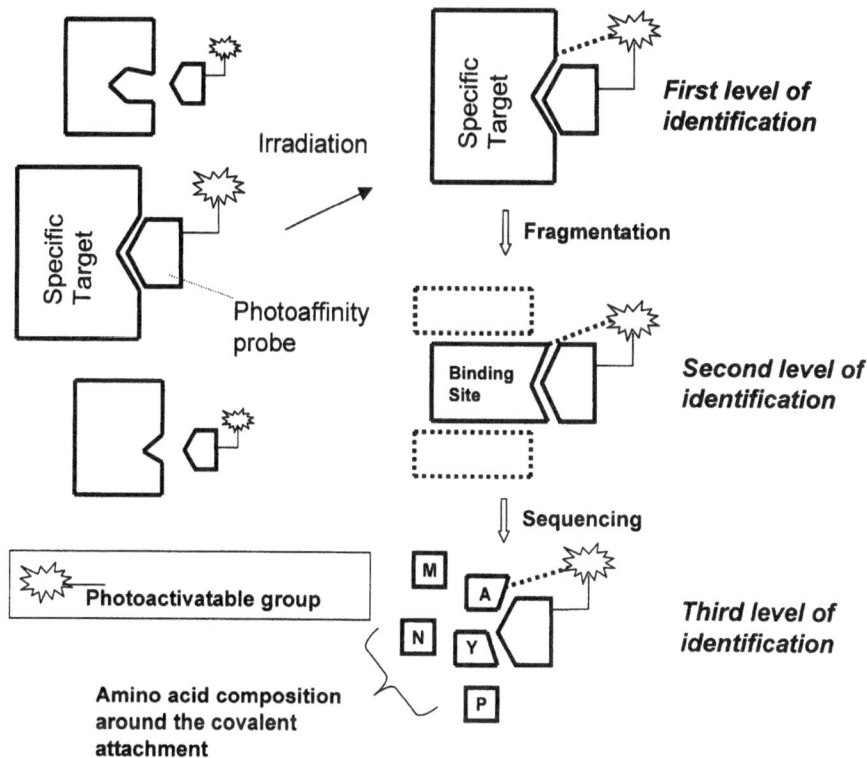

Scheme 2. The three levels of identification of biopolymers provided by photoaffinity labeling

The design of the photoprobe is based on structure-activity relationship (SAR) studies if available. Ideally the photoprobe should be bioactive over the same range as its parent compound. The next step is to synthesize the bioactive photoprobe in radiolabeled form. Similarly, non-radioactive labels (primarily biotin) can also be attached via a linker arm [18].

In the classical photoaffinity experiment a tissue homogenate is solubilized by standard detergents and incubated with the radiolabeled photoprobe for non-covalent binding prior to irradiation. By illumination at the excitation wavelength a covalent linkage is established between the ligand and receptor around the binding site as shown in Scheme 2.

After maximum radioactivity incorporation the protein is denatured and generally subjected to HPLC or gel-electrophoresis. Those methods separate the proteins from the specific tissue by size and the radioactivity distribution can be determined among the protein components. The specifically labeled bio-polymers are distinguished simply by a competition experiment performed by the addition of excess of non-labeled parent ligand. It eliminates the radio-activity incorporation.

The specifically labeled proteins are isolated and subjected to limited proteo-lytic digestion or chemical fragmentation. The highest (radio)activity fragments are either localized by immunological methods or comparison with known fragmentation patterns (analyzed by Maldie MS or MS/MS) or subjected to amino acid sequencing. The result is frequently subjected to homology search with known sequences in order to identify conserved regions and the relationship of the newly discovered biopolymer to existing protein families. The identified short sequence could also promote finding the right cDNA, which allows genetic recombination of the complete receptor.

3
Practical Overview of the Current Photoaffinity Techniques

3.1
Summary of the Most Frequent Photophores

There are several photoreactive groups and structural units ("photophores") used in PAL [3, 4, 6, 8]. The main criteria for the selection are:

1. The photoreactive groups should have sufficient stability under ambient light.
2. The life-time of the photochemically generated excited state should be shorter than the dissociation of the ligand-receptor complex, but long enough to spend sufficient time in a close proximity to a target site for covalent linkage.
3. The photophore should have an unambiguous photochemistry to provide a single covalent adduct.
4. The excited state or radical should react preferably with unreactive C-H groups yielding a uniform and stable covalent label rather than interact with nucleophiles including solvents of that type.

The activation wavelength of the photophore should be higher than the polypeptide UV absorption (> 300 nm).

Several comparative studies are available in the literature, which clearly shows that, according to the latest results, tetrafluoro-*phenyl azides* (AZ), trifluoromethyl-phenyl *diazirines* (DZR), and particularly *benzophenone* (BP) are the best choice (Scheme 3.). Diazocarbonyl compounds, which played a historically role in the evolution of PAL, can also be considered together with aryl-diazonium salts [6] (not shown). For aryl azides the more common unsubstituted photophore is presented. It should be noted that tetrafluoro-phenyl azides have an increased reactivity towards CH-groups and they do not rearrange. Other substituted phenyl azides can readily alter the excitation wavelength allowing the application of milder or differential photoactivation.

Three of the four photophores, listed in Scheme 3, eliminate nitrogen upon excitation while the biradicaloid triplet state of benzophenone can be reversibly activated, and creates covalent linkage upon excitation-relaxation cycling. This fact and their extremely low reactivity towards protic solvents make them very efficient in the majority of the cases. [7, 19, 20]. Unsaturated ketones can be activated by a similar mechanism, although the secondary processes are much more complex [21]. That mechanism is mainly utilized in steroid hormones possessing unsaturated ketones as intrinsic photophores.

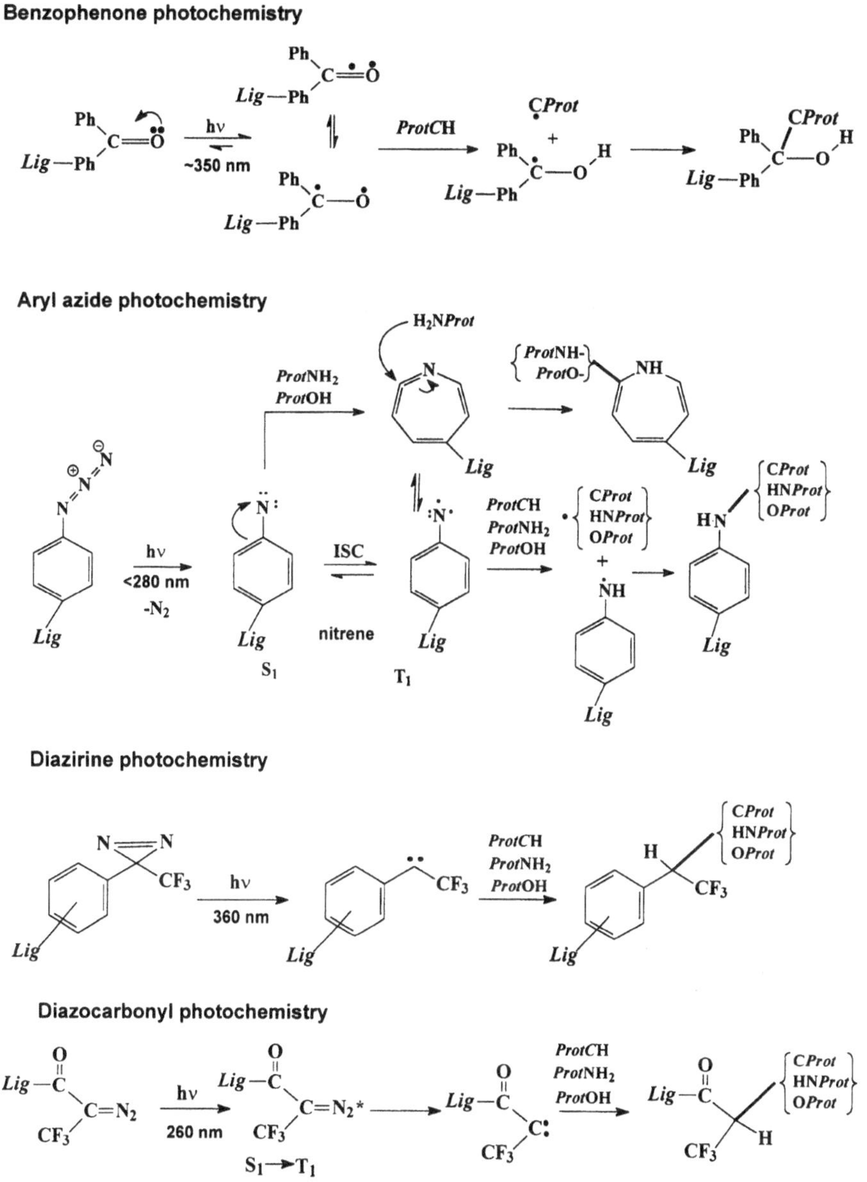

Benzophenone photochemistry

Aryl azide photochemistry

Diazirine photochemistry

Diazocarbonyl photochemistry

Scheme 3. Photochemistry of the four major "photophores"

On the other hand, aryl azides can be introduced without increasing the bulkiness of the ligand if the aryl substructure is already present, and this photophore also allows fast excitation and efficient radiolabeling, maintaining the popularity of that group among the protein target hunters.

3.2
Main Principles for the Synthesis of Photoaffinity Probes

While the majority of the chemistry strategies and synthesis design is equivalent to standard analogue producing approaches, a unique feature can be taken into account based on the relative photosensitivity of the probes. The introduction of the radio or other types of labels can also modify the reaction sequences.

The following principles can be determined:

1. During the chemical synthesis the photophore should remain intact, and therefore in many cases a precursor is constructed early in the synthetic steps, and in the final step that precursor structural unit is converted into the photolabile moiety (linear approach).
2. Most practically, the photoreactive group or its precursor can be directly attached to the parent biologically active ligand at a suitable functionality or complementing a structural unit already present (semisynthetic approach). Direct benzoylation of an aromatic ring forming a benzophenone, or azido-salicylic acid attached to hydroxyl groups can represent two examples). In that approach the photophore is generally designed to be part of the pharmaco-phore (endo-photoaffinity ligands), which allows more accurate identification of the binding site.
3. In the absence of suitable functionality for direct attachment a small linker sidearm can be introduced, which contain an appropriate functionality at the terminal position.
   Practically, a terminally protected linker can be built at any stage of the synthesis allowing the attachment of the photophore group in the final step,

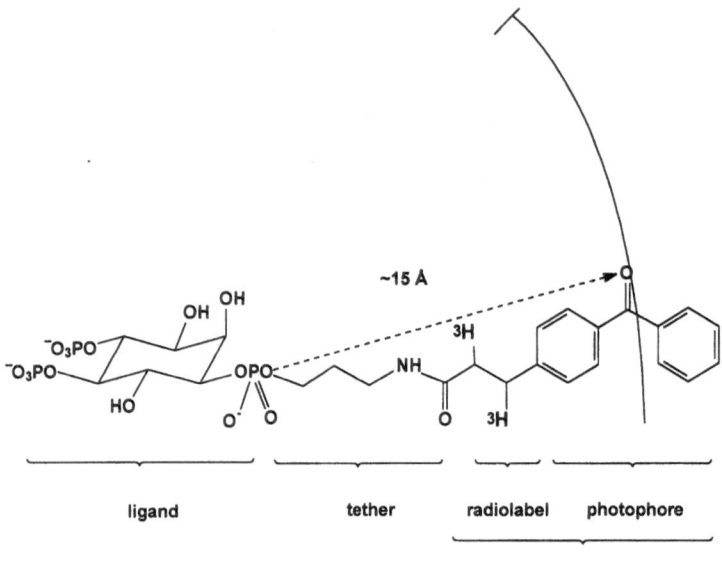

Fig. 2. A typical construction of a tethered photoaffinity probe

frequently in a radiolabeled form (tethered ligand approach). A typical example, an amino-alkyl tethered inositol-1,4,5-trisphosphate, is shown in Fig. 2, indicating the maximum span (15 Å) between the phosphate residue and the photoactivation site [22,23].

Relatively large numbers of photoreactive, heterobifunctional reagents are available commercially, and simple precursors or photophore synthons can also be utilized (the most commonly used heterobifunctional cross-linker reagents are listed in Fig. 4). The tethered ligand approach not only saves time and effort but, by using a flexible linker with optimal length, the probability of the photochemical attachment can be increased. On the other hand it may cause multiple amino acid attachment around the bound ligand within the binding site. In that approach the photophore, connected via a linker of defined length is generally placed far from the pharmacophore; therefore it does not interfere with the binding but creates covalent linkages at a defined distance (exo-photoaffinity ligands).

4. Another important principle is related to the synthetic efficiency. The ligand derivatization and tethering should be achieved with the minimal modification of the total synthesis of the parent ligand.

5. If de novo synthesis is necessary, the main principle is to use a strategy that allows one to synthesize a series of photoaffinity ligands with the photophore in multiple position around the pharmacophore. Using that approach the investigators can test a multiple set of photoprobes, which gives a better chance to find biologically active species and a definite mapping of the binding site.

The most frequent synthetic approaches, summarized in Scheme 4, are towards the primary photophores. The preparation of *aryl azide* derivatives follows the typical retro-synthetic pathway in the majority of the reported cases (Scheme 4A), and, practically, diazotation is the most commonly used procedure [24-29]. In the case of diazirines only one major synthetic sequence is repeated: ammonolysis of oximes followed by dehydrogenation (Scheme 4B) [30-32]. There are various ways of preparing diazo- or diazocarbonyl-compounds; most frequently the Forster and Bamford-Stevens reactions (Scheme 4C) are employed [33-37].

Benzophenones are usually attached as a complete photophore. Apart from the standard chemical techniques, new C-C coupling procedures extend the synthetic repertoire. However, in some cases direct benzoylation (e.g., Friedel-Crafts acylation) of aromatic or heteroaromatic rings can provide an easy access to BP or BP-like photophores (Scheme 4D) [38,39].

3.3
Photoreactive Amino Acids and Heterobifunctional Cross-Linkers

In the chemical communication many peptide-protein (e.g., peptide hormone agonist) or protein-protein signaling interaction take place. Various photoreactive amino acids, e.g., azido-phenyalanines (Apa, TFApa) [40, 41], benzoyl-phenylalanines (Bpa, p-OH-Bpa) [41,42], trifluorometil-diazirine-phenylalanine (TMDPhe) [43], were developed (Fig. 3), which can be incorporated into any places in peptide sequences by standard solid-phase synthetic techniques.

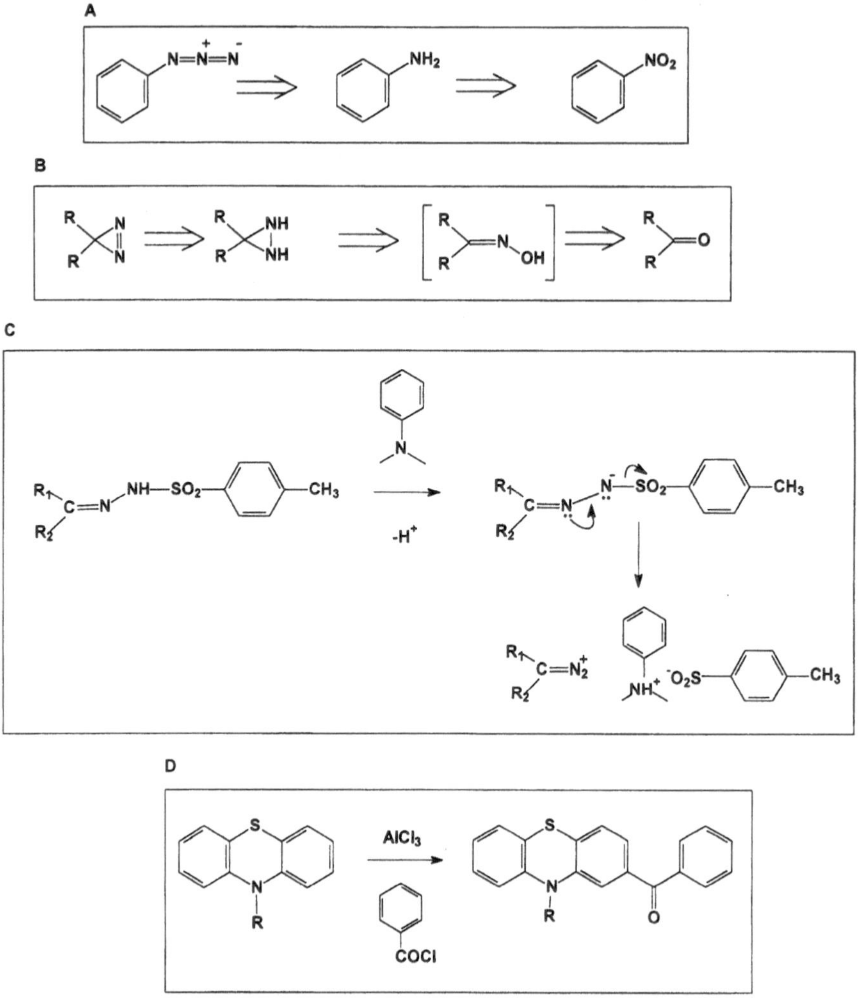

Scheme 4A–D. Selected synthetic approaches to the major "photophores"

A systematic replacement of any amino acid in the sequence for photoreactive analogues allows a photoaffinity 'scanning' of the binding interface. Since solid-phase synthesis is limited in the length of the peptide, Schultz et al. developed a sophisticated method which makes it possible to incorporate unnatural amino acids into large peptide sequences. The photoreactive amino acid was linked to *transfer* RNA, which inserted the amino acid into the required position by in vivo translation [44].

The major requirements for photoactivatable amino acids are that they should exhibit high optical purity and contain appropriate protecting groups for solid-phase synthesis; there is also a particular advantage if they contain the radiolabel. While p-azido-phenylalanines can easily be labeled with ^{125}I and

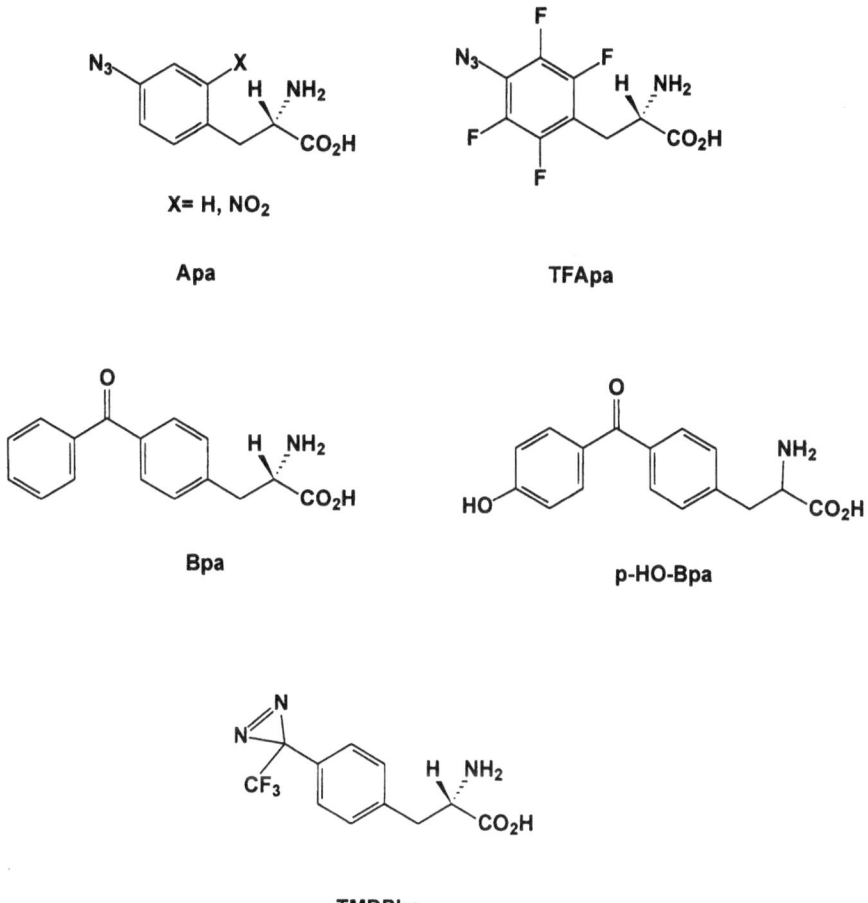

Fig. 3. Frequently used photoactivatable amino acids

tritium [40, 41], Bpa needed a multi-step synthesis to incorporate tritium [45]. More recently Maggio et al. reported the synthesis of p-HO-Bpa which allows efficient radioiodination adjacent to the hydroxyl groups [46].

Alternatively heterobifunctional cross-linker reagents [47] can be attached to specific amino acid functionalities in peptide sequences (e. g., to lysine ε-amino groups or cystein mercapto groups). These reagents contain a functional group specific anchoring unit, photoreactive reporter group with radiolabel. Some frequently used heterobifunctional crosslinkers are shown in Fig. 4, such as the amino-group specific NHS esters, the iodinatable ASA, and tritiated [³H]-BZDC [48]. The thiol group specific cross-linkers are to be attached to cystein such as MBP [20] and the cleavable ABAPS [49].

The application of heterobifunctional cross-linkers allows macromolecular PAL to probe protein-protein interactions, including subunit interactions and location, monitoring the conformational changes induced by signal transmission.

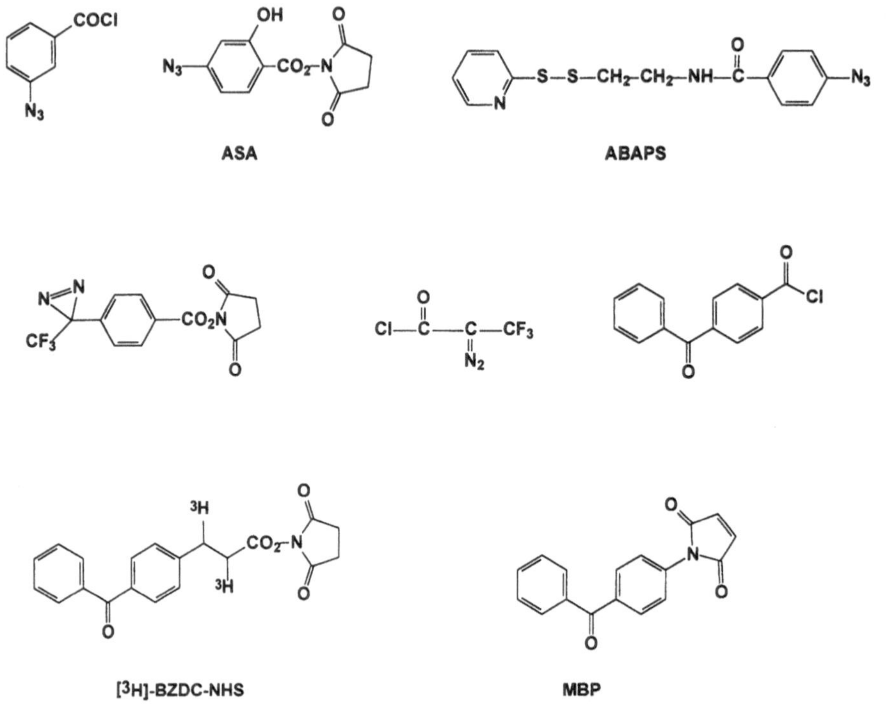

Fig. 4. Frequently used photoactivatable bifunctional cross-linker reagents

3.4
Detection/Labeling Possibilities for Cross-Linked Sites

The primary feature of the photoaffinity technique is not only to create a co-valent cross-link between the ligand and the binding protein causing permanent activation or inactivation but at the same time a detectable tag is placed at the functional domain, which allows easy localization. Even if the non-radioactive labeling technique is rapidly developing, still the majority of the recent photo-probes utilize radioactive detection [50]. There are two different ways of label-ing: in one approach the tag is joined directly to the photophore; alternatively it is placed somewhere else in the photoprobe. In the first approach, practically, the label should be placed in a precursor or heterobifunctional reagent containing both the label and photophore. The second method is commonly used in short peptide photoaffinity ligands (tyrosine residues are good targets for radioiodination).

The most common radionuclei are ^{3}H, ^{14}C, ^{125}I, and in oligonucleotide probes ^{32}P. The highest specific activity can be achieved with ^{125}I – on the other hand γ-irradiation requires more safety precautions. The ease of incorporation into aryl-azides (e.g., azidosalycilic acid [ASA] derivatives) and tyrosine residues by Chloramine T/Na^{125}I makes it particularly popular in spite of the obvious dis-advantages. In a new coupling technique organotin derivatives are replaced

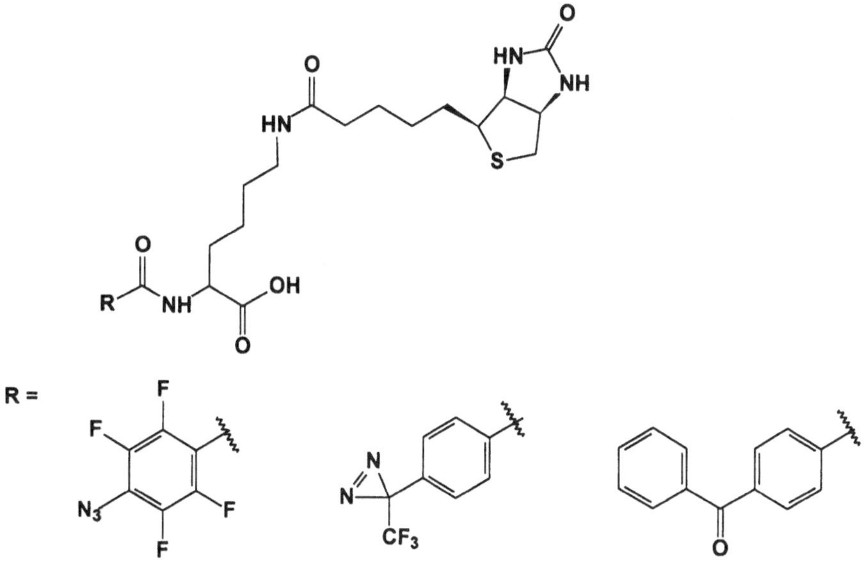

Fig. 5. A biotinylated heterobifunctional cross-linker reagent

with iodine with high efficiency. Tritium offers a relatively safe alternative, although the high-specific activity incorporation frequently needs multi-step well-designed synthesis and selective tritiation. Another problem with tritium is that in certain positions it is easily replaceable with hydrogen. Halogen-tritium exchange reaction, catalytic tritiation, or reduction with NaB^{3}H$_4$ are the choices for its introduction.

The non-radioactive labeling utilizes fluorescence, chemiluminescence, or biotin/avidin interactions. Capillary electrophoresis with laser-induced fluorescence was first employed in PAL by Miller et al. [51]. Gilbert and Rando recently reported several biotin-containing heterobifunctional reagents and used them successfully [18] (Fig. 5).

The rapid development and sensitivity of the mass spectrometric methods can be foreseen and in the near future the labeling can be more frequently eliminated. The identification of the cross-linked peptide can be detected first with immunological methods and then the digested and cleaved fragments with specific tandem MS techniques. The different photophores hold discrete MS fingerprints, which allow fast recognition of the modified sites.

4
Specific Examples of the Application of Photoaffinity Labeling in Signaling Pathways

The specific examples are classified according to the major signaling pathways and following the track of the signal from receptor binding to ultimate intracellular destination.

The account is primarily focusing on multi-protein signaling cascades, while other pathways are not discussed in detail, and only mentioned to illustrate their position in the colorful world of cell to cell communication.

4.1
Photoaffinity Labeling of Ion-Channels

Ion-channels have been extensively studied with photoaffinity labeling since the earliest years and were the subject of several excellent reviews [52, 53] and papers. For illustration some of them are listed here on calcium channels [54, 55] and sodium-channels [56, 57].

Receptor binding in neuronal synapses causes ion-release inducing very fast polarization or depolarization. The main signaling event in the ion-channel interaction is a conformational change which allows opening of the narrow pore within the protein transmembrane subunits for anions, or cations. One example is given here, which provides evidence that other associated biopolymers also participate in the channel-blocking interactions even if they are not directly involved in the channel-opening effect.

Nakanishi studied philanthotoxin (polyamine-amide) interaction with nicotinic acetylcholine ion-channel [58]. Philanthotoxin-133 (PhTX-133) is a noncompetitive channel-blocker found in venom of the wasp *Philanthus*. Nicotinic acetylcholine ion-channel is composed of five transmembrane subunits (α, α', β, γ, and δ), which forms a 270-kDa glycoprotein. The major acetylcholine binding sites are in the α,α' subunits. A 43-kDa cytoplasmic protein is associated non-covalently with the receptor, but interaction with the receptor is not essential for the channel opening (Fig. 6).

Interestingly, the aryl-azide modified PhTX-133 (1) photoprobe labeled all the subunits upon irradiation when the 43-kDa protein was not present, while the expected α-subunits were labeled only in the presence of that cytoplasmic protein. That finding provided evidence that the additional protein is necessary for orienting the toxin to the binding region. The binding site of the toxin was identified in the cytoplasmic loop of the α-subunit. In the paper a complete binding site architecture was postulated.

4.2
Photoaffinity Labeling of G-Protein Coupled Receptors and Other Protein Targets Involved in GPCR Signaling Cascade

4.2.1
G-Protein Coupled Receptors

G-protein coupled receptor family comprises most well-known cell surface receptors including the major drug targets, as previously stated. Early PAL results have been reviewed in several papers, and book chapters. For opiate, NMDA, sigma, benzodiazepine, GABA, acetyl choline, and adrenerg, serotonine receptors see [52, 59, 60], and for purinerg, histamine, and dopamine receptors see [61].

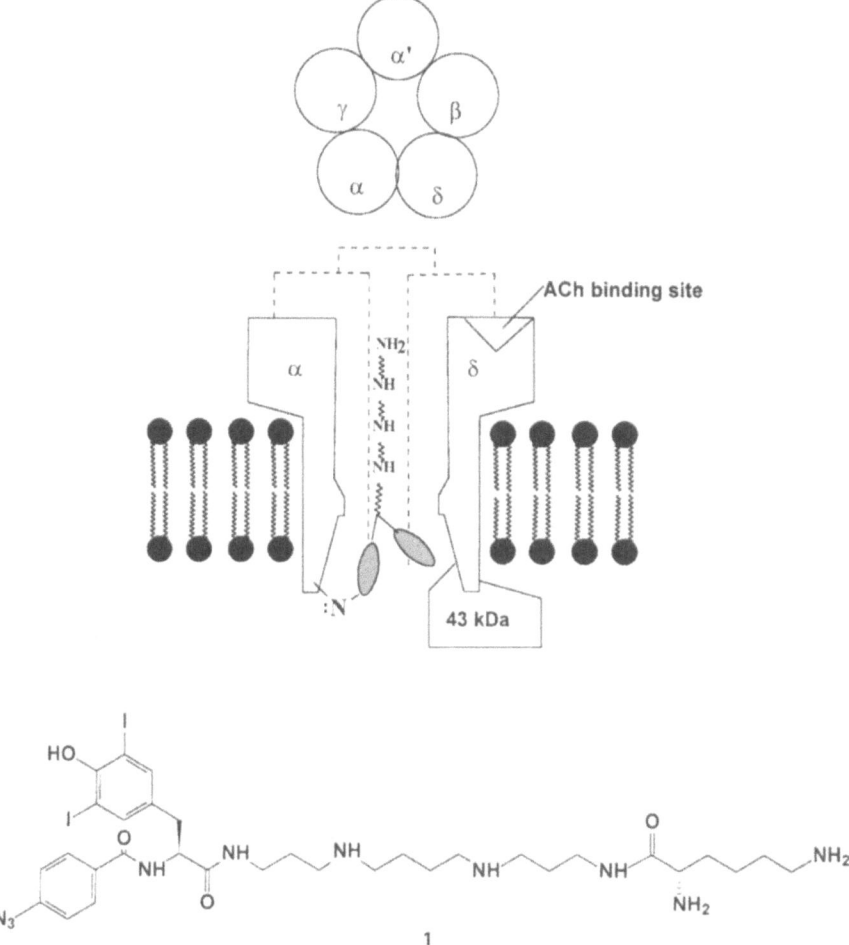

Fig. 6. Photoaffinity labeling of the nicotinic acetylcholine ion-channel with a Philanthotoxin probe

Beta-adrenerg receptors are one of the most widely studied systems in pharmacological sciences. Catechol and chatecholamines interacts primarily with the transmembrane domain-5 (TM-5) at Ser-204 and Ser-207 as well as with TM-3 at Asp-113 in β_2-adrenerg receptors (β_2-AR). Salmeterol is a long-acting β_2-AR agonist, which exhibits unusually slow dissociation from the receptor. In addition to the regular catechol interaction it has been proposed that salmeterol interacts with an additional site termed "exosite". The current model postulates that salmeterol is tethered by its long side-chain to the exosite with high affinity and that site anchors to preserve the agonist in the catechol binding site. To prove experimentally the existence of an exosite, an aryl azido derivative of salmeterol (2, Fig. 7) was prepared and used in photoaffinity experiments [62].

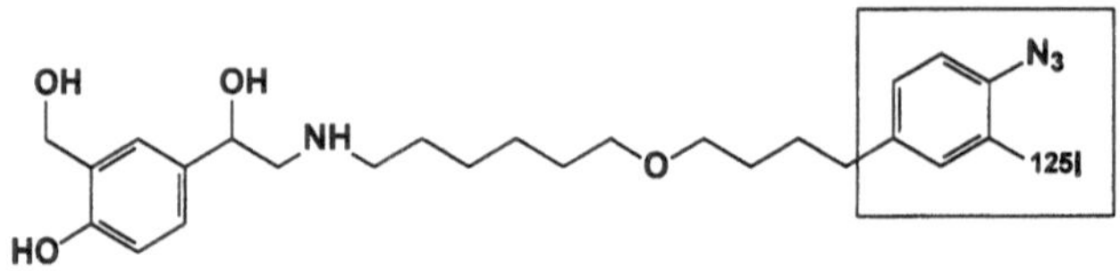

Fig. 7. Photoaffinity probes for studying G-protein coupled receptors

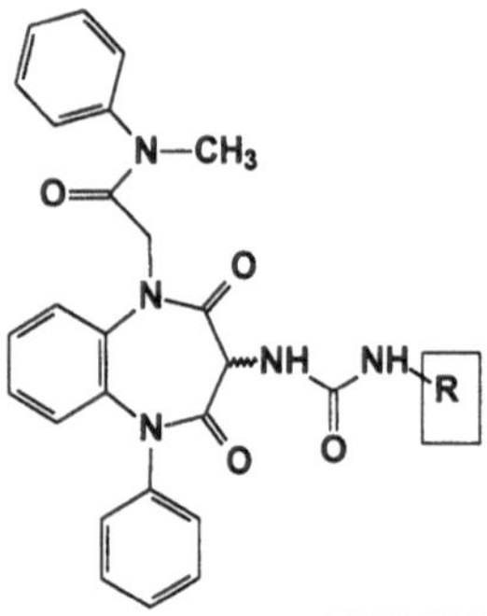

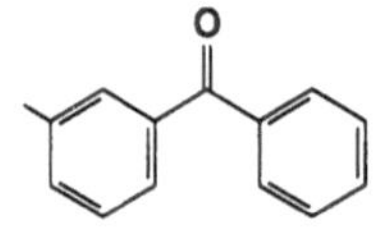

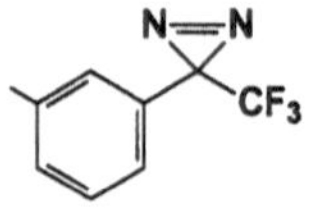

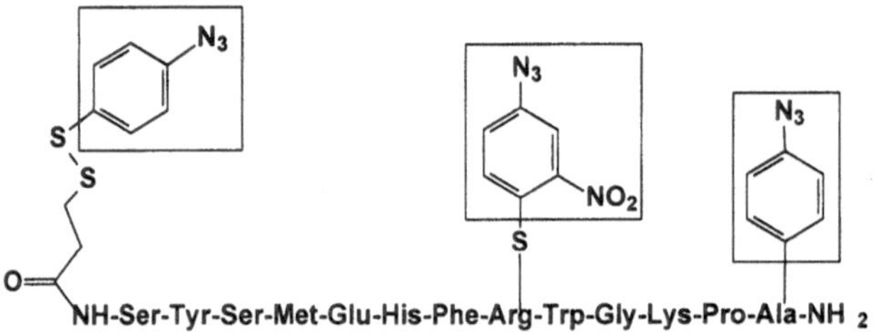

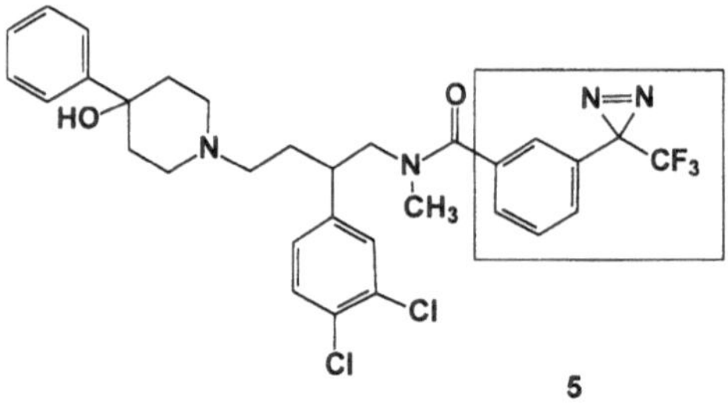

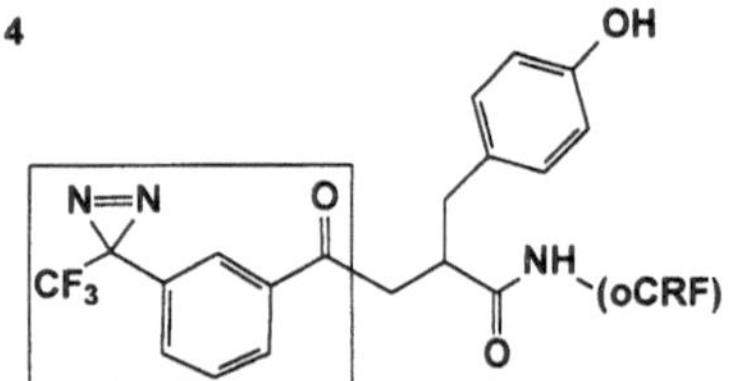

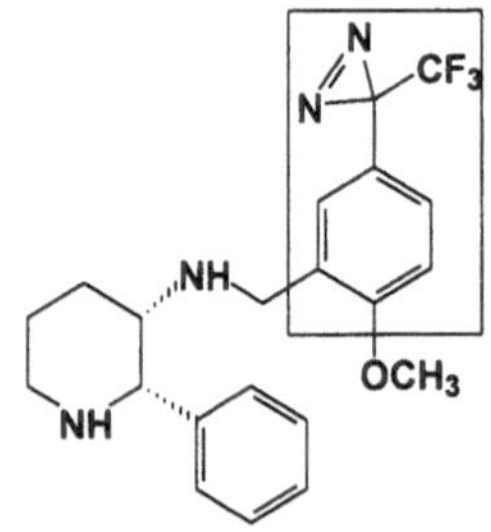

In fact, the majority of the radiolabel was recovered in the small C-terminal fragment containing TM-6 and TM-7. It delivered biochemical evidence that the aryloxyalkyl side-chain of salmeterol was positioned in that region and represented a secondary binding site, while the pharmacophore portion of the compound bound to the expected catechol site.

Calcitonin, an inhibitor of bone resorption, is a 32-amino-acid peptide hormone which acts through a G-protein coupled receptor on the target cell surface. To our knowledge, calcitonin activates not only Gα-subunits, but also other G-proteins linked to PLC, PKC enzyme activity as well as IP$_3$/Ca^{2+} releasing pathways. The receptor belongs to the same family of receptors like parathyroid hormones, secretin, vasoactive intestinal peptides or glucagon-like peptide-1. To elucidate the molecular signaling mechanism upon receptor activation an array of photoaffinity scanning analogues was prepared with altered structures. Thus, first all Lys residues were replaced for Arg in salmon calcitonin, followed by replacement of hydrophobic amino acids for photoactivatable Lys(ε-p-benzoyl-benzoyl) groups. Val-8, Leu-16, and Leu-19 were successively replaced for the photoactive structural unit. The N-terminus was also modified with the photophoric group. The photoreactive calcitonine analogue specifically labeled a 70-kDa single protein. The size of the protein was higher than expected, which might be explained by posttranslational glycosylation. Further studies will reveal the exact modification site of the photo-crosslinking [63].

Cholecystokinine (CCK) receptor was extensively studied with different research groups, which allow an excellent opportunity for comparison. CCK plays a role in CNS modulating dopamine release, and is thus involved in neurological disorders, e.g., panic attack. Both subtypes of CCK bind to G-protein coupled receptors, which causes stimulation of the phospholipase Cγ with release of IP$_3$ ultimately leading to an increase in intracellular Ca^{2+} · level. CCK has additional effects on the gastrointestinal tract. The gastrin and CCK families share a common C-terminus (Trp-Met-Asp-Phe-NH$_2$) that functions as the minimum bioactive motif. The peptide can be found in neurons as an octapeptide with the same motif as described. Baculovirus expressed receptor (CCK$_b$ subtype) was investigated with two photolabile, modified octapeptide portions of CCK, which contained photoactivatable p-benzoyl-benzoyl moieties and a p-aminobenzoyl residue instead of Met-28 and Gly-29. The main labeled band corresponded to a 45-kDa protein, which was previously confirmed by immunoprecipitation. After deglycosylation a 41-kDa protein was retained [64].

Miller reported several monomeric, photolabile CCK agonists and antagonists. The photoreactive residue (*L*-Bpa) was placed at the *N*-terminal and a fluorescent reporter group was also linked to it. The CCK receptor in the study was expressed on Chinese hamster ovary-CCKR cells. To identify the labeled domains on the receptor capillary electrophoresis was used with laser induced fluorescence detection. Separate regions were labeled with the two photoprobes, one labeled the first extracellular loop (96–121), while the other probe labeled a fragment in the second extracellular loop (174–195).

A receptor model was deduced based on the labeling pattern [51].

Several non-peptidic CCK photoaffinity probes were synthesized by Darrow et al. [65]. Benzophenone and diazirine photophores were linked to 1,5-benzodiazepine derivatives, (3a, b, Fig. 7) and the photoprobes exhibited either agonist or antagonist activity in rat amylase assay.

Secretin receptor is a member of the class II G-proteins, sharing only 12% homology with rhodopsin/β-adrenerg receptors. Secretin receptor families have long amino acid terminal domains containing six highly conserved Cys amino acids, which participate in disulfide bonds. Mutagenesis studies suggested that the referred domain had a critical role in binding. [Tyr^{10}, Bpa^{27}]-secretin-27 was a full agonist and stimulated cAMP accumulation. Upon photolysis the photolabile ligand labeled a 57–62-kDa band. After cyanogen bromide treatment the majority of the radioactivity was recovered in a 19-kDa band, which was deglycosylated into a 9-kDa polypeptide. Multiple immunoprecipitation and further proteolysis resulted in a binding site model, which located in the N-terminal region [66].

Thrombin is multifunctional serine protease involved in regulating clotting processes, wound healing, and inflammatory responses. During activation a tethered small peptide ligand is released from thrombin N-terminal by enzymatic self-slicing and acts on the G-protein-coupled thrombin receptor. After standard signaling pathways Ca^{2+} is released, which converts fibrin-stabilizing factor to fibrinoligase which strengthens the fibrin-fibrin links in the presence of Ca^{2+}. The thrombin receptor activating peptide (TRAP) is a short sequence which corresponds to the sequence cleaved from thrombin and activates the receptor itself even if not linked to the enzyme. In order to study the thrombin receptor, a photolabile Bpa-FLLRN-NH$_2$ analogue was first synthesized. That antagonist was an analogue of the minimum active sequence SFLLRN-NH$_2$. In a radiolabeled form Bpa-FLLRN-NH$_2$ specifically labeled four proteins in blood-platelets but the radioactivity incorporation was weak and prevented further studies (G. Dormán, J.T. Elliott, and G.D. Prestwich, unpublished data). In order to improve the radiolabeling efficiency Elliott prepared a series of new Bpa-photoprobes based on a recently developed receptor antagonist [67].

Alpha-melanocyte-stimulating hormone (α-MSH) is a 13-amino acid-containing peptide which is a member of the anterior pituitary peptide hormones. There are three subtypes, α, β, and γ. It has been proven that in fish and amphibia it controls pigmentation. On the other hand, in human the exact role is still unknown although it might play a part in temperature control. Eberle prepared a photoactivatable α-MSH analogue, which contained three aryl azides at 1, 9, and 13 position (4, Fig. 7). The photophore at position 1 was connected by a cleavable disulfide linkage. Photocrosslinking of the α-MSH receptor on melanophores of *Anolis carolinensis* led to almost a complete receptor inhibition, which was transferred into long-lasting receptor stimulation by exposure to a disulfidecleaving reagent disconnecting the crosslink between position 1 and the receptor. This result allows the receptor to become arrested in an activated or inactivated state and also allows a switchable change between the states. Since α-MSH receptor belongs to the G-protein coupled receptor families, this major intracellular communication pathway can be studied with the present methodology upon turning on and off the signals [68].

Angiotensin II is an octapeptide hormone that plays an important role in the regulation of the cardiovascular system (blood pressure and sustained hypertension). Angiotensin II receptors are coupled to the effector system via G-proteins, which results in regulation of Ca^{2+} current in sympathetic neurons. Primarily, stimulation of angiotensin II receptors induces the $PLC\gamma/IP_3$ signaling sequence, and ultimately results in an activation of the Ca^{2+}/calmodulin system. A hexapeptide portion of angiotensin II (Phe-Pro-His-Ile-Tyr-Val) exhibits its activity on a unique binding site, as recently discovered. Bernier studied the AT-4 receptor by PAL. The N-terminal amino acid was replaced for Bpa and azido-phenylalanine (Apa) forming two different types of photoprobes. Both photoligands performed over 60% covalent labeling. The primarily labeled 186-kDa protein was deglycosylated giving a 129-kDa biopolymer. Mild trypsin treatment released a large fragment, which constituted the majority of the protein, revealing that AT-4 the receptor is a single transmembrane protein resembling the family of growth factor or cytokine receptors (see later) rather than a G-protein coupled receptor as the AT-2 or AT-3 receptor [69, 70].

Substance P is a neuropeptide and member of the tachykinin family, released in nociceptive sensory or enteric neurons and in many regions of the brain. SP remained a pharmacological curiosity with implication in several disease states (including pain modulation and inflammation) which derives from wide range of physiological responses in many cell types. SP receptor binding activates the PLC $\gamma/IP_3/Ca^{2+}$ intracellular signaling pathways via a G-protein. The widely distributed SP related neurokinin (NK) receptors were extensively studied by photoaffinity probes in order to identify specific binding sites, which could be a starting point for efficient antagonist design. First a Bpa-modified undecapeptide was synthesized by Boyd et al. where the Phe-8 in the C-terminal was substituted with Bpa (Bpa^8-SP). Upon irradiation the photoprobe labeled two polypeptides (46 kDa and 53 kDa) and further studies revealed that the lower protein was a proteolysis product [71]. The very efficient crosslinking (63%) allowed further investigation using SDS-PAGE and MALDI MS analysis. This led to the identification of the exact attachment site as a methionine residue (Met-181) in the second extracellular loop (178–183) [72]. Maggio et al. studied murine SP receptor with two photolabile analogues ($SP-Bpa^3$ and $SP-Bpa^8$). The initially identified 75-kDa band was shifted to 42 kDa followed by deglycosylation, which was the weight expected from cDNA sequence. Interestingly $SP-Bpa^3$ interacted with the N-terminal extracellular tail, whereas $SP-Bpa^8$ labeled a region in the second extracellular loop [73]. The same group developed for SP receptor labeling a new radioiodinatable benzophenone containing amino acid (*p*-hydroxybenzoylphenyl alanine, p-OH-Bpa) (see Fig. 3) allowing one to unite the photophore and the radiolabel in the same structural unit [46].

Non-peptidic tachykinine antagonists were converted to photoprobe ligands by Ward. First, a piperidine derivative, CP-99,994 (Glaxo) was appended with a diazirine photophore (6, Fig. 7) to study SP (NK1) receptors [74]. A similar modification on a neurokinin A antagonist, SR 48968 (Sanofi) produced a photoligand (5, Fig. 7) in order to investigate NK2 receptor proteins [75].

N-formyl-peptides (FP) are G-protein coupled receptors and members of the phagocyte chemotactic receptor family, which are involved in inflammatory processes. *N*-formyl-peptides are indicators of the presence of bacteria or damage to host cells in mitochondria. Receptor binding is giving a signal for infection activating chemotaxis and defense system. The exact binding site of the receptor was identified by using an agonist analogue, N-formyl-peptide photoprobe (Formyl-Met-*Bpa*-Phe-Tyr-Lys-ε-N-fluorescein). The photoaffinity labeled recombinant human phagocyte FP receptor was cleaved with CNBr and the fluorescent fragments were isolated on an antifluorescent immunomatrix. MALDIE MS identified a major fragment with a mass corresponded to the ligand crosslinked to a CNBr fragment of FPR residues (83–87). Further cleavage resulted in the identification of the crosslink to Val-Arg-Lys (amino acids: 83–85). These residues lie within the putative second transmembrane spanning region of the receptor near the extracellular surface [76].

Neuropeptide Y (NPY), a 36-amino acid peptide, is distributed widely in CNS and is known as a potent vasoconstrictor. NPY is a neuronal and endocrine messenger that acts on several receptor subtypes. Sequence comparison indicates that Y1, Y4, and Y6 share more similarity with each other forming one distinct family. The other two (Y2 and Y5) represent a different subtype. Y2 is located primarily in the brain and particularly in the hippocampus. Two photoaffinity analogues were prepared containing 3-trifluoromethyl-3H-diazirin-3-ylphenylalanine (TMDPhe). The photoreactive amino acid was inserted into the 36 (TMDPhe^{36}) and 27 position (TMDPhe^{27}) and used in labeling recombinant receptors. A radiolabel was introduced via Lys ε-amino groups as tritiated propionates. After photolysis the cross-linked peptides were detected by immunological methods. Two proteins were detected for each cell line: a 58-kDa and a 50-kDa. From cDNA estimation a 42-kDa protein was expected, and therefore the difference revealed posttranslational glycosylation. Experiments with endoglycosidase confirmed the presumption. The two identified proteins represent different glycosylation level of the Y2-receptor [77].

Corticotropin-releasing factor (CRF) is a 41-amino acid polypeptide which is thought to synchronize physiological responses to stress. CRF exhibits its activity through G-protein-coupled receptors mainly found in pituitary and brain. Beside the two known subtypes, a 37-kDa binding protein was also identified, not homologous to the known receptors. The contact domain of the receptors (CRFR-1 and CRFR-2) and other CRF binding proteins were studied by Spiess et al. with photoaffinity probes. A diazirine photoreactive group was connected to the N-terminus of the bioactive peptide together with a tyrosine residue for radiolabeling (7). Upon irradiation the photoactivatable CRF labeled a 75-kDa protein (CRFR-1) from rat brain which, after deglycosylation, resulted in a 45-kDa polypeptide. That size was in agreement with the estimated size for rat CRFR-1 from cDNA coding [78].

Parathyroid hormone (PTH) regulates calcium levels in blood and bone remodeling. The 'activation domain' of that 84-amino acid polypeptide locates around the N-terminal (1–34 amino acids). Parathyroid hormone receptor is a typical G-protein coupled receptor, which is coupled to both adenyl cyclase/cAMP and PLCγ/IP$_3$/cytosolic Ca^{2+} intracellular signaling pathways. In order to identify the structural elements involved in the peptide hormone binding and signal initiation, Chorev et al. employed a photoaffinity scanning approach. The N-terminal amino acids were successively deleted or modified and the new N-terminus was replaced for photoreactive Bpa. The most active peptide analogue, Bpa[1], (*norleucine*)[8,18], Arg1[3,26,27], *L*-2-(*naphtyl-alanine*)[23], Tyr[34]*b*PTH-(1–34)NH$_2$ specifically labeled Met-425 within the receptor putative contact domain (409–437 amino acids) located in the transmembrane helix 6 [79].

Vasopressin receptor (VP) is a typical member of the G-protein coupled receptors. The knowledge of the hormone binding site is a particular interest. V$_2$ activates an adenylate cyclase pathway, which mediates the antidiuretic action. A radiolabeled photoreactive VP agonist [^{3}H][Mpa[1], Lys(Apam)[8]]VP (Apam = azidophenylamidino) was used in a PAL experiment and specifically labeled the second extracellular domain. The cross-linked amino acid was identified as Thr-102 and Arg-106. The sequence homology of the second extracellular domain is in fact highly conserved among the VP receptor subtypes, which most likely confirm the presence of the binding site in that region [80].

4.2.2
Light-Induced and Chemosensory Signaling Pathways

Rhodopsin, the receptor of light, consists of a chromophore (11-*cis*-retinal) that is covalently bound to apoprotein opsin (37–46 kDa) via a Schiff-base to a lysine residue. Light initiates the visual process by the photoisomerization of the chromophore from an 11-*cis* to *trans* configuration. This transformation alters the conformation of the surrounding opsin moiety. In that altered state, called metarhodopsin, the conformational change is transmitted to the cytoplasmic surface, enabling interaction of metarhodopsin with the G-protein transducin and a kinase that phosphorylates metarhodopsin. Activated transducin stimulates the exchange of GDP to GTP at the α-subunit. The GTP-bound α-subunit in turn activates cGMP-specific phosphodiesterases, which reduces the level of cGMP, causing the cGMP-activated Na$^+$-channel to close. The other G-protein induced pathway is the PLCγ/IP$_3$/Ca^{2+} signaling cascade in different species. The termination of the signal can be achieved by the light-activated GTPase activity and alternatively a soluble 48-kDa protein binds to the phosphorylated metarhodopsin, and interchanges phosphorylation. That light-induced signal transduction was extensively studied by PAL. Since only activated G-proteins are able to bind GTP, photocrosslinking can selectively distinguish the light-dependent transducin α-subunit from other G-protein guanosine phosphate-binding subunits not involved directly in the transmission. In an early study a photoreactive, enzymatically non-cleavable GTP analogue, azidoanilido-GTP (**8**, Fig. 8), was used as an efficient photolabel for the G-protein [81]. Preillumination with a blue light, which rapidly initiates the conversion of rhodopsin to metarhodopsin, was followed by irradiation with an intense UV light. Two proteins were identified, a 41-kDa and a 39-kDa protein. When the preillumination was carried out with red light, suppressing the photoisomerization of the retinal only a 39-kDa G-protein subunit was detected, clearly showing the direct involvement of the 41-kDa polypeptide in the photosignaling process.

GTP-bound α-subunit activates a cGMP-specific phosphodiesterase as described briefly earlier. A benzophenone derivative of cGMP was prepared and found to inhibit several types of cyclic nucleotide phosphodiesterases at low concentrations. In rod outer segment preparations, the [α-^{32}P]-tagged probe

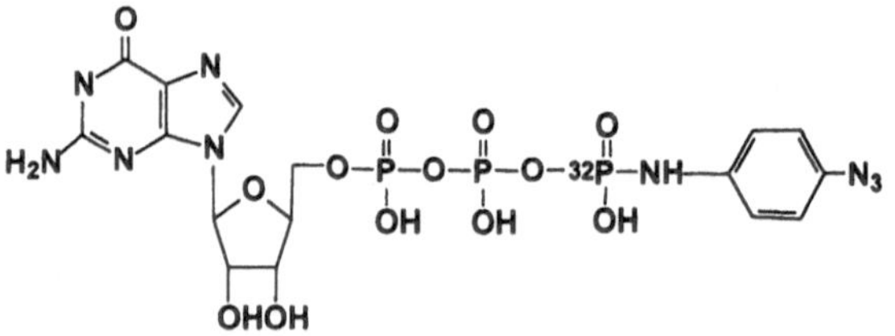

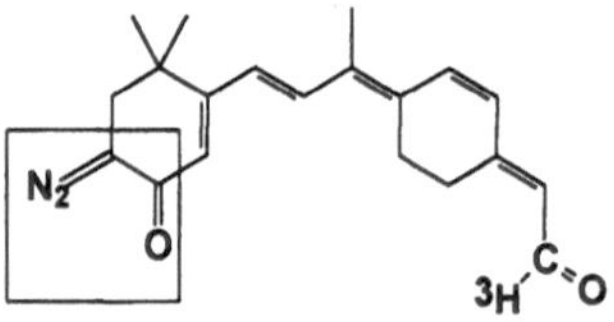

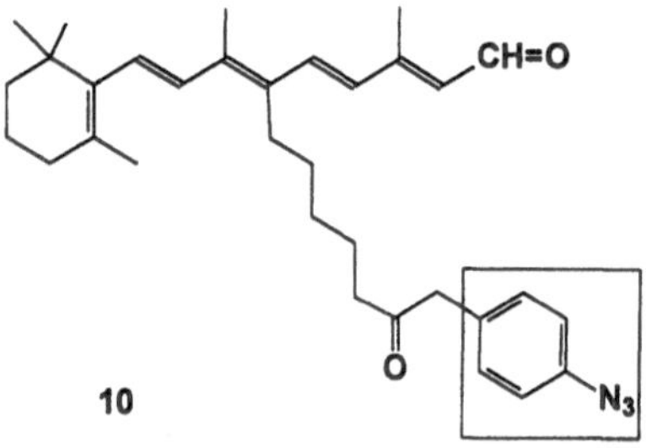

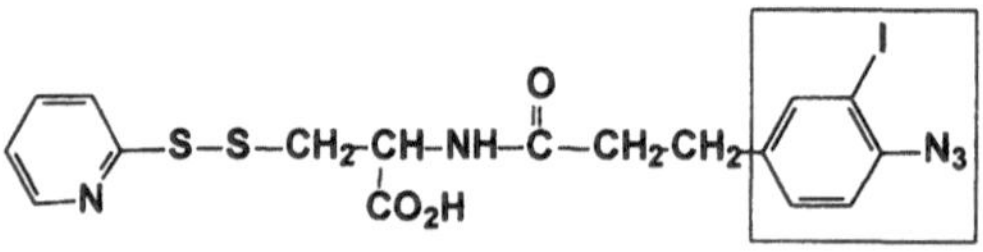

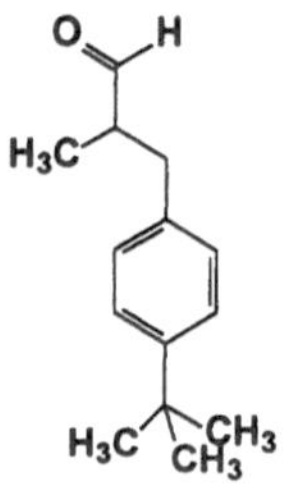

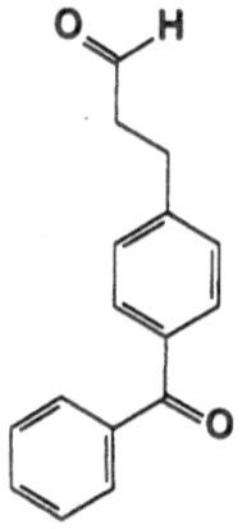

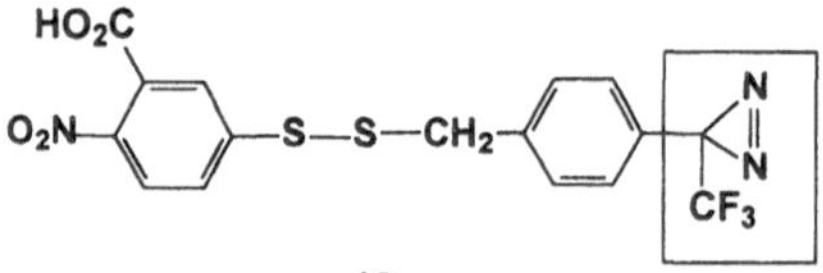

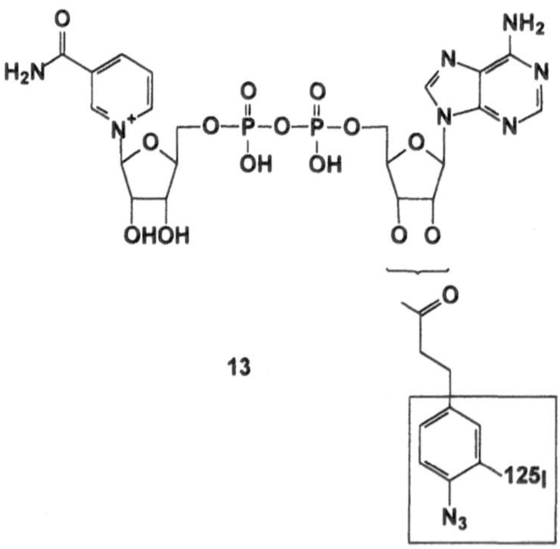

Fig. 8. Photoaffinity probes for studying light-induced and chemosensory signaling pathways

specifically labeled the light-stimulated cGMP phosphodiesterase, opening the possibility of using this probe for identification and active site mapping of cyclic nucleotide-binding protein or enzymes [82].

A tritiated, photoreactive, diazo ketone analogue of 11-*cis*-retinal (9, Fig. 8) was used to study the location and orientation of the chromophore in rhodopsin. The 11-*cis*-double bond was trapped by building it into a six-membered ring in order to prevent light directed isomerization. Experimentally first the retinal analogue was incubated with bovine opsin for 24 h, and allowed to form rhodopsin in a ratio of 1:0.65 to avoid random labeling of the non-bound retinal. The photoaffinity labeling was carried out for 10 min at 254 nm, and then the Schiff-base linkage of retinal was chemically cleaved. The released retinal-derived radioactive tag was identified to modify Trp-265 and Leu-266 in the α-helix. According to the current receptor model (Fig. 9), retinal is linked to Lys-296 in the seventh (G) α-helix in the middle of the lipid bilayer. The present photoaffinity results confirmed that the entire chromophore resides near the center of the bilayer, with the polyene's long axis angled only slightly relative to

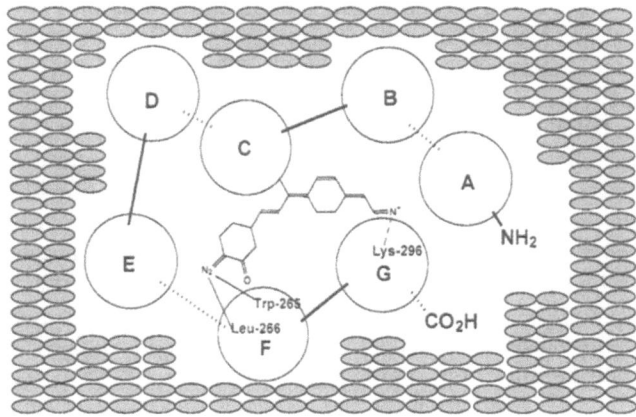

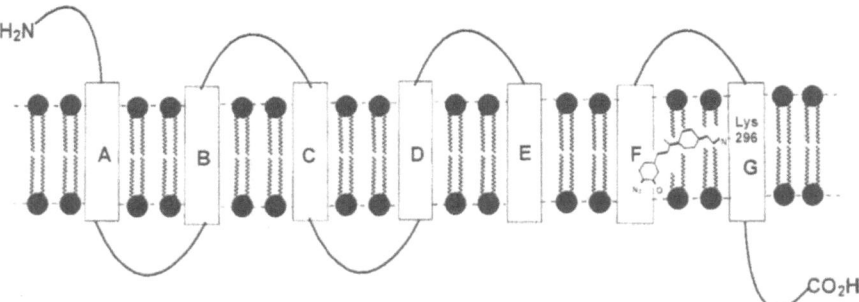

Fig. 9. Photocovalent attachment to rhodopsin

the membrane plane. This position leaves the chromophore accessible to absorb the light easily [83, 84]. In a previous study when photoisomerizable photo-probes were used, both the C and F transmembrane α-helix were labeled. The comparison of the two labeled regions gives a clear picture of the motion of the photosensor small molecule in rhodopsin [85].

Nakanishi also studied bacteriorhodopsin, which is a 26-kDa hydrophobic protein found in the outer membrane of *Halobacterium halobium*. It functions as a light-driven proton pump. It is composed of a 248-amino acid protein and an all-*trans* retinal is anchored to Lys-216 through a Schiff-base. Upon light-absorption the chromophore undergoes a *trans* → *cis* isomerization about the C-13, C-14 double bond and deprotonation/reprotonation of the Schiff base. The latter event is directly involved in the proton pumping. The orientation of the chromophore can provide information, to which direction the proton transfer takes place beginning at the Schiff-base. The determination of the orientation of C-9, C-13 methyl groups, which are opposite to the location of the proton studied, can solve the problem. A C-10 tethered (13 Å long) photoaffinity ana-logue (**10**, Fig. 8) was synthesized with a tritiated, terminal aryl azide. The major crosslinking site was identified as Arg-175 and Asn-176 of α-helix F, which revealed that the C-9 methyl group is oriented towards the extracellular space, the β-ionone ring was tilted at 27° and the chromophore's plane is closely per-pendicular to the plane of the membrane [83].

Transducin is a typical α-subunit of the light-activated G-protein containing four functional sites: rhodopsin binding, GTP binding, cGMP phosphodiesterase activation sites, and $\beta\gamma$-subunit contact sites. The structural organization of these functional sites in the tertiary structure was identified by photoaffinity cross-linking. A photoactivatable heterobifunctional cross-linking reagent N-(3-iodo-4-azidophenylpropionamido)-S-(2-thiopyridyl)cystein (**11**, Fig. 8) was attached to cystein units in transducin. Within the α-subunit only two cysteins (in different domains: Cys-210 or Cys-347) were conjugated with the crosslinking reagent but, separately, bis-conjugated product was not detected. Any of the modification inhibited the rhodopsin catalyzed activation of that G-protein. After irradiation the Cys-347 azide derivative was able to cross-link to the 12-kDa intramolecular domain in the α-subunit, which otherwise con-tains the second cystein (Cys-210). The modification prevented the replacement of GDP to GTP. The result indicates that conformational changes within the α-subunit could be readily transmitted between the two functional domains [86]. In a more recent study, Resek prepared rhodopsin mutants that preserved only a single reactive cystein residue per rhodopsin molecule at 65, 140, 240, or 316 position on the cytoplasmic face; all the other cysteins were replaced for serine. A carbene generating diazirine was connected via a cleavable disulfide group (**12**, Fig. 8) to each of the rhodopsin mutants. Visible light activation formed metarhodopsin II, which caused rhodopsin to bind to transducin. After UV irradiation (355 nm) only the Cys-240-linked photophore (in the third cyto-plasmic loop) was able to connect covalently the α-subunit to transducin. All the other crosslinks were found to be intramolecular interactions [87].

NAD$^+$ is a substrate for pertussis toxin, allowing the transfer of an ADP-ribose derivatives to G-proteins. Ruoho used [^{125}I]-iodoazidophenylpropionyl-NAD$^+$

(13, Fig. 8) and 2-azido-NAD$^+$ to probe G-proteins using that toxin induced transfer methodology.

The aryl azido derivatives of [^{32}P]NAD$^+$ were employed to site-specifically incorporate photoactivatable ADP-riboses to transducin at Cys-347 by means of pertussis toxin, and the conjugation was followed by irradiation. Photocross-linking revealed three major bands after irradiation which corresponded to α-γ (47 kDa), α-α (83 kDa), and α-α-α (105 kDa) subunit contact domains [88].

Vertebrate olfactory receptors also belong to the family of G-protein coupled receptors. Olfactory receptor families have the capacity to recognize several hundreds of different odorants. Lilial (14, Fig. 8) is one of the general odorant ligands which activates a specific receptor subtype and induces an intracellular signal, which leads to Ca^{2+} mobilization via IP$_3$ second messenger release. The olfactory receptor was overexpressed in bacterial systems. Since isolated recombinant receptors have no biochemical or cellular responses, these proteins can be easily characterized by PAL. A photoreactive lilial analogue (15, Fig. 8) was prepared and the photoprobe specifically labeled the recombinant receptors (31 kDa) with high efficiency confirming the presence of the correct polypeptide [89].

4.2.3
Studying Interactions at G-Protein Level

If a receptor is coupled to multicellular processes, a drug which would act at the receptor/G-protein interface would deliver a remarkable specificity – that is why so much importance is attached to studying such interaction at the molecular level.

Photoreactive GTP analogues have proven beneficial for studies of G-proteins, since using such probes can not only lead to the right G-protein being identified, activated by the agonist, but also information about the environment of the GTP binding site can be assessed.

$G\alpha_0$-*proteins* are widely distributed in CNS. Three subtypes $G\alpha_{01}$, $G\alpha_{02}$, and $G\alpha_{03}$ are involved in the muscarinic receptor Ca^{2+} channel signaling machinery. Anis studied that system in rat brain with [α^{32}P]GTP-azidoanilide (8) described earlier.

Considerable evidence has been gathered showing that, in synaptoneurosome membrane, depolarization induces the activation of G-proteins. This finding was confirmed by PAL, when synaptoneurosomes in depolarized and repolarized states were incubated and then photolized separately with [α^{32}P]GTP-azidoanilide. Two subtypes, $G\alpha_{01}$ and $G\alpha_{03}$, showed five- to sevenfold enhanced photocovalent labeling in the depolarized state relative to the polarized. This indicated an enhanced exchange of GDP to GTP (or its analogues) [90].

The membrane-bound α_2-adrenerg receptor is coupled to its cytoplasmic effector enzyme, through G-proteins. Site-directed mutagenesis studies confirmed that the third or second cytoplasmic loops of the transmembrane receptor and the C-terminal tail are important in contacting to G-proteins and transmitting the signal. In order to determine the receptor contact site of the G-protein responsible for receiving the signal, Taylor designed a photoaffinity probe that contained a 14-amino acid portion from the receptor. This *"Peptide Q"* *(G-protein activator peptide)* derived from the third cytoplasmic loop (362–373) and contained an additional cystein C-terminal residue. Peptide Q itself proved an efficient inhibitor of the signaling system when α_2-adrenerg agonists were employed. On the other hand Peptide Q activated GTPase in purified G-proteins, confirming its receptor mimic action. A photoreactive diazoketone was conjugated to the cystein C-terminal of Peptide Q and the photoprobe (16, Fig. 10) was used as a photoaffinity label for the G-protein. The diazoketone derivative of "Peptide Q" specifically labeled the G_0-subunit of the G-protein [91].

4.2.4
The PLC/IP$_3$/Ca^{2+} Signaling Cascade

The phospholipases (PLC) isozymes cleave the phosphodiester bond in phosphatidyl-inositol-4,5-bisphosphate (PIP$_2$) releasing two second messenger molecules inositol 1,4,5-trisphosphate (IP$_3$) and diacylglycerol (DAG) as shown before. The β-isozyme are controlled by the G_α or $G_{\beta,\gamma}$ subunits of the heterotrimeric G-proteins coupled to surface receptors. The γ-isozymes are substrates for tyrosine kinases, such as growth factors.

The recently discovered PLC-δ_1 is activated by a new class of G-proteins which are also linked to cell surface receptors. The PIP$_2$ binding site of the enzyme was studied by a photoaffinity analogue of IP$_3$ ([^{3}H](benzoyldihydrocinnamyl)-aminopropyl-IP$_3$, 17a, Fig. 10) where the benzophenone-containing unit efficiently imitated the apolar acyl chain of PIP2. After succesful photolabeling limited digestion separated the major domains. Careful radioactivity analysis revealed that the majority of the radiolabel was recovered in the pleckstrin homology (PH) domain instead of the catalytic region as initially anticipated. PH domain is a highly conserved sequence (130–140 amino acids) found originally in platelet protein pleckstrin and in several signaling proteins. They form a characteristic pseudo-β-sandwich containing a carboxy terminal β-helix. The labeling pattern showed that PIP$_2$ binds the PH domain, which orientates the substrate to the catalytic site in the best conformation for the efficient binding [92].

The PLC-γ-produced inositol 1,4,5-trisphosphate is a common second messenger, which plays an important role in the intracellular Ca^{2+} mobilization.

IP$_3$ receptors are located in the membrane of endoplasmic reticulum as a homotetramer forming a Ca^{2+} channel. The 260-kDa monomer contains eight transmembrane regions and the majority of the amino acids are cytoplasmic. That outer portion contains the putative IP$_3$ binding site at the N-terminal tail, while additional binding sites are also identified: two for ATP binding and another two for cAMP-dependent protein kinase interaction [22]. Studies on localization IP$_3$ binding sites were allowed by the above-mentioned photoactivatable analogue of IP$_3$ (17a, Fig. 10). PAL helped to identify a short sequence in the putative binding region, previously localized to the cytoplasmic N-terminal 1–788 region by mutagenesis studies.

Ryanodine receptors have a high-degree of similarity to IP$_3$ receptors and they are responsible for Ca^{2+} influx from sarcoplasmic reticulum of skeletal and cardiac muscles. The ryanodine receptor also forms a very large homotetramer,

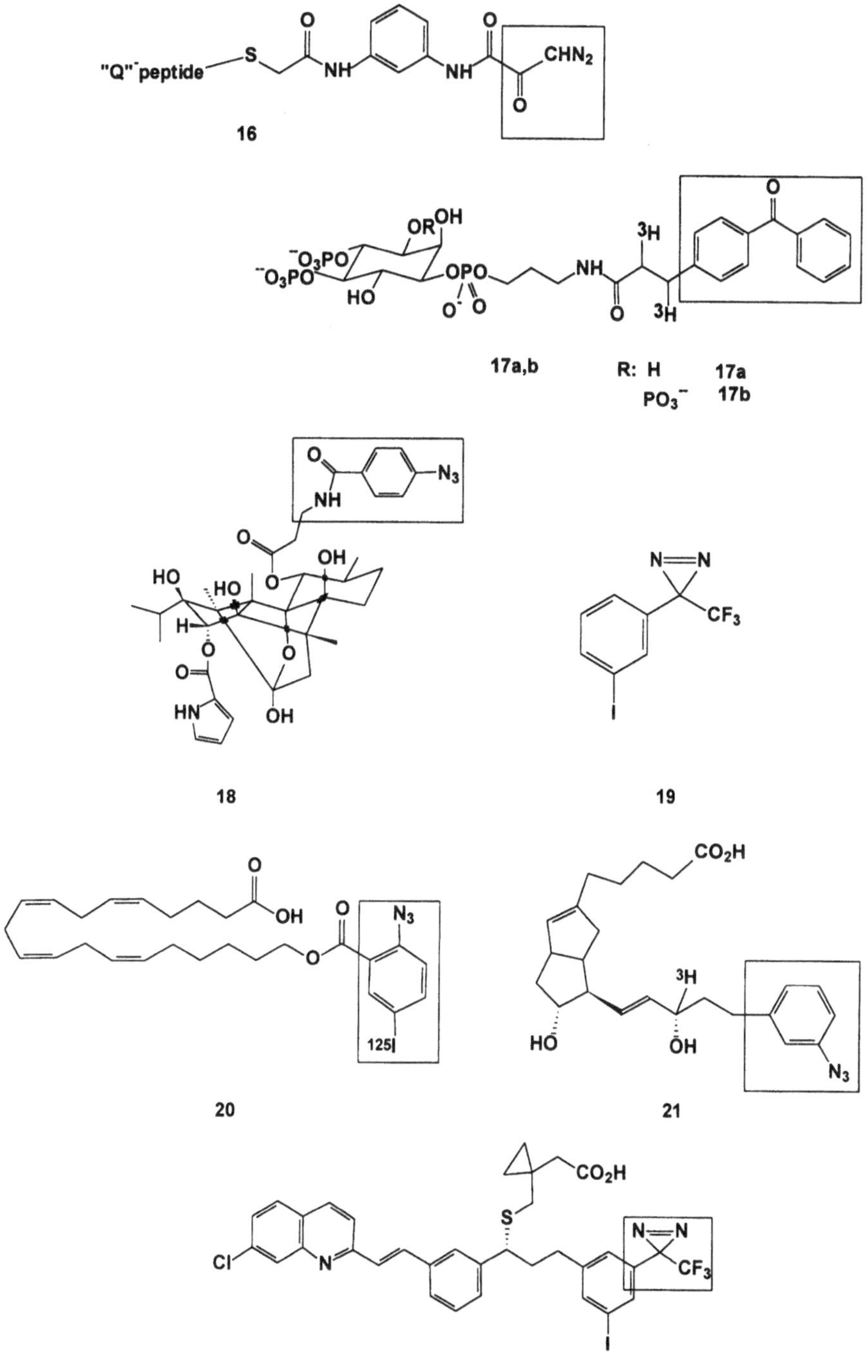

Fig. 10. Photoaffinity probes for studying G-protein interactions; PLC/IP$_3$/Ca^{2+} signaling pathways, and the "arachidonic acid cascade"

with a monomer subunit of 565 kDa. A tethered photoaffinity analogue ([³H]-10-O-[3-(4-azidobenzamido)propionyl]ryanodine, **18**, Fig. 10) efficiently labeled the 565-kDa protein that was previously identified. The labeled 76-kDa tryptic fragment was recognized with an antibody raised against the COOH-terminal. It confirmed the exact location of the high-affinity binding site for ryanodine [93]. Another group described the synthesis of a similar probe with a radioiodinated form [94].

4.2.5
PLA Pathways: Eicosanoids and their Targets

In an other signaling pathway, which also utilizes PIP_2, membrane-bound PLA cleaves the ester groups between the acyl moieties and most frequently arachidonic acid is released. That 20-carbon and four double-bond-containing fatty acid results in about 20 major and almost 100 different metabolites with strikingly different biological activities. The series of events are collectively called the 'arachidonic acid cascade' and the bioactive products of the metabolic pathways are local hormones called eicosanoids. The major subclasses of eicosanoids are prostaglandins, leukotrienes, thromboxanes, lipoxines. These *local* hormones are released near the membrane interface, allowing these molecules to migrate to neighboring cells and bind primarily to G-protein coupled receptors. Recently it was also shown that these compounds can also act on intracellular receptors by translocating the 5-LO enzyme from the membrane interface.

The biosynthesis of eicosanoids utilizes several enzymes till the ultimate bioactive ligand is obtained. The literature is very rich in PAL studies on these enzymes and the local hormone receptors; therefore only some major results on the key biotransformations and receptors are represented here.

Prostaglandin Endoperoxide Synthase-1 (or cyclooxygenase, COX-1) converts arachidonic acid to prostaglandin endoperoxides. The enzymatically unstable intermediate is the precursor of prostaglandins, including the inflammatory factor PGEs, and prostacyclin/thromboxane components, which regulate blood platelet aggregation. In addition, PGES-1 (COX-1) enzyme is the common molecular target for the non-steroid anti-inflammatory drugs (NSAIDs) like aspirin, ibuprofen, etc. Structurally, COX-1 is an integral membrane protein; however, it lacks transmembrane domains. The topology of the membrane associated region, which is an amphipathic helix, was postulated to contain the catalytic site. That region was identified by an apolar diazirin probe (3-trifluoro-3-[*m*-iodophenyl]-diazirine, TID, **19**, Fig. 10). The apolar membrane photoprobe caused broad labeling on the protein region embedded into the membrane and it was eliminated by standard NSAIDs as anticipated [95].

The other major arachidonic acid (AA) converting enzyme is an integral binding protein, *5-lipoxygenase*, which is responsible for the initial transformation in a cascade of events towards the biosynthesis of leukotrienes. Leukotrienes are major mediators of numerous biological processes, including chemotaxis, and are implicated in hypersensitivity disorders like asthma. It was discovered in the early 1990s that another protein is necessary for the cellular synthesis of leukotrienes, called 5-lipoxygenase activating protein (FLAP). A radiolabeled arachidonic acid analogue [^{125}I]-L739,059 (20, Fig. 10), was synthesized by Wang et al. [96] and used in PAL studies of human recombinant FLAP. The AA analogue efficiently labeled the 17-kDa FLAP, and the labeling was eliminated by preincubation with AA and a known leukotriene biosynthesis inhibitor, MK-886. These results suggest that FLAP may activate 5-LO by specifically binding AA and transferring that substrate to the enzyme catakytic site [97].

Prostaglandin receptors, including prostacyclin (PGI$_2$), are all G-protein coupled receptors, which activate cAMP cyclase, resulting in cAMP. PGI$_2$ receptors were identified in mouse mastocytoma P-815 cells (43 kDa) by a photolabile analogue, [15-^{3}H]-19-(3-azidophenyl)-20-norisocarbacyclin (21, Fig. 10). The photoprobe also labeled a 45-kDa band in membranes of procine platelets [98].

Leukotriene B$_4$ has been implicated in inflammatory processes and chemotaxis. The G-protein-coupled receptor in human myeloid cells was postulated to possess two distinct binding sites. Photoaffinity studies using unmodified tritiated LTB$_4$ as photoreactive species labeled two different proteins in the presence (53 kDa) and in the absence of a non-cleavable GTP analogue (56 kDa). This means that stabilization of the G-protein in the GTP-bound state resulted in an interconversion of the first high affinity binding site to an alternative low-affinity binding site [99].

Cysteinyl leukotrienes (LTC$_4$, LTD$_4$, LTE$_4$) are derived from the biotransformation of an unstable epoxide precursor (LTA$_4$), which is produced from arachidonic acid in a 5-LO catalyzed process in the presence of FLAP. LTD$_4$ receptors also belong to the family of G-protein coupled receptors. First, an azido-LTD$_4$ derivative was used in PAL experiments [100], and then a diazirine-linked MK-0476 antagonist analogue photoprobe (22, Fig. 10) was applied in further studies. In both cases a 43-kDa protein was labeled; however, the diazirine group was better suited for the G-protein coupled receptors, since these proteins are sensitive to wavelengths lower than 360 nm [101].

4.2.6
Studying Unusual G-Proteins

Microtubules have a key role in mitosis and cell-proliferation. They are dynamic assemblies of heterodimers of α and β tubullin. In the cell-reproduction cascade tubulin polymerizes fast and subsequently depolymerizes. Tubulin dimers are unusual guanyl nucleotide binding (G) proteins, which bind GTP reversibly at a site in the β-tubulin. GTP irreversibly hydrolyzes to GDP during polymerization.

Rajagopalan reported a bidentate GTP photoaffinity probe to investigate GTP binding site on tubulin, combining two mechanistically different photoreactive groups, aryl azides and benzophenones (23, Fig. 11). Sequential photochemical activation of the noncovalent complexes gave cross-linking patterns different from any of the monodentate probes. If the GTP probe-tubulin complex was briefly irradiated at 254 nm to activate aryl azides, the covalent attachment was detected in the β-subunit. On the other hand, prolonged photolysis at 360 nm led to incorporation of benzophenone into the α-subunit [102].

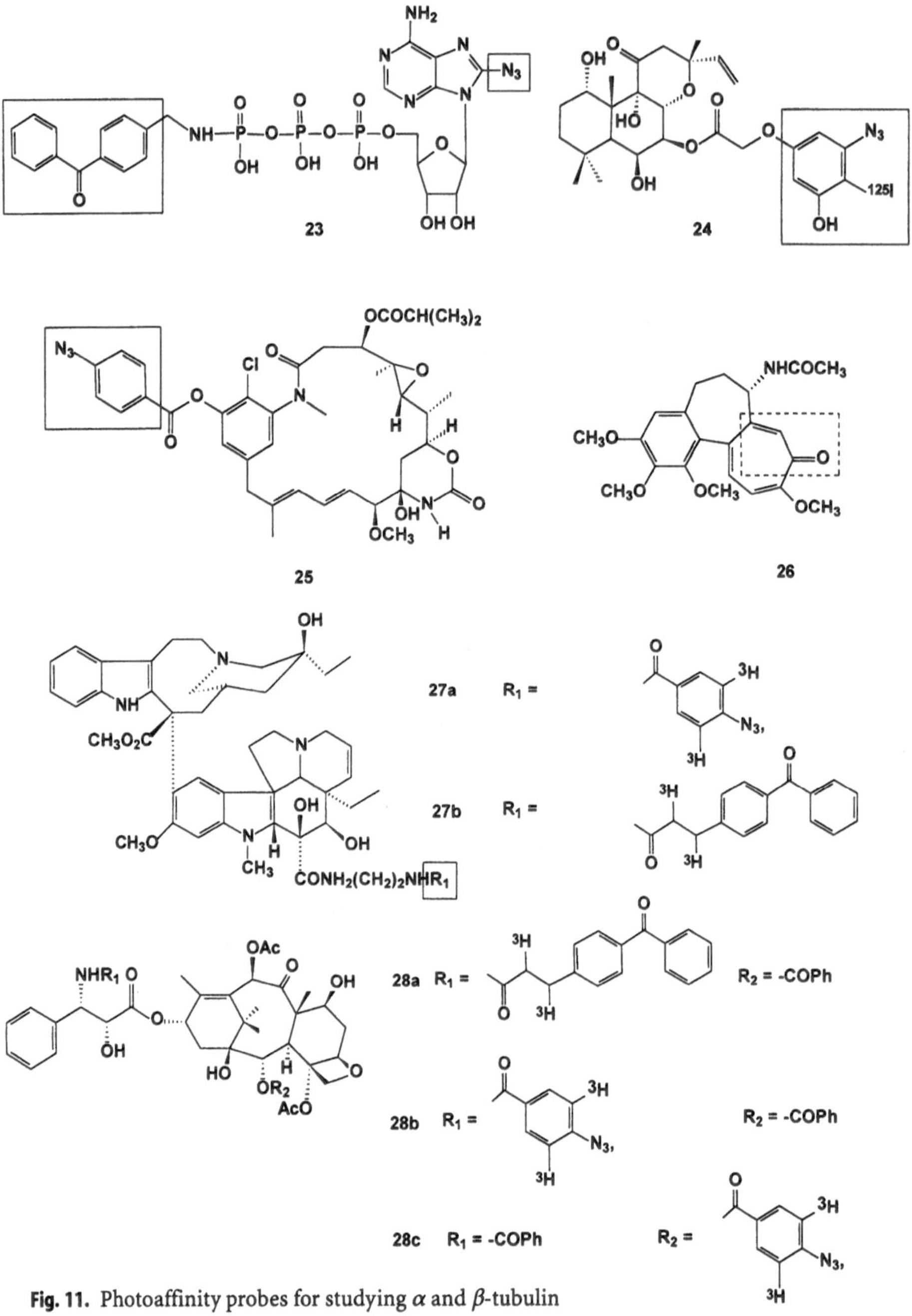

Fig. 11. Photoaffinity probes for studying α and β-tubulin

The dynamic instability of the microtubules was blocked by several anti-mitotic agents, but according to the mechanism of action two major classes can be distinguished. The first type of antimitotic agents suppresses the polymerization of microtubules, while others like Paclitaxel inhibit depolymerization and freeze the tubulin in the polymerized state. Several groups prepared photoaffinity analogues of paclitaxel. In earlier studies forskolin (24, Fig. 11), maytansine (25, Fig. 11), and colchicine (26, Fig. 11) affinity probes labeled both α and β-tubulins (55 kDa and 52 kDa) with a labeling preference to the β-sub-unit [103–105]. Azidophenyl [106] vinca-alkaloid probes (27a, b, Fig. 11) were also developed and it was found that they largely retained the biological activity as tubulin polymerization inhibitors. A similar benzophenone probe was also prepared (G. Dormán, I. Ujváry, G. D. Prestwich, unpublished results) and exhibited good inhibitory activity. Horwitz first reported the identification of an amino acid sequence from paclitaxel binding domain in β-tubulin. When the azidophenyl group was placed in the isoserine sidechain in paclitaxel (28b, Fig. 11) a 32-amino acid sequence was labeled (1–31) in the N-terminus [107].

A benzophenone analogue in the same position (28a, Fig. 11) efficiently labeled both subunits; however, that probe found more relevant use in identifying paclitaxel binding domains in glycoprotein-P [108]. Another photoprobe containing the azidophenyl group in the 2-position of the baccatine core (28c, Fig. 11) labeled the 217–231-amino acid portion in β-tubulin [109]. Very recently Ojima et al. reported paclitaxel photoprobes where the benzophenone-moiety was attached to the 7- and 10-position. The 7-substituted analogue showed particularly good activity in inhibiting tubulin depolymerization [110].

Finally a French group developed a novel taxoid photoprobe [111], which allowed the first identification of an amino-acid sequence in the α-subunit (281–304) together with expected labeling in the β-subunit (217–229).

4.2.7
Protein Kinases Involved in G-Protein Coupled Pathways

The covalent modification of cellular proteins by phosphorylation of serine/threonine and tyrosine residues provides an efficient molecular switch for altering cellular responses.

Phosphorylation plays an important role in several processes of the cell life-cycle including cell proliferation, division, and apoptosis.

Reversible protein phosphorylation occurs by means of protein kinases. Eukaryotic protein kinases comprise a 250-amino acid, conserved catalytic domain. They are widely exploited in the regulation of physiological processes; according to estimates the mammalian genome contains over 2000 PK genes.

Protein kinases have a general architecture: catalytic domain, a binding domain that orients the substrate to the catalytic site, and phosphate donor binding site which donates the γ-phosphate to the acceptor hydroxyl residues.

There are several subfamilies of PKs, which either require Ca^{2+}, diacyl glycerol or phorbol esters or extracellular agonists for activation. Four major groups of PKs are distinguished. Basic amino acid-directed kinases (PKA, PKB, PKC) phophorylate serine/threonine around such amino acid residues. PKA is activated by cAMP upon stimulation of G-protein-coupled receptors and comprised of heterotetrameric complexes of two catalytic and two regulatory subunits. PKC_μ needs membrane-bound phospholipids for activation.

CaMK kinases are Ca^{2+}/calmodulin regulated kinases, while CMGC Pks collect cyclic-dependent, mitogene-activated peptide (MAP) kinases, GSK3 and Clk kinases, that phosphorylate substrates in proline-rich domains. The last group are PTKs (protein tyrosine kinases), which include both receptor and non-receptor kinases, which phosphorylates tyrosine residues (this will be discussed in Sect. 4.3).

Protein kinases, in cooperation with other proteins, form multiprotein complexes which are susceptible to activation upon external agonist stimuli. According to different functions in cell-cycle regulation, the conformational changes are initiated by autophosphorylation and dimerization transmitted by the previously discussed second messengers: cAMP, cGMP, IP_3, PIP_3, AA and DAG.

Intracellular signaling cascades provide complex networks and interconnect protein systems for cell function regulation relies on phosphorylation/ dephosphorylation of specific proteins. Since large numbers of ligands, receptor agonists, and messengers are involved in signaling processes using protein kinase-mediated pathways as transmitters, several very different probes were investigated.

Vitamin E (tocopherol), which is a major lipid-soluble antioxidant, inhibits the differentiation of smooth muscle presumably by modulating the activity of protein kinase C. PKC inhibition by tocopherol has also been implied in human platelets.

A different set of photoaffinity probes (**29**, Fig. 12) was designed and synthesized by Atkinson et al. The photosensitive 4-azido-2,3,5,6-tetrafluorobenzyloxy group was placed at the terminus of the phytyl side chain [112].

It has been found that the catalytic activity of PKC is enhanced by a lipid component of the cell membrane, namely phosphatidylserine. This activity is further stimulated by *sn*-1,2-diacylglycerol. Oleic acid also activates the enzyme in the presence of 1,2-diacylglycerol, and thus it is presumed to mimic phosphatidylserine. In order to identify that modulating binding site for oleic acid on PKC, a photoaffinity analogue was devised. A carbene generating photophore, diazirine was placed in the apolar terminus of the unsaturated fatty acid ligand (**30**, Fig. 12). The synthesis and the photochemical activation properties were reported by Rühmann and Wentrup [113].

Phorbol esters are known as powerful tumor promoters and activators of PKC. The enzyme also needs physiologically 1,2-diacylglycerol for activation. Shibasaki et al. reported the preparation of a photolabile phorbol derivative that contained the photophore at the C-13 position (**31**, Fig. 12) and a fluorescent label. The acyl groups of phorbol esters are considered to be critical in their binding activity and mimicking DAG. In fact, the specific photolabeling localized the enzyme (82 kDa) and showed that the C-13 ester is in close proximity to the target protein in the bound state [114]. Wender et al. synthesized another phorbol ester probe, placing the photophore at the C-3 position. The diazo ester probe (**32**, Fig. 12) bound to PKC with high affinity [115]. Other tumor promoters

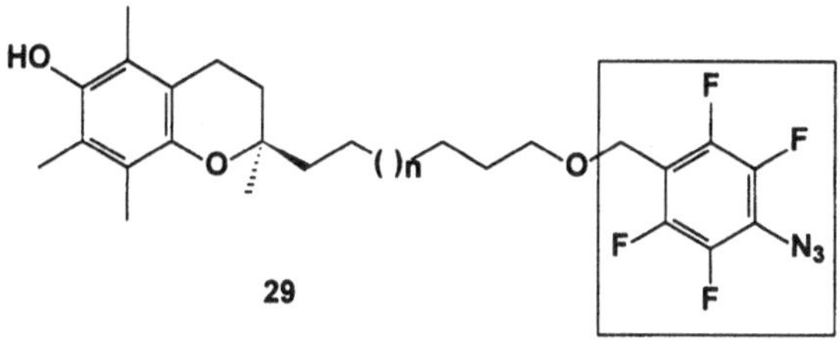

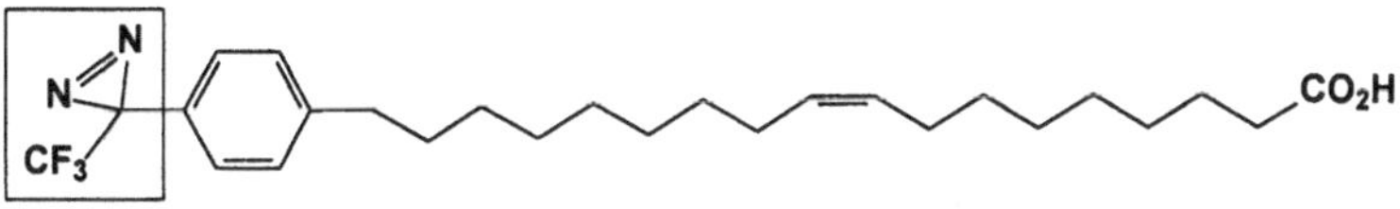

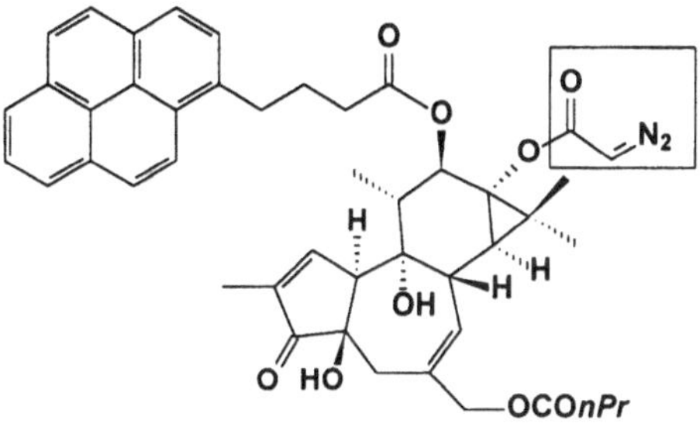

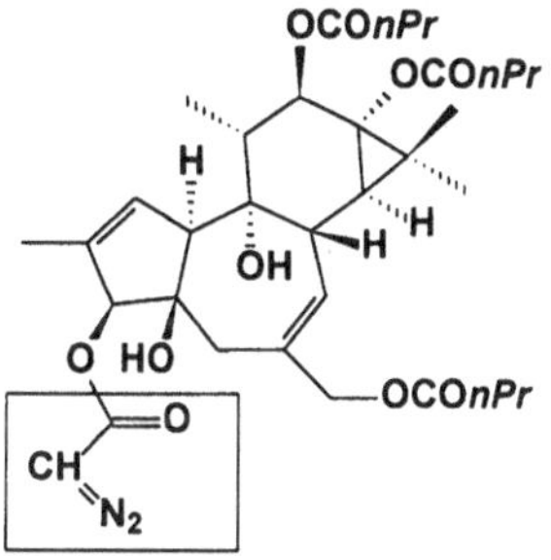

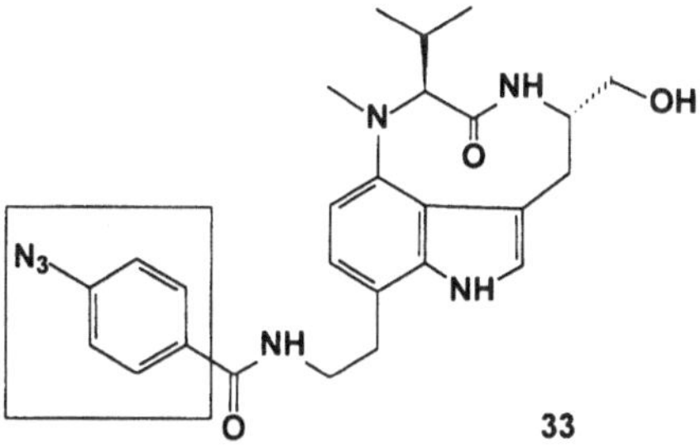

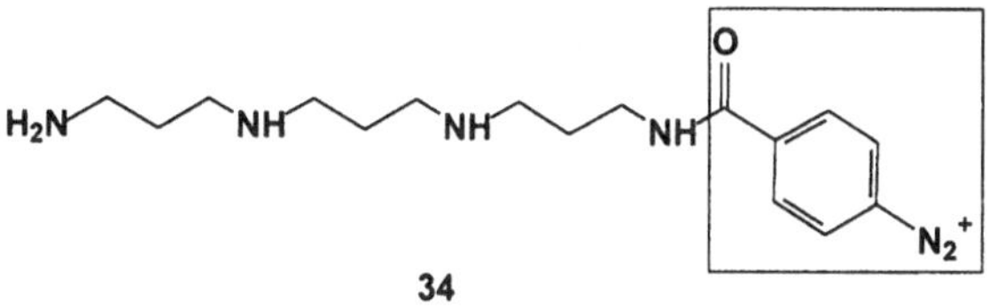

Fig. 12. Photoaffinity probes for studying protein kinases

are also believed to act via activating the PKC pathway. A series of arylazide teleocidin photoprobes (**33**, Fig. 12) were designed and synthesized by Irie et al. The biological activity of the photoligands was in the same range as the unmodified parent compound [116].

Protein kinase casein kinase (CK2) is found in cytoplasm and the nucleus of eucaryotic cells and is elevated in many types of cancer. Unlike most serine/ threonine kinases they are not activated by specific stimuli, but are essentially constitutively active. They autophosphorylate, which causes inactivation, and various protein phosphatases are able to recover their activity. It forms a heterotetramer, with an α-subunit (35–44 kDa) and β-subunit (24–29 kDa). The α-subunit displays the catalytic activity while the β-subunit contains a binding site for the substrate which is self-phosphorylated by the α-subunit. That is required for optimal CK2 activity of the α-subunit. A number of protein substrates are phosphorylated by CK2, including several nuclear oncoproteins including p53. CK2 and p53 associate in a tight complex that involves the β-subunit of the enzyme. The regulation of this enzyme in these processes is still poorly understood. Interestingly, CK2 is increasingly activated by polycationic systems, and spermine was found to be the most active. A photoaffinity analogue of spermine (**34**, Fig. 12) efficiently labeled the β-subunit as a primary binding site, but a minor labeling was detected in the α-subunit (catalytic site). The close proximity of the subunits has been proved and it was also confirmed that the binding site in the β-subunit functioned as orientating the substrate towards the catalytic site [117]. The binding site on the β-subunit was mapped and the location of the covalent attachment was identified as Thr-72, while in the α-subunit the probe labeled Leu-220.

4.2.8
Transcription Factors

As was mentioned earlier, DAG activates protein kinase C, which phosphorylates transcription factors like NFκB: nuclear transcription factor. NFκB forms a multisubunit complex with an inhibitory subunit which is phosphorylated by PKC. The complex disintegrates and what is released translocates to the nucleus and initiates gene transcription. NFκB is a heterodimer, with two distinct DNA-binding subunits: 50 kDa and 65 kDa, both being members of the Rel transcription factor family. These proteins have an important role in the signaling cascade of the cellular defense system, and activate numerous genes in response to pathogens or inflammatory cytokines.

Rel proteins contain a 280-amino acid homology region, which mediates both DNA binding and dimerization. DNA contact regions are located in the aminoterminal half and the residues responsible for the dimerization reside in the carboxy terminal half.

A photoreactive nucleotide, 8-bromo-2'-deoxy-adenosine, was incorporated post-synthetically into a DNA sequence (**35**, Fig. 13) within the previously determined DNA contact site for the transcription factor. Upon irradiation the resulting nitrene cross-linked primarily to the 50-kDa subunit and covalently modified Lys-244. On the other hand, that amino acid residue appeared not

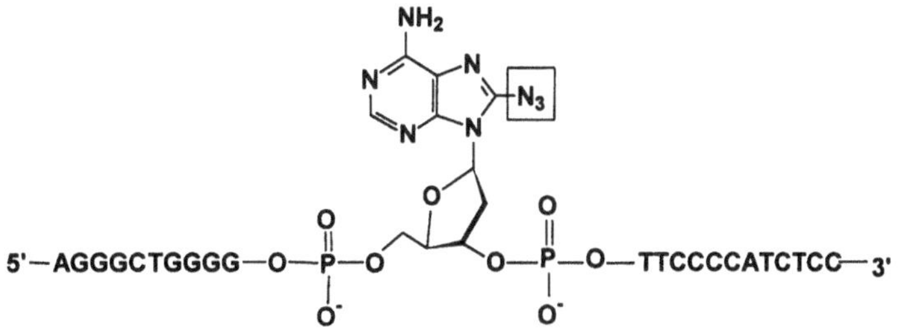

35

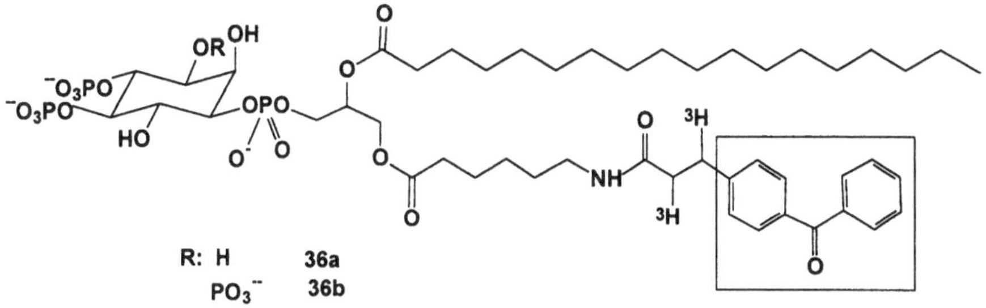

R: H   **36a**
PO$_3^-$ **36b**

Ac-Asp-Ala-Asp-Glu-Tyr992(PO$_3$H$_2$)-*Bpa*-Ile-Pro-Gln-Gln-Gly   **36a**

Ac-Asp-Ala-Asp-*Bpa*-Tyr992(PO$_3$H$_2$)-*Bpa*-Leu-Pro-Gln-Gln-Gly   **36b**

Ac-Phe-Leu-Pro-Val-Pro-Glu-Tyr1068(PO$_3$H$_2$)-*Bpa*-Asn-Gln-Ser-Val   **36c**

Bpa-Asn-Gly-Asp-<u>Tyr(PO$_3$H$_2$)</u>-Met-Pro-Met-Ser-Pro-Lys-Ser

Gly-*Bpa*-Gly-Asp-<u>Tyr(PO$_3$H$_2$)</u>-Met-Pro-Met-Ser-Pro-Lys-Ser

Gly-Asn-*Bpa*-Asp-<u>Tyr(PO$_3$H$_2$)</u>-Met-Pro-Met-Ser-Pro-Lys-Ser

Gly-Asn-Gly-*Bpa*-<u>Tyr(PO$_3$H$_2$)</u>-Met-Pro-Met-Ser-Pro-Lys-Ser

Gly-Asn-Gly-Asp-<u>Tyr(PO$_3$H$_2$)</u>-*Bpa*-Pro-Met-Ser-Pro-Lys-Ser

Gly-Asn-Gly-Asp-<u>Tyr(PO$_3$H$_2$)</u>-Met-*Bpa*-Met-Ser-Pro-Lys-Ser

Gly-Asn-Gly-Asp-<u>Tyr(PO$_3$H$_2$)</u>-Met-Pro-*Bpa*-Ser-Pro-Lys-Ser

Gly-Asn-Gly-Asp-<u>Tyr(PO$_3$H$_2$)</u>-Met-Pro-Met-*Bpa*-Pro-Lys-Ser

Gly-Asn-Gly-Asp-<u>Tyr(PO$_3$H$_2$)</u>-Met-Pro-Met-Ser-*Bpa*-Lys-Ser

Gly-Asn-Gly-Asp-<u>Tyr(PO$_3$H$_2$)</u>-Met-Pro-Met-Ser-Pro-*Bpa*-Ser

Gly-Asn-Gly-Asp-<u>Tyr(PO$_3$H$_2$)</u>-Met-Pro-Met-Ser-Pro-Lys-*Bpa*

37a-37k

Fig. 13. Photoaffinity probes for studying transcription factors, inositol-polyphosphate binding proteins, and biopolymers containing an SH2 domain

to contact with a DNA base, but rather a phosphate moiety that lays in close proximity [118].

4.2.9
Inositol-Polyphosphate Binding Proteins

Profillin is a 14-kDa protein which binds actin monomers, proline rich sequences, and PIP_2. The exact consequence of these bindings at cellular level is still mainly unresolved. It has been suggested that profillin possess a regulatory role in the signaling pathways: it links transmembrane signal transduction to the reorganization of the actin cytoskeleton required for cell motility, and it also participates in the PIP_2 cleavage pathway. Profillin binds to the inner side of the cell membrane in contact with PIP_2 residues. Receptor activation, which makes PIP_2 fall apart and leave the membrane, mediates the translocation of profillin to the cytosolic fraction. On the other hand, profillin is able to inhibit PLC-γ1. Benzophenone-based photoaffinity analogues of PIP_2 (35 a, b, Fig. 13) were used to localize the binding site on profillin. Upon irradiation, quite interestingly, Ala-4 was labeled in the N-terminus helix. Based on that result the bis-phosphate binding domain was modeled at Arg-135 and Arg-136 close to the carboxy-terminal helix. In fact PIP_2 seemed to stretch out between the terminal region. remaining in a fairly open position in the periphery of the protein [119].

4.3
Photoaffinity Labeling of Receptor Protein Tyrosine Kinases (PTK)
and Biomolecular Targets Involved in the Signaling Pathways

PTKs are monomeric in the active form, and consist of an extracellular binding domain, single membrane-associated α-helix and cytosolic domain with the kinase activity. Ligand binding induces dimerization and autophosphorylation that enables binding to adaptor molecules. These molecules connect the receptor to the signal transmitting biopolymers.

Growth factor activation of PTKs induces binding the receptor to Grb2 (growth factor receptor binding protein 2), containing SH2 (sarcoma homology) and 2 SH3 domains. In turn it localizes SOS protein to the plasma membrane. SOS activates Ras by exchange of GTP to GDP. Ras-GTP binds directly to serine/threonine kinases and activates the MAP kinase signaling machinery leading to increased gene expression. Normal Ras proteins play an important role in growth factor initiated tyrosine kinase receptor signaling and are considered as central switches used by cells in cell-proliferation, cell-differentiation and other genetic programming. Ras gene and the encoded Ras G-protein are closely associated with carcinogenesis. Mutations of the *ras* gene contribute to 20 – 30 % of all human cancers. In cells in which the *ras* gene is mutated, the G-proteins can be perpetually "turned on," continuously giving the signal for cell division even in the absence of growth factors.

4.3.1
Studying Biopolymers Containing an SH2 Domain

Sarcoma homology 2 domains (SH2 domains) are discrete phosphotyrosine binding modules (composed of about 100 amino acids) found in a wide variety of cytoplasmic signaling proteins, mediating the physical association. Two classes of SH2 domains exist with a major difference in the neighboring amino acids next to the phosphotyrosine-binding residue. SH2 domains receive signals from the autophosphorylated tyrosine kinase receptors activated by agonist binding. SH2 domains transmit the signals leading to activation of key enzymes in the signaling cascade.

Another representative of PTK signaling is the *insulin receptor* that is a membrane-spanning glycoprotein linked tyrosine kinase and present on the surface of insulin-responsive tissues. The receptor autophosphorylates and also phosphorylates insulin receptor substrate-1 (IRS-1), which serves as a docking site for proteins with SH2 domains. In order to localize the interaction between insulin and its receptor, Shoelson envisaged an insulin agonist photoprobe that would cross-link to the receptor with high efficiency and activate permanently. Based on previous studies, which suggested that Phe-25 (in B-chain) is crucial in receptor recognition and *para*-substituents of this amino acid were well tolerated, he designed a photoprobe with modified C-terminal octapeptide of the B chain. This new sequence contained a radio- and biotin-label and a photoreactive amino acid, Bpa. This photoaffinity ligand labeled a single 135-kDa peptide with an average 80% cross-linking efficiency. The covalent receptor-ligand complex resulted in tyrosine kinase activation in vitro and in intact cells insulin receptor phosphorylation and internalization were both activated [120].

Miller studied the interface between *tyrosine-kinase type of EGFR (epidermal growth factor receptor)* autophosphorylation sites (Tyr-992, 1068, 1086, 1148, 1173) and phospholipase Cγ1 N-terminal SH2 domain. PLCγ1 has two SH2 domains at the N-terminal and at the C-terminal. Upon interaction with phosphopeptide sites of the SH2 domains the tyrosine kinase acts on the enzyme phosphorylating it, which finally turns on the enzyme activity and results in the release of two second messengers, IP$_3$ and diacylglycerol. The initial peptide-peptide interaction was examined by PAL, using the benzophenone-based photoactivatable amino acid, Bpa. The photoreactive amino acid was incorporated into different phosphotyrosine-containing 12-mer peptides (**36a – c**, Fig. 13) derived from the major autophosphorylation sites of EGFR (Tyr-992 and Tyr-1068). The photoaffinity experiment resulted in efficient cross-linking to the SH2 domain with the small peptides: Tyr-992-derived peptide was covalently attached to Arg-562 and the Tyr-1068-derived peptide cross-linked to Leu-653. Based on these results a three-dimensional model was constructed predicting the entire contact site between the SH2 domain and the model phosphopeptides [121].

A similar study was conducted by Shoelson. Phosphatidyl inositol 3-kinase is activated by contact with autophosphorylated tyrosine kinase receptors followed by agonist binding to the cell-surface receptor. The truncated regulatory subunit of the enzyme (p85), which contains an SH2 domain, retained its full binding activity. Williams and Shoelson studied the phosphotyrosine-binding region of the enzyme by photoaffinity scanning [122]. A series of IRS-1-related (insuline receptor subtrate-1) 12-mer phosphopeptides was prepared in such a way that each amino acid in the sequence was successively replaced by the photoactivatable Bpa (**37a-l**, Fig. 13). When the replaced amino acid was adjacent to the phosphotyrosine residue, efficient cross-linking was observed at Gln-83 of the SH2 domain within the α-helix-I, while Bpa+4 substitution led to specific cross-linking at Asn-85 within the flexible loop C-terminal to α-helix-II.

The *phosphatidyl inositol 3-kinase* hold a heterodimer regulatory domain (85 kDa) including SH2 sequence and a catalytic domain (110 kDa). Its function has been widely investigated in the signaling processes and, particularly, the resulting second messenger PIP$_3$ was the subject of several studies. The major putative targets were reported as serine/threonine kinase Akt/PKB, protein kinase C, etc. In order to promote the identification of those targets a photo-activatable analogue was prepared by Prestwich et al. and are currently under investigation with several targets [123]. In rat cerebellar membrane using an inositol 1,3,4,5-tetrakisphosphate photoprobe, [^{3}H](benzoyldihydrocinnamyl)-aminopropyl-inositol 1,3,4,5-tetrakisphosphate ([3H]BZDC-IP$_4$, **17b**, Fig. 10) a 42-kDa protein was identified and named centaurin. In the region where the photoprobe cross-linked, a high-affinity PIP$_3$ binding site was identified by photoaffinity competition assays [124].

4.3.2
Studying Protein Prenylation

In normal cells, the GDP/GTP-binding proteins, after protein synthesis, move to the cell membrane to which they become "hooked" by a hydrophobic farnesyl group. The γ-subunit is anchored in the membrane by a post-translational modification of the C-terminal CAAX sequence (C – cystein, AA – aliphatic amino acids, X – methionine). This protein is first enzymatically farnesylated by a specific farnesyltransferase, then the AAX part is cleaved by specific proteases and finally the cystein residue is converted to a methyl ester.

Inhibition of the lipid modification cascade provides an alternative way to block aberrant signaling pathways and that opportunity can be exploited in anticancer therapy. As part of the growth factor, signaling of the false activation is transmitted by the mutated *ras* gene encoded proteins (Ras) and ultimately leads to uncontrolled cell growth. These typical GTP binding proteins are also subject to membrane anchoring and the biosynthesis of those Ras proteins can be blocked at the posttranslational prenylation step.

The molecular elements of that pathway were mapped with photoaffinity labeling by different investigators. Farnesyltransferase contains α and β hetero-dimer subunits, and binds to both protein and farnesyl diphosphate. The main recognition elements for the protein is the C-terminal CAAX motif. Coleman et al. attached two benzophenones to the recognition sequence and the resulting photoprobe (**38**, Fig. 14) specifically labeled both subunits [125].

In recent years several FPTase inhibitors were developed as peptidomimetics of the key tetrapeptide fragment for use as potential anticancer agents. Miller at

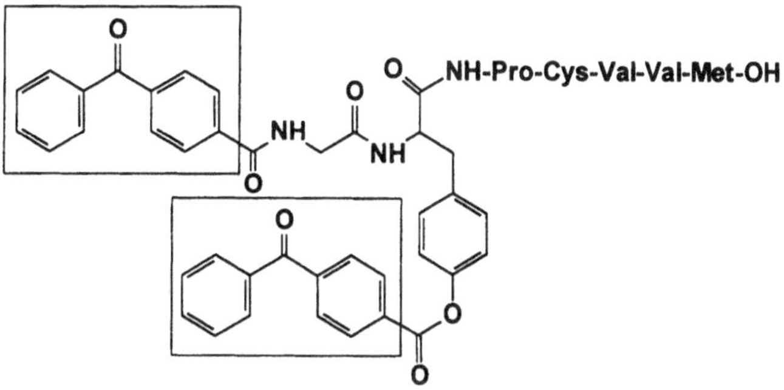

38

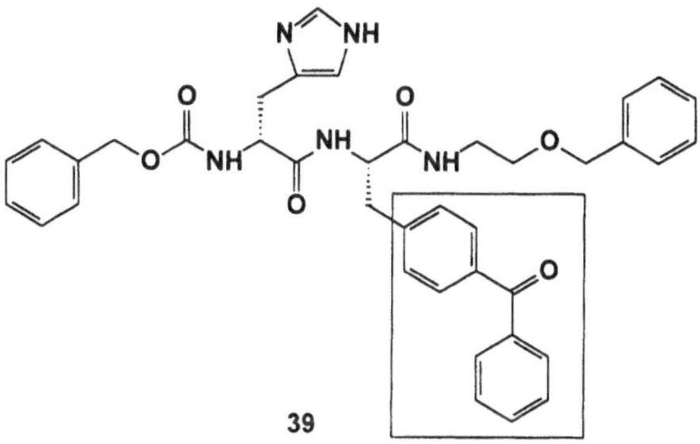

39

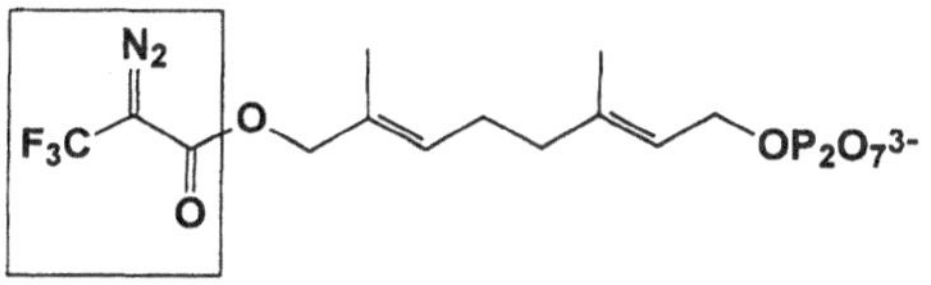

40

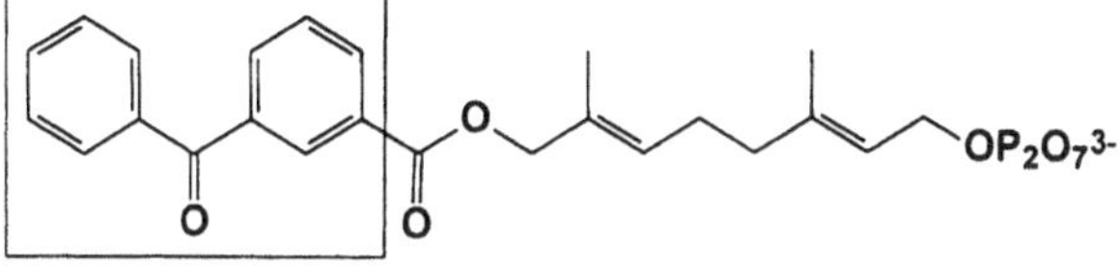

41

Fig. 14. Photoaffinity probes studying biomolecular targets in protein prenylation

Stony Brook synthesized a photoreactive benzophenone containing inhibitor (39, Fig. 14) which efficiently labeled the active site of the enzyme. The photo-inhibition was prevented by adding native Ras to the reaction mixture. That competition indicated that the labeling was specific at the active site. Peptide mapping of the labeled enzyme by HPLC, Edman sequencing and MALDI-MS allowed the identification of key amino acids in the substrate binding, as Asp-110 and Asp-112 in the α-subunit [126].

Other research groups attempted to identify the farnesyl diphosphate binding site of the enzyme. Allen reported a trifluoromethyl-substituted diazoester probe (40, Fig. 14), where the photoreactive group was introduced at the apolar end of farnesyl diphosphate. The [^{32}P] labeled probe upon irradiation specifically cross-linked to the β-subunit (45 kDa) both in the case of farnesyl transferase and geranylgeranyl transferase [127]. DiStefano et al. synthesized and evaluated other photoreactive analogues of farnesyl or geranylgeranyl pyrophosphate containing apolar benzophenone (41, Fig. 14), which replaced one geranyl unit. The photoprobes, which partially mimicked the geranyl-geranyl isoprenoid unit, preferably modified the β-subunit of the transferase enzyme [128].

4.3.3
Discovery of New Protein Targets Involved in Tyrosine Kinase Signaling

Benzoquinoid ansamycin antibiotics are extensively studied as antitumor agents and as inhibitors of tyrosine kinases. Much evidence was collected showing that the enzyme activity was rapidly depleted after treatment with ansamycin antibiotics. Since the exact molecular mechanism of the tyrosine kinase inhibition was unclear, an aryl azide photoaffinity analogue (42, Fig. 15) was prepared. The erbB2 encoded p185 tyrosine kinase was effectively depleted by the ansamycin analogue hebimycin. The photoaffinity labeling result was unexpected, the specific radioactivity incorporation was detected in a 100 kDa protein rather than in the p185. The same protein was labeled in two fibroblast lines transfected with the oncogenes erbB2 and v-src. The detailed study confirmed that binding of p100 presumably indirectly activates the degradation of p185 and it was postulated that the newly discovered protein was an important regulator for tyrosine kinases [129].

The mitogen-activated protein kinases (MAPK) are proline-directed Ser/Thr kinases that activate their substrates by dual-phosphorylation. These kinases are activated in growth factor or cytokine stimulated, as well as stress reaction or UV. light induced signaling pathways. Activated MAPKs after nuclear translocation induce phosphorylation of nuclear transcription factors. MAPKs are involved in many cellular processes such as cell-proliferation, oncogenesis, and cell cycling. They are also implicated in a series of protein-protein interactions, which altogether regulate distinct cellular out-puts. One of the diverse cascades was discovered using PAL techniques by researchers at SKB. Proinflammatory cytokines like interleukin-1 or tumor necrosis factors were successfully inhibited with a new series of pyridinyl-imidazoles. Photoaffinity experiments revealed that the target (cytokine suppressive drug binding protein – CSBP) for the in-

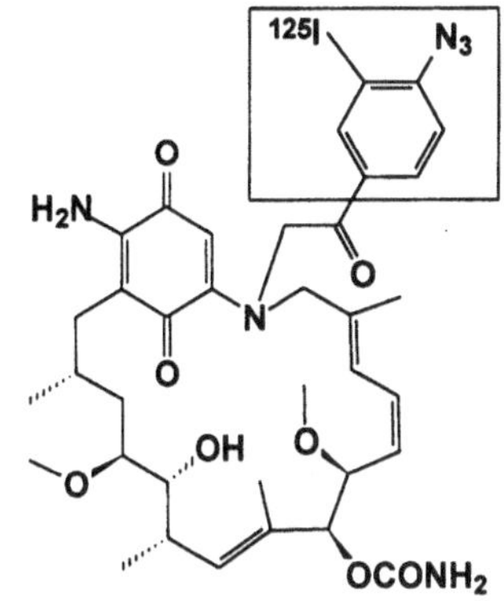

42

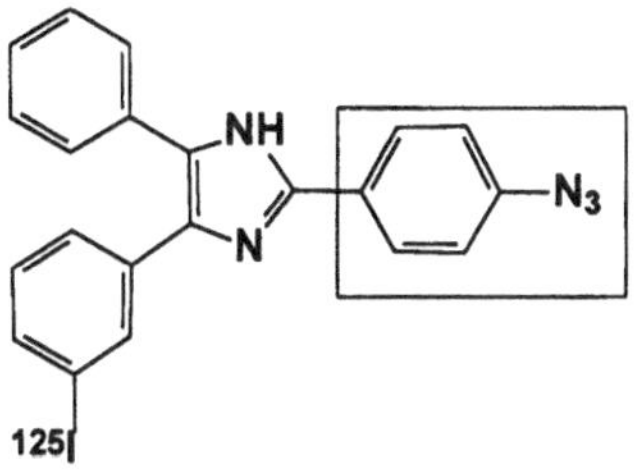

43

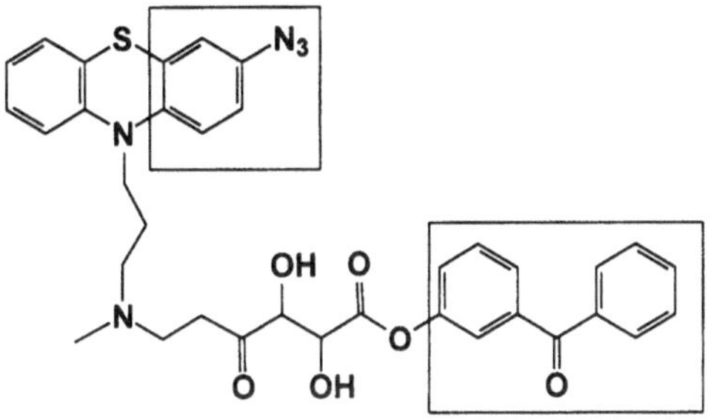

44

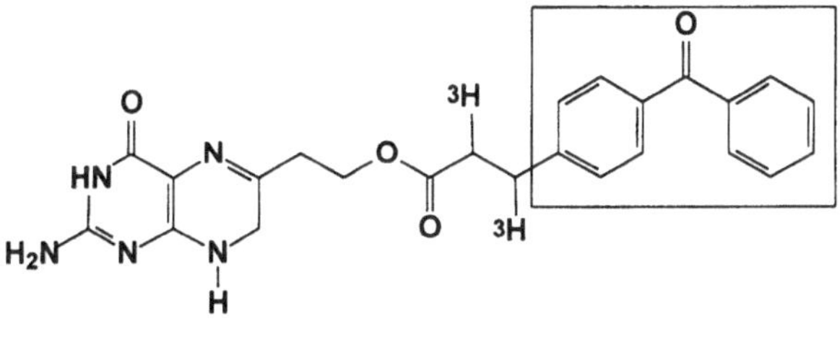

45

Fig. 15. Miscellaneous photoaffinity probes for studying interactions in tyrosine kinase, calmodulin and nitric oxide related signaling

hibitors was actually a protein kinase. Several initial studies pointed in the same direction, the inhibition was at the translational level and particularly at the initiation stage. Finally, by means of a PAL the involvement of a protein kinase was unambiguously identified. The azidophenyl appended pyridinyl-imidazoles photoprobe (**43**, Fig. 15) labeled an approximately 45-kDa protein. After limited digestion with trypsin and cyanogen bromide both the labeled and unlabeled fragments were subjected to sequence homology search. A 15-amino acid tryptic fragment showed 85% homology to a C-terminal portion of a protein belonging to a MAPK family. Another sequence derived from an 8-kDa CNBr cleavage site that contained the majority of the radiolabel. From sequence reading a unique MAP kinase was identified. Based on the sequence a reverse-translated oligonucleotide portion was synthesized and used in screening the cDNA library from human monocyte. Surprisingly, two distinct cDNAs were isolated. The two cDNAs as well as the resulting recombinant CSBPs shared only 43% identity, while their sizes were almost identical (42 kDa), close to the detected size for the natural enzyme. The anti-inflammatory agents inhibited both kinases, confirming the presence of the right cloned polypeptides. It is still unclear why the kinase exists in two nearly identical forms in monocyte cells exhibiting the same cellular function [130].

4.4
Photoaffinity Labeling of Protein Targets in Additional Signaling Pathways

4.4.1
Calmodulin Signaling Pathways

This pathway is discussed separately since the general Ca^{2+}-ion signaling participates in several cascades.

Calmodulin (CaM), the principal intracellular calcium receptor, is implicated in several signaling pathways. Since CaM pathways can be directly linked to nitric oxide generation and immunomodulation, it was rational to discuss them earlier. CaM is an acidic protein composed of 148 amino acids and two hand-like subunits connected by a tether. This structural arrangement provides sufficient flexibility when binding to several different targets. Upon Ca^{2+}-ion binding CaM interacts with the target proteins and induces a conformational change on the targets, which results in their activation. The structural features of CaM activation have been studied for more than a decade with photoactivatable peptides containing Bpa. First DeGrado investigated the Ca^{2+}-dependent conformational changes and binding of CaM to model peptide sequences, which derived from CaM activated proteins [131].

Phenothiazines bind to calmodulin and stimulates the interaction with target proteins. In order to map the conformational changes (subunit shifts) in CaM upon phenothiazine binding, Watt devised a bidentate photoaffinity ligand (44, Fig. 15), which contained a "fast reacting" photophore (aryl azide) and a "slowly reacting" photophore (benzophenone). Using different wavelengths upon irradiation at 260–280 nm the phenothiazine residues was covalently linked to its binding site which resulted in permanent conformational change. The second photolytic activation at 360 nm allowed mapping of the activated conformation. The photoaffinity reagent also contained a cleavable tartrate linkage, which allowed the removal of the primary ligand phenothiazine residue from the covalent complex [38].

4.4.2
Guanylate Cyclase Receptors and Biomolecular Targets Involved in that Signaling

Guanylate cyclase signaling pathway also utilizes an enzyme-linked receptor model. The effector enzyme, guanylate cyclase, converts GTP to cGMP, which in turn activates cGMP dependent protein kinase or phosphodiesterases.

Atrial natriuretic peptide (ANP) is a 28-amino acid endocrine hormone. It has several effects: it increases the Na^+ and water excretion by the kidney and the vascular permeability, relaxes vascular smooth muscle, and inhibits several other hormones. The major actions of ANF are mediated through membrane-bound receptors, stimulating guanylate cyclase catalytic activity that is a fundamental part of the receptor. In order to map the ANF binding site in the receptor a photoaffinity analogue was prepared, in which Arg at the C-terminal was replaced with the photoreactive Bpa. A Gln residue was changed to iodinatable Tyr. This residue together with Bpa and the Bpa-adjacent Phe form a hydrophobic core deep inside the peptide binding domain. The photoprobe selectively labeled a 130-kDa glycoprotein with high efficiency [132].

A special guanylate cyclase receptor can be found in the nitric oxide (NO) signaling pathway. The activation of the sequence of events in that pathway results in smooth muscle relaxation. This pathway is directly linked to other cascades by receiving a Ca^{2+} signal and utilizing calmodulin (CaM) as transmitter protein.

Nitric oxide (NO), also termed endothelial-derived relaxing factor – EDRF – is a gaseous second messenger, which is released in blood vessels, neurones, macrophages, etc., by numerous extracellular stimuli and displays important biological functions. It has an extremely short half-life of 10 s. NO is released in endothelial cells, and causes primarily relaxation in neighboring smooth muscle cells. NO binds to the heme group of the soluble guanylate cyclase receptor within the cell and causes an immediate increase in the level of cGMP. The primary step in the NO signaling pathway is the event when NO-synthase (NOS) converts L-arginine to NO. That process requires Ca^{2+}-bound calmodulin for activating NOS and a cofactor, tetrahidro-L-biopterin (H4Bip). An intimate H4Bip-heme-arginine interaction is a critical component in the catalytic activity, and it was presumed that H4Bip facilitates an electron transfer during the reduction of the oxyferroheme complex. BP-linked photolabile antagonists (45, Fig. 15) were designed to identify the pterin contact region. PAL experiments allowed identification of a 341-amino acid sequence in the oxygenase/dimerization domain of NOS. This finding will prove that H4Bip may act simply by stabilizing the dimer during L-arginine turnover without any implication in the electron transfer process [133].

4.4.3
Signaling Pathways Involved in Immunomodulation

T-cell antigen receptors belong to a separate family, which is composed of eight noncovalently bound trans-membrane subunits. The central ligand-binding element is a disulfide linked α,β-heterodimer linked to the membrane by a short transmembrane sequence. The dimeric, variable subunits are the cellular equivalents of the antigen-specific immunoglobulin G antigen binding fragment (IgG Fab). The remaining "complementary determining" (CD2 and 3) chains consist of γ and δ-type glycoproteins and four other nonglycosylated subunits ($\varepsilon, \varepsilon', \zeta, \zeta'$). When a macrophage interaction delivers an antigen fragment to the receptor by the assistance of the major histocompatibility complex (MHC), the fragment is recognized and bound to these major subunits. Upon binding, the eight individual subunits closely associate, resulting in a tyrosine phosphorylation in the cytoplasmic portions by a coupled kinase. The phosphorylated tyrosines in the ζ-subunit are the recognition element in the signaling pathway for the SH-2 domain of a protein kinase, which stimulates PLC γ activity. The known hydrolysis pathway produces IP_3 and then Ca^{2+} as already described. Luescher et al. studied extensively the T-cell receptor antigen ligand interactions with PAL [134]. Upon ligand binding, the MHC-protein-antigen ligand complex slowly dissociates and leaves the T-cell receptor. In order to control the ligand-receptor interaction, first the ligand was covalently linked to the MHC molecules, then cross-linked to the receptor. These two events were directed by a bidentate photoaffinity ligand using two aryl-azide photophores (46, Fig. 16), allowing a sequential activation at a different wavelength. First, the N-terminal of the antigen fragment, derived from *Plasmodium berghei* circumsporozoite (YIPSAEKI), was linked with iodo-hydroxyl-aryl azides and upon irradiation at >350 nm covalently attached to the K_d-restrictive peptide (K_d-Q10). That irradiation left unreacted the other photophore (unsubstituted aryl azide) attached to the Lys-259 residue in the short peptide. It was excited at < 350 nm and cross-links were created efficiently to the T-cell receptor. The labeling was inhibited by K_d-Q10 specific antibodies which confirmed the binding at the anticipated site. Very recently the same group reported the above photoprobe labeled T-cell receptor residues (Tyr-48 and/or Tyr-50 in the CD2 region [135].)

As explained, the immunomodulation signal starts with binding the antigen on T-cell receptor, which ultimately results in Ca^{2+} release. Ca^{2+} induces calmodulin activation, which in turn activates calcineurin, a Ca^{2+}-dependent phosphatase. This event leads to translocation of the cytoplasmic component of the transcription factor, which is required for IL-2 gene expression and T-cell activation.

Cyclosporin A (CsA) and FK506 are immunosuppressive drugs, which allow the patient to overcome the normal immune-based rejection of foreign organs in organ transplantation. *Immunophylin* is a collective term for peptidyl-proline *cis-trans* isomerase proteins which bind with high affinity to immunosuppressant agents, such as cyclophylines that are CsA-binding proteins, and FKBP, that stands for FK506 binding proteins. CsA and FK506 inhibit T-cell proliferation at an early stage by preventing transcription of primarily interleukin-2 (IL-2). The CsA/cyclophylin and FK506/FKBP complexes bind to calcineurin and inhibit its activity, thus stopping the immune-signal. The *immunophylin-calcineurin* interface was mapped by Husi et al. with PAL [136].

Calcineurin is a heterodimer consisting of a large calmodulin-binding catalytic subunit 'A' (61 kDa) and a Ca^{2+}-dependent subunit 'B' (19 kDa). A diazo-group containing photoreactive CsA (47, Fig. 16) was prepared after binding to cyclophylin and after incubation with calcineurin. Irradiation resulted in specific cross-linking mainly to the subunit 'B', with some minor interaction to subunit "A".

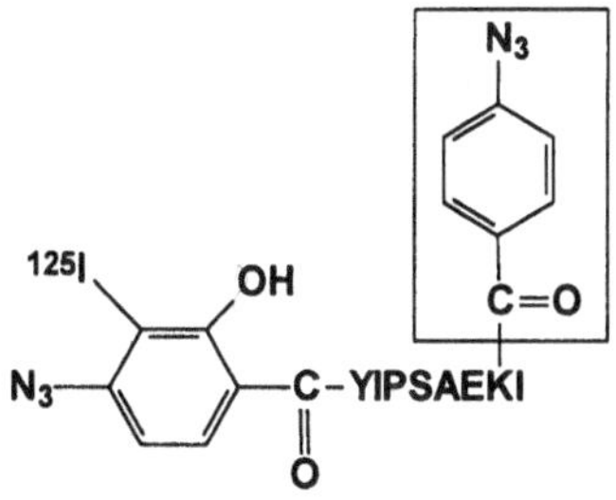

46

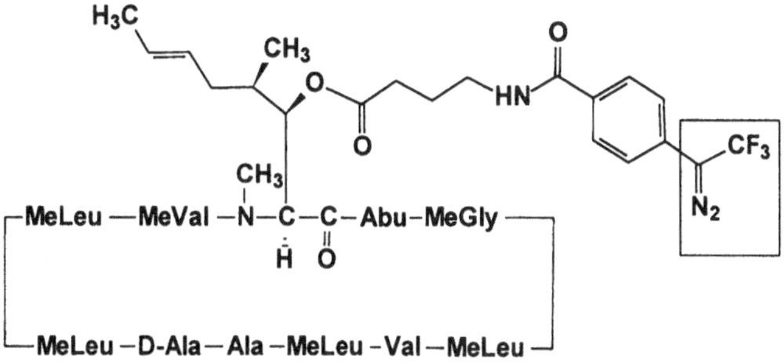

47

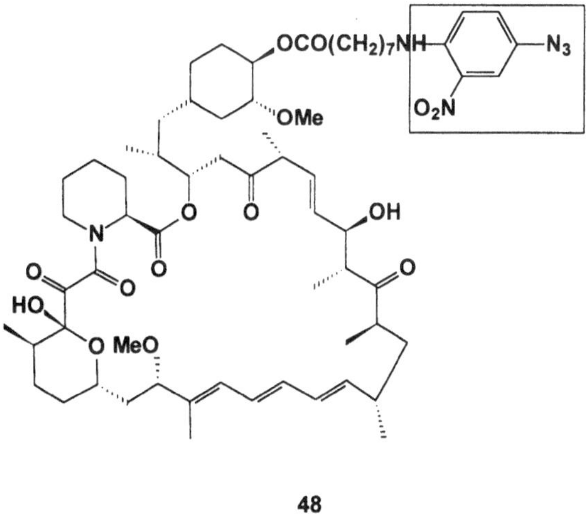

48

Fig. 16. Photoaffinity probes for studying signaling pathways in immunomodulation

Another class of immunosuppressant agents, e.g., rapamycin has a different mechanism. Although rapamycin binds to FKBP12, the complex does not interact with calcineurin. Alternatively, the rapamycin-FKBP complex binds to a specific rapamycin (sirolimus) effector protein (SEP). Since that SEP protein possesses homology to phosphatidyl inositol kinases, it has been suggested that rapamycin acts as an inhibitor of the kinase activity through the complex blocking further incompletely resolved signaling pathways. In order to map the ternary complex rapamycin-FKBP12 and the effector enzyme photoactivatable rapamycin analogues were prepared (C-31 and C-42 esters [**48**, Fig. 16]), which are currently under investigation. The C-42 modified analogues exhibited T-cell antiproliferative activity almost equipotent to the parent compound [137].

4.4.4
Integrin Signaling

Cell-adhesion molecules [138] are multifunctional molecules which are involved in a number of cell-regulatory processes, including growth, differentiation, proliferation, and programmed cell-death. Adhesion proteins share a similar

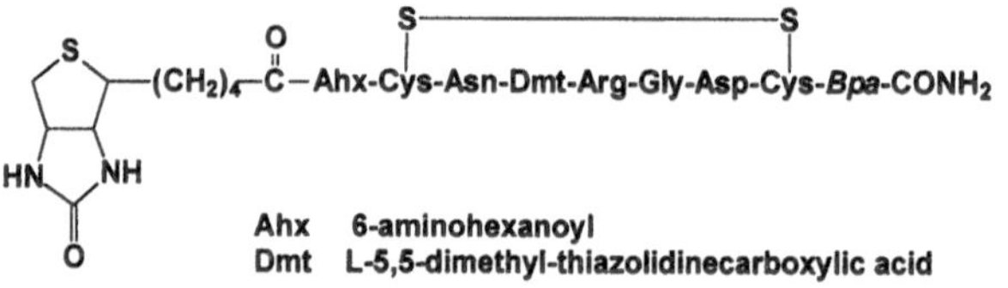

$(CH_2)_4$–C–Ahx-Cys-Asn-Dmt-Arg-Gly-Asp-Cys-*Bpa*-$CONH_2$
Ahx 6-aminohexanoyl
Dmt L-5,5-dimethyl-thiazolidinecarboxylic acid

49

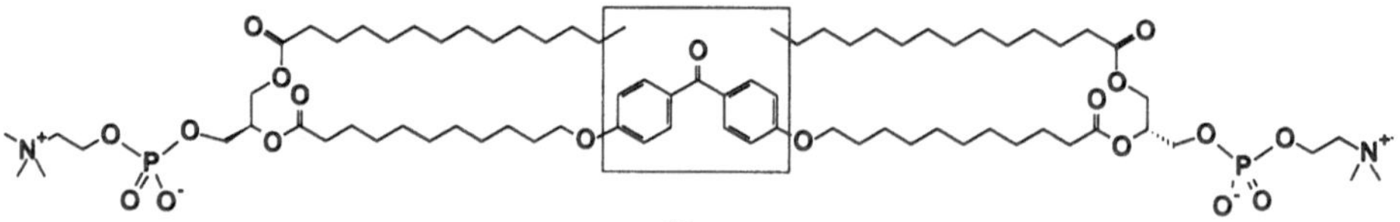

50

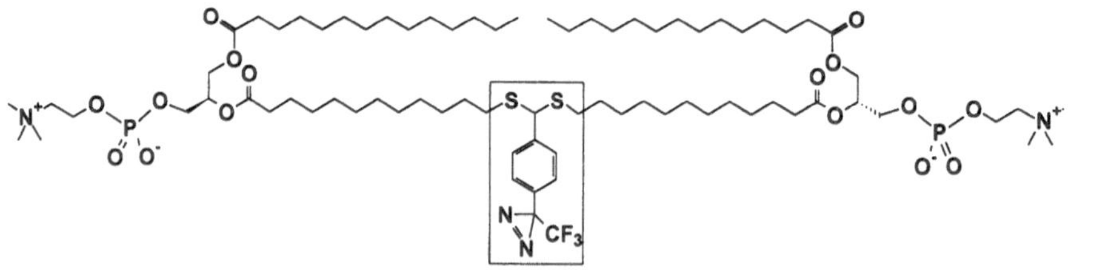

51

Fig. 17. A photoaffinity probe for integrin signaling and photoactivatable membrane probes

structure and contain a conserved region called adhesion domain (AD). Cell-extracellular (ECM) interactions occur by involving cell-surface adhesion receptors, like integrin, which also function as signaling molecules. Focal adhesion biopolymers are rich in tyrosine phosphorylated proteins that are associated with receptors and signaling pathways. Ligand binding to integrin initiates focal adhesion, which transmits the signal by tyrosine phosphorylation to internal signaling molecules. Integrin is composed of two subunits, α and β, both possessing a large N-terminal extracellular region and a short intracellular C-terminal sequence. There are a large number of combinations between 17 different α and 8 β-subunits. According to the current receptor model, both subunits are necessary for the activation and the presence of Ca^{2+} and Mg^{2+} is also essential. The metal binding domains are localized in the N-terminal part of the β-subunit. The most common feature of integrins is that small peptides containing a defined sequence (RGD motif) bind to the receptor with high affinity. One family of these receptors is $\alpha_v\beta_3$ integrin (vironectin receptor), which plays an important role in angiogenesis, apoptosis, and particularly in bone resorption. In order to map the integrin-ligand interface conformationally constraint photoaffinity analogues were designed, where Bpa was placed close to the RGD motif at the C-terminus. Both radiolabeled and biotinylated (49, Fig. 17) probes were synthesized. A 100-kDa polypeptide was specifically labeled, which corresponded to the β_3 subunit. The cross-linking was dependent on the presence of the Ca^{2+} and Mg^{2+} ions. Limited digestion resulted in a 20-amino acid labeled fragment which was identified as a 19-amino acid portion of the β_3 (99–118) subunit, located as the primary contact domain [139].

4.5
Photoaffinity Labeling of Steroid Hormone Receptors and Membrane Probes

As recently discussed, the majority of the cell surface receptor interactions starts a signal within the cell, which induces several intracellular responses and the signal ends at the nucleus initiating transcription/gene expression. Steroids partition in the cell membrane and readily pass through into the cytoplasm. They bind directly to nuclear receptors. Nuclear steroid receptors consist of four major domains: *steroid binding domain* is at the C-terminus; *hinge domain* is involved in nuclear localization; *DNA-binding domain* consists of two "zinc fingers," which wrap around the DNA helix, and finally *the regulatory domain* activates gene-specific transcription factors.

First, there are several photoaffinity approaches, which aim to identify ligands in the membrane phospholipid bilayer. Several groups reported lipid monolayer [140] or bilayer mimic probes with different photophores. In bilayer imitating probes the photophores (benzophenone – 50 [141], diazirine – 51 [142], Fig. 17), are placed in bridged position in the center. Using a monolayer photoprobe, Ourisson et al. [143] reported creating crosslinks to steroids (cholesterol analogues) while partitioning in the deep core of the membrane. It provided evidence that they can be readily transported to the cytoplasm without binding to surface receptors or using surface proteins for the internalization.

Binding to the nuclear receptors by steroid hormones have been illustrated with some examples.

Katzenellenbogen studied several steroid receptors [21, 144]. Estrogen receptor is one of the most frequently investigated receptors by PAL. Hexestrol is a non-steroidal agonist, which was modified with a diazirine photophore via a sulfide linkage (52, Fig. 18). The photoprobe efficiently (30%) labeled the

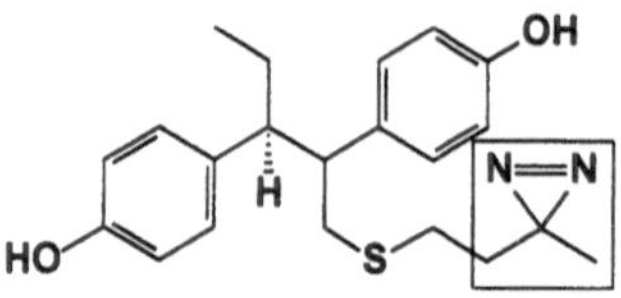

52

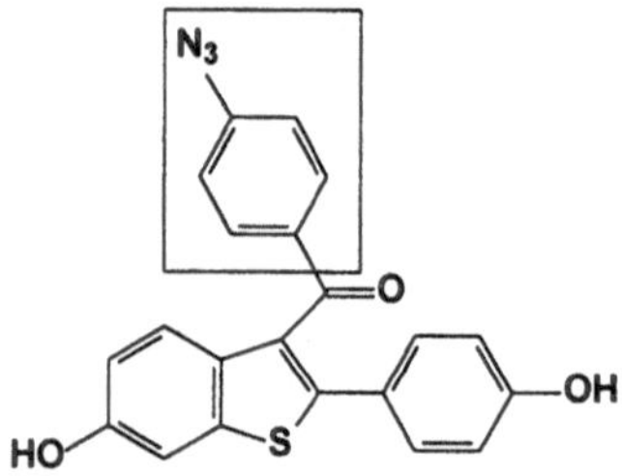

53

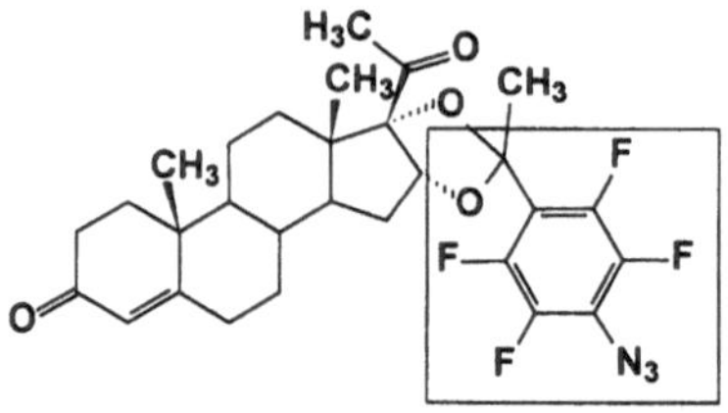

54

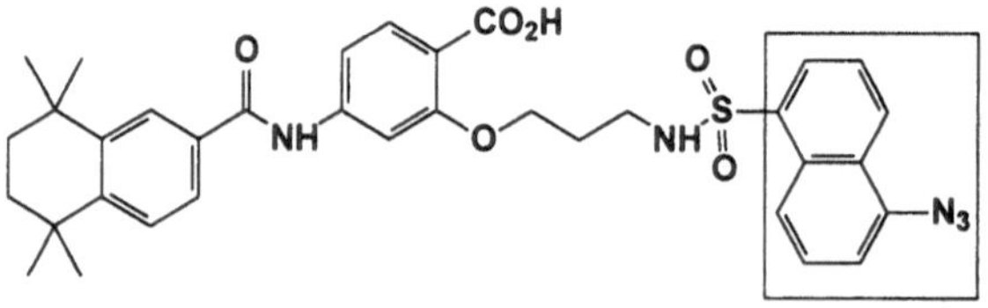

55

Fig. 18. Photoaffinity probes for studying steroid receptors

receptor protein (60 kDa) [145]. Alternatively an anti-estrogen photoprobe was designed based on the structure of LY117018 (53, Fig. 18) [146].

Katzenellenbogen has also studied progesterone receptor with an aryl azide derivative of the hormone (54, Fig. 18). The extremely efficient (60%) photocovalent attachment was identified in two subunits (B: 109 kDa, A: 87 kDa) in a ratio of 3.3:1 [147].

In another recent example, Hashimoto reported photoaffinity experiments on retinoic acid receptors (RAR). Retinoic acid plays a critical role in cell proliferation and differentiation. RARs belong to the superfamily of nuclear/thyroid hormone receptors. They consist of six transmembrane domains (A–F) which is a general feature of these receptors. The A/B domains have an autonomous transactivation function while the C-domain contains the Zn-finger, which binds to DNA. The large E-domain participates in ligand binding, dimerization, and ligand dependent transactivation. Finally, D- and F-domains help the orientation and stabilization of the E-domain.

Photoreactive aryl azide linked retinoic acid derivative (termed ADAM-3, 55, Fig. 18) labeled efficiently the recombinant E-subunit and the covalent attachment was identified within residues 492–510 and 589–594, which correspond to similar sequences found in the human retinoic receptor [148].

5
Summary and Future Directions

In the fine regulation of the cellular functions, the recognition sites of related molecules and their binding affinities are different, allowing them to induce selective responses. On the other hand many of the second messengers, adaptor molecules or effector enzymes, interact with several different protein targets, which have multiple binding sites on regulatory proteins. As recently depicted, the "tangled web" of the signal transduction provides diverse opportunities for PAL techniques. Photochemistry provides mild, remotely controllable and selective tools for the discovery scientist. PAL techniques both help to identify multiple binding sites for a specific ligand, and provide valuable information about location, binding region and contact architecture of the interaction. They contribute to the discovery of the complete signaling machinery, which helps to develop specific drugs for restoring the normal cell function in aberrant signaling. Photoaffinity labeling is very efficient in complementing other techniques like site-directed mutagenesis. In the future new innovative applications are expected to evolve based on the experience gathered up to now in photochemistry and in molecular biology. It will also be accelerated with new sensitive analytical and separation techniques.

Acknowledgements. The author is grateful to Prof. Glenn D. Prestwich (University of Utah) for multiple opportunities in the area of photoaffinity labeling and for his continuing support, as well as his encouragement in writing the present account. The author thanks Dr. Ferenc Darvas (ComGenex) for his continuing support during the preparation of the manuscript.

6
References

1. Chowdhry V, Westheimer FH (1979) Ann Rev Biochem 48:293
2. Knowles JR (1972) Acc Chem. Res 5:155
3. Bayley H (1983) Photogenerated reagents in biochemistry and molecular biology. Elsevier, Amsterdam
4. Fleming SA (1995) Tetrahedron 51:12,479
5. Brunner J (1993) Ann Rev Biochem 62:483
6. Kotzyba-Hibert F, Kapfer I, Goeldner M (1995) Angew Chem Int Ed Engl 34:1296
7. Prestwich GD, Dormán G, Elliott JT, Marecak DM, Chaudhary A (1997) Photochem Photobiol 65:222
8. Dormán G, Prestwich GD (2000) Trends Biotech (in press)
9. Hatanaka Y, Naskayama H, Kanaoka Y (1996) Reviews on Heteroatom Chemistry 14:213
10. Eyster KM (1998) Biochem Pharm 55:1927
11. Hinterding K, Alonso-Diaz D, Waldmann H (1998) Angew Chem Int Ed Engl 37:688
12. Rang HP, Dale MM, Ritter JM, Gardner P (eds) (1995) Pharmacology. Churchill Livingstone, New York
13. Meyers RA (ed) (1995) Molecular biology and biotechnology; a comprehensive desk reference. Chemie, Weinheim
14. Grutter T, Goeldner M, Kotzyba-Hibert F (1999) Biochemistry 38:7476
15. Trentham J, Corrie T (1993) In: Morrison H (ed) Biological applications of photochemical switches. Wiley, New York
16. Dormán G, Ceruso M, Prestwich GD (1994) ChemTracts 7:322
17. Fang K, Hashimoto M, Jockusch S, Turro NJ, Nakanishi K (1998) J Am Chem Soc 120:8543
18. Gilbert BA, Rando R (1995) J Am Chem Soc 117:8061
19. Dormán G, Prestwich GD (1993) ChemTracts 6:131
20. Dormán G, Prestwich GD (1994) Biochemistry 33:5661
21. Katzenellenbogen JA, Katzenellenbogen BS (1984) Affinity labeling of receptors for steroid and thyroid hormones. Academic Press, New York
22. Mourey RJ, Estevez VA, Marecek JM, Barrow RK, Prestwich GD, Snyder SH (1993) Biochemistry 32:1719
23. Dormán G, Chen J, Prestwich GD (1995) Tetrahedron Lett 36:8719
24. Patai S (ed) (1972) Chemistry of the azido group. The chemistry of the functional groups. Interscience, New York
25. Neumeyer JL, Baindur N, Yuan J, Booth G, Seeman P, Niznik HB (1990) J Med Chem 33:521
26. Staretz M, Hastie SB (1993) J Med Chem 58:1589
27. Kapfer I, Jaqueces P, Toubal H, Goeldner MP (1995) Bioconjugate Chem 6:109
28. Suzuki M, Koyano H, Noyori R (1992) Tetrahedron 48:2635
29. Thurkauf A, de Costa B, Berger P, Paul S, Rice KC (1991) J Labelled Comp Radiopharm 29:125
30. Hassner A (ed) (1984) Azetidines, β-lactams, diazetidines, and diazirines. Small ring heterocycles. Part 2, The chemistry of heterocyclic compounds. Interscience, New York, p 589
31. Hatanaka Y, Hashimoto M, Kurihara H, Nakayama H, Kanaoka Y (1994) J Org Chem 59:383
32. Chen FQ, Zheng JL, Hirano T, Niwa H, Ohmiya Y, Ohashi M (1995) J Chem Soc Perkin Trans 1 2129
33. Patai S (ed) (1972) Chemistry of the diazo groups. Chemistry of the functional groups. Interscience, New York
34. Wender PA, Irie K, Miller K (1993) J Org Chem 58:4180
35. Liu J, Stipanovic RD, Benedict CR (1996) J Labelled Comp Radiopharm 38:139
36. Ok H, Caldwell C, Schroeder DR, Singh AK, Nakanishi K (1988) Tet Lett 29:2275

37. Kapfer I, Hawkinson JE, Casida JE, Goeldner MP (1994) J Med Chem 37:133
38. DeLaLuz PJ, Golinski M, Watt DS, Vanaman TC (1995) Bioconjugate Chem 6:555
39. Holt DA, Yamashita DS, Konialian-Beck AL, Luengo JI, Abell AD, Bergsma DJ, Brandt M, Levy MA (1995) J Med Chem 38:13
40. Eberle A, Schwyzer R (1976) Helv Chim Acta 59:2421
41. Escher E (1988) Pharmacol Ther 37:37
42. Kauer JC, Erickson-Viitanen S, DeGrado WF (1986) J Biol Chem 261:10,695
43. Nassal M (1984) J Am Chem Soc 106:7540
44. Cornish VW, Benson DR, Altenbach CA, Hideg K, Hubell WL, Schultz PG (1994) Proc Natl Acad Sci USA 91:2910
45. Dormán G, Olszewski JD, Hong Y, Ahern DG, Prestwich GD (1995) J Org Chem 60:2292
46. Wilson CJ, Husain SS, Stimson ER, Dangott LJ, Miller KW, Maggio JE (1997) Biochemistry 36:4542
47. Wong SS (1991) Chemistry of protein conjugation and cross-linking. CRC Press, Boca Raton, Florida
48. Olszewski JD, Dormán G, Elliott JT, Prestwich GD, Ahern DG, Hong Y (1995) Bioconjugate Chem 6:395
49. Chen Y, Ebright Y, Ebright R (1994) Science 265:90
50. Prestwich GD (1991) In: Buncel E, Jones JR (eds) Isotopes in the physical and biomedical sciences. Elsevier, Amsterdam, pp 137–166
51. Dong M, Ding X-Q, Pinon DI, Hadac EM, Oda RP, Landers JP, Miller LJ (1999) J Biol Chem 274:4778
52. Newman AH (1991) Ann Rep Med Chem 25:271
53. Changeux JP, Galzi JL, Devillers-Thiery A, Bertrand D (1992) Quat Rev Biophys 25:395
54. Kalasz H, Watanabe T, Yabana H, Itagaki K, Naito K, Nakayama H, Schwartz A, Vaghy PL (1993) FEBS Lett 331:177
55. Nakayama H, Taki M, Striessnig J, Glossmann H, Catterall WA, Kanaoka Y (1991) Proc Natl Acad Sci USA 88:9203
56. Trainer VL, Baden DG, Catterall WA (1994) J Biol Chem 269:19,904
57. Peng L, Alcaraz ML, Klotz PF, Kotzyba-Hibert F, Goeldner M (1994) FEBS Lett 346:127
58. Choi SK, Kalivretenos AG, Usherwood PNR, Nakanishi K (1995) Chemistry & Biology 2:23
59. Eberle AN, DeGraan PNE (1985) Methods Enzymol 109:129
60. Schuster DI, Probst WC, Ehrlich GK, Singh G (1989) Photochem Photobiol 49:785
61. Hansch C (ed) (1990) Comprehensive medicinal chemistry, vol 6. Pergamon Press, Oxford, pp 633, 336, 242
62. Rong Y, Arbabian M, Thiriot DS, Seibold A, Clark RB, Ruoho AE (1999) Biochemistry 38:11,278
63. Suva LJ, Flannery MS, Caulfield MP, Findlay DM, Jüppner H, Goldring SR, Rosenblatt M, Chorev M (1997) J Pharm Exp Ther 283:876
64. Gimpl G, Anders J, Thiele C, Fahrenholz F (1997) Eur J Biochem 237:768
65. Darrow JW, Hadac EM, Miller LJ, Sugg EE (1998) Bioorg Med Chem Lett 8:3127
66. Dong M, Wang Y, Pinon DI, Hadac EM, Miller LJ (1999) J Biol Chem 274:903
67. Elliott JT, Hoekstra WJ, Maryanoff BE, Prestwich GD (1999) Bioorg Med Chem Lett 9:279
68. Eberle AN (1995) J Mol Recog 8:47
69. Escher E (1988) Pharmacol Ther 37:37
70. Bernier SG, Bellemare JML, Escher E, Guillemette G (1998) Biochemistry 37:4280
71. Kage NR, Leeman SE, Boyd ND (1993) J Neurochem 60:347
72. Kage NR, Leeman SE, Krause JE, Costello CE, Boyd ND (1996) J Biol Chem 271:25,797
73. Li Y, Marnerakis M, Stimson ER, Maggio JE (1995) J Biol Chem 270:1213
74. Kersey ID, Fishwick CW, FindlayBC, Ward P (1995) Bioorg Med Chem Lett 5:1271
75. Kersey ID, Bhogal N, Donnely D, Fishwick CW, Findlay BC, Ward P (1996) Bioorg Med Chem Lett 6:605
76. Mills JS, Miettinen HM, Barnidge D, Vlases MJ, Wiemer-Mackin S, Dratz EA, Sunner J, Jesaitis AJ (1998) J Biol Chem 273:10,428

77. Ingenhoven N, Eckard CP, Gehlert DG, Beck-Sickinger AG (1999) Biochemistry 38:6897
78. Rühmann A, Köpke AKE, Dautzenberg FM, Spiess J (1996) Proc Natl Acad Sci USA 93: 10,609
79. Bisello A, Adams AE, Mierke DF, Pellegrini M, Rosenblatt M, Suva LJ, Chorev M (1998) J Biol Chem 273:22,498
80. Kojro E, Eich P, Gimpl G, Fahrenholz F (1993) Biochemistry 32:13,537
81. Rasenick MM, Talluri M, Dunn WJ (1994) Methods Enzymol 237:100
82. Clack JW, Stein PJ (1994) J Med Chem 37:2406
83. Zhang H, Lerro KA, Takekuma S, Baek DJ, Moquin-Pattey C, Boehm MF, Nakanishi K (1994) J Am Chem Soc 116:6823
84. Zhang H, Lerro KA, Yamamoto T, Lien TH, Sastry L, Gawinowicz A, Nakanishi K (1994) J Am Chem Soc 116:10,165
85. Nakayama TA, Khorana HG (1990) J Biol Chem 265:15,762
86. Dhanasekaran N, Wessling-Resnick M, Kelleher DJ, Johnson GL, Ruoho A (1988) J Biol Chem 263:17,942
87. Resek JF, Farrens D, Khorana HG (1994) Proc Natl Acad Sci USA 91:7643
88. Vaillancourt R, Dhanasekaran N, Ruoho A (1994) Methods Enzymol 237:70
89. Kiefer H, Krieger J, Olszewski JD, von Heinje G, Prestwich GD, Breer H (1996) Biochemistry 35:16,077
90. Anis Y, Nürnberg B, Visochek T, Nachum R, Naor Z, Cohen-Armon M (1999) J Biol Chem 274:7431
91. Taylor JM, Jacob-Mosier GG, Lawton RG, Neubig RR (1994) Peptides 15:829
92. Tall E, Dormán G, Garcia P, Runnels L, Shah S, Chen J, Profit AA, Gu Q-M, Chaudhary A, Prestwich GD, Rebecchi MJ (1997) Biochemistry 36:7239
93. Witcher DR, McPherson PS, Kahl SD, Lewis T, Bentley P, Mullinix MJ, Windass JD, Cambell KP (1994) J Biol Chem 269:13,076
94. Bidasee KR, Besch HR Jr, Kwon S, Emmick JT, Besch KT, Gerzon K, Humerickhouse RA (1993) J Labelled Comp Radiopharm 34:33
95. Otto JC, Smith WL (1996) J Biol Chem 271:9906
96. Perrier H, Prasit P, Wang Z (1994) Tetrahedron Lett 35:1501
97. Mancini JA, Abramovitz M, Cox ME, Wong E, Charleson S, Perrier H, Wang Z, Prasit P, Vickers PJ (1993) FEBS Lett 318:277
98. Ito S, Hashimoto H, Negishi M, Suzuki M, Koyano H, Noyori R, Ichikawa A (1992) J Biol Chem 267:20,326
99. Slipetz DM, Scoggan KA, Nicholson DW, Metters KM (1993) Eur J Pharm Mol Pharm 244:161
100. Metters KM, Zamboni RJ (1993) J Biol Chem 268:6487
101. Gallant M, Sawyer N, Metters KM, Zamboni RJ (1998) Bioorg Med Chem 6:63
102. Rajagopalan KN, Chavan AJ, Haley BE, Watt DS (1993) J Biol Chem 268:14,230
103. Chavan AJ, Richardson SK, Kim H, Haley BE, Watt DS (1993) Bioconjugate Chem 4:268
104. Sawada T, Kato Y, Kobayashi H, Hashimoto Y, Watanabe T, Sugiyama Y, Iwasaki S (1993) Bioconjugate Chem 4:284
105. Wolff J, Knipling L (1995) J Biol Chem 270:16,809
106. Safa AR, Felsted RL (1987) J Biol Chem 262:1261
107. Rao S, Krauss NE, Heerding JM, Swindell CS, Ringel I, Orr GA, Horwitz SB (1994) J Biol Chem 269:3132
108. Ojima I, Duclos O, Dormán G, Simonot B, Prestwich GD, Rao S, Lerro KA, Horwitz SB (1995) J Med Chem 38:3891
109. Rao S, Orr GA, Chaudhary AG, Kingston DGI, Horwitz SB (1995) J Biol Chem 270:20,235
110. Ojima I, Bounaud P-Y, Ahern DG (1999) Bioorg Med Chem Lett 9:1189
111. Loeb C, Combeau C, Ehret-Sabatier L, Breton-Gilet A, Faucher D, Rousseau B, Commerçon A, Goeldner M (1997) Biochemistry 36:3820
112. Lei H, Marks V, Pasqule T, Atkinson JK (1998) Bioorg Med Chem Lett 8:3453
113. Rühmann A, Wentrup C (1994) Tetrahedron 50:3783
114. Sodeoka M, Uotsu K, Shibasaki M (1995) Tetrahedron Lett 48:8795

115. Wender PA, Irie K, Miller K (1993) J Org Chem 58:4180
116. Irie K, Okuno S, Koizumi F, Koshimizu K, Nishino H, Iwashima A (1993) Tetrahedron 49:10,817
117. Leroy D, Schmid N, Behr J-P, Filhol O, Pares S, Garin J, Bourgardt J-J, Chambaz E, Cochet C (1995) J Biol Chem 270:17,400
118. Liu J, Fan QR, Sodeoka M, Lane W, Verdine GL (1994) Chem Biol 1:47
119. Chaudhary A, Chen J, Gu Q-M, Witke W, Kwiatkowski DJ, Prestwich GD (1998) Chem Biol 5:273
120. Shoelson S, Lee J, Lynch CS, Backer JM, Pilch PF (1993) J Biol Chem 268:4085
121. Gergel JR, McNamara DJ, Dobrusin EM, Zhu G, Saltiel AR, Miller WT (1994) Biochemistry 33:14,671
122. Williams KP, Shoelson SE (1993) J Biol Chem 268:5361
123. Chen J, Profit AA, Prestwich GD (1996) J Org Chem 61:6305
124. Hammond-Odie LP, Jackson TR, Bladder IJ, Profit AA, Turck C, Prestwich GD, Theibert AB (1996) J Biol Chem 271:18,859
125. Ying W, Sepp-Lorenzino L, Cai K, Aloise P, Coleman PS (1994) J Biol Chem 269:470
126. Pellicena P, Scholten JD, Zimmermann K, Creswell M, Huang CC, Miller WT (1996) Biochemistry 35:13,494
127. Bukhtiyarov Y, Omer CA, Allen CM (1995) J Biol Chem 270:19,035
128. Gaon I, Turek TC, Weller VA, Edelstein R, Singh SK, DiStefano MD (1996) J Org Chem 61:7738
129. Miller P, Schnur RC, Barbacci E, Moyer MP, Moyer JD (1994) Biochem Biophys Res Comm 201:1313
130. Lee JC, Laydon JT, McDonnell PC, Gallagher TF, Kumar S, Green D, McNulty D, Blumenthal MJ, Heys JR, Landwater SW, Strickler JE, McLaughlin MM, Siemens IR, Fisher SM, Livi GP, White JR, Adams JL, Young PR (1994) Nature 372:739
131. O'Neill KT, DeGrado WF (1989) Proteins 6:284
132. McNicoll N, Escher E, Wilkes B, Schiller PW, Ong H, De Lean A (1992) Biochemistry 31:4487
133. Bömmel HM, Reif A, Fröhlich LG, Frey A, Hoffman H, Marecak DM, Groehn V, Kotsonis P, La M, Köstner S, Meinecke M, Bernhardt M, Weeger M, Ghisla S, Prestwich GD, Pfleiderer W, Schmidt HHW (1998) J Biol Chem 273:33,142
134. Luescher IF, Cerottini JC, Romero P (1994) J Biol Chem 269:5574
135. Kessler B, Michelin O, Apostolou I, Delarbre C, Gachelin G, Gregoire C, Malissen B, Cerottini JC, Wurm F, Karplus M, Luescher IF (1999) J Biol Chem 274:3622
136. Husi H, Luyten MA, Zurini MGM (1994) J Biol Chem 269:14,199
137. Chen Y, Nakanishi K, Merrill D, Eng CP, Molnar-Kimber KL, Faillin A, Caggiano TJ (1995) Bioorg Med Chem Lett 5:1355
138. Mohan C (1999) Cell adhesion molecules in signal transduction. In: Signal transduction catalogue and technical handbook. Calbiochem, San Diego, CA
139. Bitan G, Scheibler L, Greenberg Z, Rosenblatt M, Chorev M (1999) Biochemistry 38:3414
140. Alcaraz ML, Peng L, Klotz P, Goeldner M (1996) J Org Chem 61:192
141. Yamamoto M, Dollé V, Warnock W, Diyizou Y, Yamada M, Nakatani Y, Ourisson G (1994) Bull Soc Chim Fr 112:317
142. Delfino JM, Schreiber SL, Richards F (1993) J Am Chem Soc 115:3458
143. Yamamoto M, Warnock W, Milon A, Nakatani Y, Ourisson G (1993) Angew Chem Int Ed Engl 32:259
144. Pinney KG, Katzenellenbogen JA (1991) J Org Chem 56:3125
145. Bergmann KE, Carlson KE, Katzenellenbogen JA (1994) Bioconjugate Chem 5:141
146. Kym PR, Anstead GM, Pinney KG, Wilson SR, Katzenellenbogen JA (1993) J Med Chem 36:3910
147. Bergmann KE, Carlson KE, Katzenellenbogen JA (1995) Bioconjugate Chem 6:115
148. Sasaki T, Shimazawa R, Sawada T, Ijiwa T, Fukasawa H, Shudo K, Hashimoto Y, Iwasaki S (1996) Biol Pharm Bull 19:659

Author Index Volume 201–211

The volume numbers are printed in italics